FUNCTIONAL FOODS AND NUTRACEUTICALS FOR HUMAN HEALTH

Advancements in Natural Wellness and Disease Prevention

FUNCTIONAL FOODS AND NUTRACEUTICALS FOR HUMAN HEALTH

Advancements in Natural Wellness and Disease Prevention

Edited by
Cristóbal Noé Aguilar, PhD
A. K. Haghi, PhD

First edition published 2022

Apple Academic Press Inc.
1265 Goldenrod Circle, NE,
Palm Bay, FL 32905 USA

4164 Lakeshore Road, Burlington,
ON, L7L 1A4 Canada

CRC Press
6000 Broken Sound Parkway NW,
Suite 300, Boca Raton, FL 33487-2742 USA

2 Park Square, Milton Park,
Abingdon, Oxon, OX14 4RN UK

© 2022 Apple Academic Press, Inc.

Apple Academic Press exclusively co-publishes with CRC Press, an imprint of Taylor & Francis Group, LLC

Reasonable efforts have been made to publish reliable data and information, but the authors, editors, and publisher cannot assume responsibility for the validity of all materials or the consequences of their use. The authors, editors, and publishers have attempted to trace the copyright holders of all material reproduced in this publication and apologize to copyright holders if permission to publish in this form has not been obtained. If any copyright material has not been acknowledged, please write and let us know so we may rectify in any future reprint.

Except as permitted under U.S. Copyright Law, no part of this book may be reprinted, reproduced, transmitted, or utilized in any form by any electronic, mechanical, or other means, now known or hereafter invented, including photocopying, microfilming, and recording, or in any information storage or retrieval system, without written permission from the publishers.

For permission to photocopy or use material electronically from this work, access www.copyright.com or contact the Copyright Clearance Center, Inc. (CCC), 222 Rosewood Drive, Danvers, MA 01923, 978-750-8400. For works that are not available on CCC please contact mpkbookspermissions@tandf.co.uk

Trademark notice: Product or corporate names may be trademarks or registered trademarks and are used only for identification and explanation without intent to infringe.

Library and Archives Canada Cataloguing in Publication

Title: Functional foods and nutraceuticals for human health advancements in natural wellness and disease prevention / edited by Cristóbal Noé Aguilar, PhD, A.K. Haghi, PhD.

Names: Aguilar, Cristóbal Noé, editor. | Haghi, A. K., editor.

Description: First edition. | Includes bibliographical references and index.

Identifiers: Canadiana (print) 20210241209 | Canadiana (ebook) 20210241292 | ISBN 9781771889469 (hardcover) | ISBN 9781774638163 (softcover) | ISBN 9781003097358 (ebook)

Subjects: LCSH: Functional foods. | LCSH: Health. | LCSH: Medicine, Preventive.

Classification: LCC QP144.F85 F86 2022 | DDC 613.2—dc23

Library of Congress Cataloging-in-Publication Data

Names: Aguilar, Cristóbal Noé, editor. | Haghi, A. K., editor.

Title: Functional foods and nutraceuticals for human health : advancements in natural wellness and disease prevention / edited by Cristóbal Noé Aguilar, PhD, A.K. Haghi, PhD.

Description: First edition. | Palm Bay, FL : Apple Academic Press, 2022. | Includes bibliographical references and index. | Summary: "This new volume, Functional Foods and Nutraceuticals for Human Health: Advancements in Natural Wellness and Disease Prevention, provides important information on potential applications and new developments in functional health foods and nutraceuticals. The volume looks at health-promoting properties in functional foods and beverages as well as nutraceuticals. Some health issues that are considered in conjunction with these foods and nutraceuticals include oxidative stress, obesity, pharyngitis, low cognitive concentration among others. Research topics include the antioxidant properties of certain products, the development of functional and medicinal beverages, nutraceuticals and functional foods for alternative therapies, and more. This volume is a valuable reference for professionals involved in product development and researchers focusing on food and nutraceutical products"-- Provided by publisher.

Identifiers: LCCN 2021027381 (print) | LCCN 2021027382 (ebook) | ISBN 9781771889469 (hardback) | ISBN 9781774638163 (paperback) | ISBN 9781003097358 (ebook)

Subjects: LCSH: Functional foods.

Classification: LCC QP144.F85 F8638 2022 (print) | LCC QP144.F85 (ebook) | DDC 613.2--dc23

LC record available at https://lccn.loc.gov/2021027381

LC ebook record available at https://lccn.loc.gov/2021027382

ISBN: 978-1-77188-946-9 (hbk)
ISBN: 978-1-77463-816-3 (pbk)
ISBN: 978-1-00309-735-8 (ebk)

About the Editors

Cristóbal Noé Aguilar, PhD
Director of Research and Postgraduate Programs at Universidad Autonoma de Coahuila, Mexico

Cristóbal Noé Aguilar, PhD, is Director of Research and Postgraduate Programs at the Universidad Autonoma de Coahuila, Mexico. Dr Aguilar has published more than 160 papers in indexed journals, more than 40 articles in Mexican journals, and 250 contributions in scientific meetings. He has also published many book chapters, several Mexican books, four editions of international books, and more. He has been awarded several prizes and awards, the most important of which are the National Prize of Research 2010 from the Mexican Academy of Sciences; the Prize "Carlos Casas Campillo 2008" from the Mexican Society of Biotechnology and Bioengineering; National Prize AgroBio—2005; and the Mexican Prize in Food Science and Technology. Dr. Aguilar is a member of the Mexican Academy of Science, the International Bioprocessing Association, Mexican Academy of Sciences, Mexican Society for Biotechnology and Bioengineering, and the Mexican Association for Food Science and Biotechnology. He has developed more than 21 research projects, including six international exchange projects. His PhD in Fermentation Biotechnology was awarded by the Universidad Autónoma Metropolitana, Mexico.

A. K. Haghi, PhD
Professor Emeritus of Engineering Sciences, Former Editor-in-Chief, International Journal of Chemoinformatics and Chemical Engineering and Polymers Research Journal; Member, Canadian Research and Development Center of Sciences and Culture

A. K. Haghi, PhD, is the author and editor of 200 books, as well as 1000 published papers in various journals and conference proceedings. Dr Haghi has received several grants, consulted for a number of major corporations, and is a frequent speaker to national and international audiences. Since 1983, he served as professor at several universities. He is former Editor-in-Chief of the *International Journal of Chemoinformatics and Chemical*

Engineering and *Polymers Research Journal* and is on the editorial boards of many international journals. He is also a member of the Canadian Research and Development Center of Sciences and Cultures (CRDCSC), Montreal, Quebec, Canada.

Contents

Contributors .... *ix*

Abbreviations .... *xv*

Preface .... *xix*

PART I: Health-Promoting Properties in Functional Foods and Beverages 1

1. **Anxiolytic and Concentration-Stimulating Properties of *Valeriana* and *Passiflora*: Chemical Composition and Antioxidant Activity** 3
   Daniela Ibarra-Flores, Gabriela Perez-Sanchez, Liz Elena Rodriguez-Romo, Anayelli Demeneghi-Rivero, Víctor Olver-García, Paola I. Angulo-Bejarano, Victor Manuel Rodriguez-García, and Anaberta Cardador-Martinez

2. ***Tremella fuciformis* and Its Polysaccharides as an Alternative Therapy Against Oxidative Stress** 45
   Rubén F. González-Laredo, Angelica Valdez-Villarreal, Nuria Elizabeth Rocha-Guzmán, Martha Rocío Moreno-Jiménez, and José Alberto Gallegos-Infante

3. **Natural Extracts and Compounds as Inhibitors of Amylase for Diabetes Treatment and Prevention** 69
   Cynthia Selene Vasquez-Ramos, Melany Guadalupe Garcia-Moreno, Daniel García-García, Gloria Alicia Martínez-Medina, Sujey Abigail Niño-Herrera, Hugo Luna-García, Sandra Paciós-Michelena, Anna Ilyina, E. Patricia Segura-Ceniceros, Mónica L. Chávez-González, Mayela Govea-Salas, José L. Martínez-Hernández, S. Yesenia Silva-Belmares, Radik A. Zaynullin, Raikhana V. Kunakova, Rodolfo Ramos-González, and Roberto Arredondo-Valdés

4. **The Implication of Rambutan, Ginseng, and Green Tea in Energy Drinks Through Their Genetic Characterization, Total Phenolics, and Antioxidant Activity** 109
   Axel Flores-Cuéllar, Octavio Torres-Hernández, Miriam Vázquez-Segoviano, Paola Isabel Angulo-Bejarano, and Víctor Olvera-García

5. **Innovation and Challenges in the Development of Functional and Medicinal Beverages** 137
   Dayang Norulfairuz Abang Zaidel, Ida Idayu Muhamad, Zanariah Hashim, Yanti Maslina Mohd Jusoh, and Eraricar Salleh

PART II: Nutraceuticals 199

6. **Herbal Antiobesity Products and Their Function in the Gut-Brain Axis** 201

   Nuria-Elizabeth Rocha-Guzmán, Paola Flores-Rodríguez, Guadalupe Montiel-Ramírez, Martha-Rocío Moreno-Jiménez, José Alberto Gallegos-Infante, Rubén-Francisco González-Laredo, and Carlos-Alonso Salas-Ramírez

7. **Plants as a Potential Source of Acetylcholinesterase Inhibitors for Nutraceutical Therapy Disease** 229

   Yesenia Estrada-Nieto, Daniel García-García, Alejandra I. Vargas-Segura, Roberto Arredondo-Valdés, Radik A. Zaynullin, Raikhana V. Kunakova, Mónica Chávez-González, Rodolfo Ramos-González, José L. Martínez–Hernández, Mayela Govea-Salas, Anna Ilyina, and E. Patricia Segura-Ceniceros

8. **Characterization of *Heliopsis longipes* and the Potential of Its Ethanolic Extract as an Adjuvant in the Treatment of Pharyngitis** 243

   Mariela Correa-Delgado, Víctor Manuel Orozco-González, Ana Paola Aldama-Núñez, Victor Olvera-García, and Víctor Manuel Rodríguez-García

9. **Antidiabetic Natural Products: Mechanisms of Action** 291

   Erick P. Gutiérrez-Grijalva, Laura A. Contreras-Angulo, Marilyn S. Criollo-Mendoza, Sara Avilés-Gaxiola, and J. Basilio Heredia

10. **Ent-Kaurenes: Natural Agents with Potential for the Pharmaceutical Industry** 317

    Carlos Camacho-González, Francisco Fabián Razura-Carmona, Mayra Herrera-Martínez, Efigenia Montalvo-González, Sonia Sáyago-Ayerdi, and Jorge Alberto Sánchez-Burgos

11. **Anti-Inflammatory, Antioxidant, and Antitumoral Potential of Plants of the Asteraceae Family** 331

    Miguel Ángel Alfaro-Jiménez, Alejandro Zugasti-Cruz, Sonia Yesenia Silva-Belmares, Juan Alberto Ascacio-Valdés, and Crystel Aleyvick Sierra-Rivera

Index *381*

Contributors

Ana Paola Aldama-Núñez
Tecnologico de Monterrey, Escuelade Ingeniería y Ciencias, San Pablo, Querétaro 76130, Mexico

Miguel Ángel Alfaro-Jiménez
Laboratory of Immunology and Toxicology, Faculty of Chemistry, Autonomous University of Coahuila, Saltillo, Mexico

Paola I. Angulo-Bejarano
Tecnologico de Monterrey, Escuela de Ingeniería y Ciencias, Departamento de Bioingenierías, Epigmenio González 500, San Pablo, Querétaro 76130, México

Roberto Arredondo-Valdés
Nanobioscience Group, Chemistry School, Autonomous University of Coahuila, Blvd. V. Carranza e Ing. J. Cardenas V., 25280 Saltillo, Coahuila, Mexico

Juan Alberto Ascacio-Valdés
Food Research Department, Faculty of Chemistry, Autonomous University of Coahuila, Saltillo, Mexico

Sara Avilés-Gaxiola
Research Center for Food and Development, Culiacán, Sinaloa 80110, México

Carlos Camacho-González
Teccnológico Nacional de México/Instituto Tecnológico de Tepic, Av. Tecnológico # 2595, Col. Lagos del Country, 63175 Tepic, México

Anaberta Cardador-Martinez
Tecnologico de Monterrey, Escuela de Ingeniería y Ciencias, Departamento de Bioingenierías, Epigmenio González 500, San Pablo, Querétaro 76130, México

Mónica L. Chávez-González
Nanobioscience Group, Chemistry School, Autonomous University of Coahuila, Blvd. V. Carranza e Ing. J. Cardenas V., 25280 Saltillo, Coahuila, Mexico

Laura A. Contreras-Angulo
Research Center for Food and Development, Culiacán, Sinaloa 80110, México

Mariela Correa-Delgado
Tecnologico de Monterrey, Escuelade Ingeniería y Ciencias, San Pablo, Querétaro 76130, Mexico

Marilyn S. Criollo-Mendoza
Research Center for Food and Development, Culiacán, Sinaloa 80110, México

Anayelli Demeneghi-Rivero
Tecnologico de Monterrey, Escuela de Ingeniería y Ciencias, Departamento de Bioingenierías, Epigmenio González 500, San Pablo, Querétaro 76130, México

Yesenia Estrada-Nieto
Nanobioscience Group, Chemistry School, Autonomous University of Coahuila, Blvd. V. Carranza e Ing. J. Cardenas V., Saltillo 25280, Mexico

Axel Flores-Cuéllar
Tecnologico de Monterrey, Escuela de Ingeniería y Ciencias, Campus Queretaro,
Av. Epigmenio González 500, Fracc. San Pablo, CP 76130 Queretaro, Mexico

Paola Flores-Rodríguez
Centro Interdisciplinario de Investigación para el Desarrollo Integral Regional
(CIIDIR-IPN Unidad Durango). Sigma 119, Fracc. 20 de Noviembre II, 34234, Durango, Dgo. México

José Alberto Gallegos-Infante
Tecnológico Nacional de México/Instituto Tecnológico de Durango, Unidad de Posgrado,
Investigación y Desarrollo Tecnológico (UPIDET), Blvd. Felipe Pescador 1830 Ote.,
Col. Nueva Vizcaya, 34080 Durango, Dgo., México

Daniel García-García
Nanobioscience Group, Chemistry School, Autonomous University of Coahuila,
Blvd. V. Carranza e Ing. J. Cardenas V., 25280 Saltillo, Coahuila, Mexico

Melany Guadalupe Garcia-Moreno
Nanobioscience Group, Chemistry School, Autonomous University of Coahuila,
Blvd. V. Carranza e Ing. J. Cardenas V., 25280 Saltillo, Coahuila, Mexico

Rubén-Francisco González-Laredo
Tecnológico Nacional de México/Instituto Tecnológico de Durango, Unidad de Posgrado,
Investigación y Desarrollo Tecnológico (UPIDET), Blvd. Felipe Pescador 1830 Ote.,
Col. Nueva Vizcaya, 34080 Durango, Dgo., México

Mayela Govea-Salas
Nanobioscience Group, Chemistry School, Autonomous University of Coahuila,
Blvd. V. Carranza e Ing. J. Cardenas V., 25280 Saltillo, Coahuila, Mexico

Erick P. Gutiérrez-Grijalva
Cátedras CONACYT-Research Center for Food and Development, Culiacán, Sinaloa 80110, México

Zanariah Hashim
Food and Biomaterial Engineering Research Group, School of Chemical and Energy Engineering,
Faculty of Engineering, Universiti Teknologi Malaysia, 81310 Johor Bahru, Malaysia

J. Basilio Heredia
Research Center for Food and Development, Culiacán, Sinaloa 80110, México

Mayra Herrera-Martínez
Instituto de Farmacobiología, Universidad de la Cañada, Carretera Teotitlán-San Antonio Nanahuatipán
Km 1.7, Paraje Titlacuatitla, Teotitlán de Flores Magón, 68540 Oaxaca, México

Daniela Ibarra-Flores
Tecnologico de Monterrey, Escuela de Ingeniería y Ciencias, Departamento de Bioingenierías,
Epigmenio González 500, San Pablo, Querétaro 76130, México

Anna Ilyina
Nanobioscience Group, Chemistry School, Autonomous University of Coahuila,
Blvd. V. Carranza e Ing. J. Cardenas V., 25280 Saltillo, Coahuila, Mexico

Yanti Maslina Mohd Jusoh
Food and Biomaterial Engineering Research Group, School of Chemical and Energy Engineering,
Faculty of Engineering, Universiti Teknologi Malaysia, 81310 Johor Bahru, Malaysia

Raikhana V. Kunakova
Ufa State Petroleum Technological University, 1 Kosmonavtov Str., 450062 Ufa, Bashkortostan, Russia

Contributors

Hugo Luna-García
Nanobioscience Group, Chemistry School, Autonomous University of Coahuila, Blvd. V. Carranza e Ing. J. Cardenas V., 25280 Saltillo, Coahuila, Mexico

José L. Martínez-Hernández
Nanobioscience Group, Chemistry School, Autonomous University of Coahuila, Blvd. V. Carranza e Ing. J. Cardenas V., 25280 Saltillo, Coahuila, Mexico

Gloria Alicia Martínez-Medina
Nanobioscience Group, Chemistry School, Autonomous University of Coahuila, Blvd. V. Carranza e Ing. J. Cardenas V., 25280 Saltillo, Coahuila, Mexico

Efigenia Montalvo-González
Teccnológico Nacional de México/Instituto Tecnológico de Tepic, Av. Tecnológico # 2595, Col. Lagos del Country, 63175 Tepic, México

Guadalupe Montiel-Ramírez
Department of Chemical and Biochemical Engineering of the National Technology of Mexico/Technological Institute of Durango (TecNM/IT Durango)

Martha-Rocío Moreno-Jiménez
Tecnológico Nacional de México/Instituto Tecnológico de Durango, Unidad de Posgrado, Investigación y Desarrollo Tecnológico (UPIDET), Blvd. Felipe Pescador 1830 Ote., Col. Nueva Vizcaya, 34080 Durango, Dgo., México

Ida Idayu Muhamad
Food and Biomaterial Engineering Research Group, School of Chemical and Energy Engineering, Faculty of Engineering, Universiti Teknologi Malaysia, 81310 Johor Bahru, Malaysia

Sujey Abigail Niño-Herrera
Nanobioscience Group, Chemistry School, Autonomous University of Coahuila, Blvd. V. Carranza e Ing. J. Cardenas V., 25280 Saltillo, Coahuila, Mexico

Víctor Olvera-García
Tecnologico de Monterrey, Escuela de Ingeniería y Ciencias, Campus Queretaro, Av. Epigmenio González 500, Fracc. San Pablo, CP 76130 Queretaro, Mexico

Víctor Manuel Orozco-González
Tecnologico de Monterrey, Escuelade Ingeniería y Ciencias, San Pablo, Querétaro 76130, Mexico

Sandra Paciós-Michelena
Nanobioscience Group, Chemistry School, Autonomous University of Coahuila, Blvd. V. Carranza e Ing. J. Cardenas V., 25280 Saltillo, Coahuila, Mexico

Gabriela Perez-Sanchez
Tecnologico de Monterrey, Escuela de Ingeniería y Ciencias, Departamento de Bioingenierías, Epigmenio González 500, San Pablo, Querétaro 76130, México

Rodolfo Ramos-González
CONACYT—Autonomous University of Coahuila, 25280 Saltillo, Coahuila, México Saltillo, COAH, México. Blvd. V. Carranza e Ing. J. Cardenas V., Saltillo, Coahuila, Mexico, CP 25280

Francisco Fabián Razura-Carmona
Teccnológico Nacional de México/Instituto Tecnológico de Tepic, Av. Tecnológico # 2595, Col. Lagos del Country, 63175 Tepic, México

Nuria-Elizabeth Rocha-Guzmán
Tecnológico Nacional de México/Instituto Tecnológico de Durango, Unidad de Posgrado, Investigación y Desarrollo Tecnológico (UPIDET), Blvd. Felipe Pescador 1830 Ote., Col. Nueva Vizcaya, 34080 Durango, Dgo., México

Victor Manuel Rodriguez-García
Tecnologico de Monterrey, Escuela de Ingeniería y Ciencias, Departamento de Bioingenierías, Epigmenio González 500, San Pablo, Querétaro 76130, México

Liz Elena Rodriguez-Romo
Tecnologico de Monterrey, Escuela de Ingeniería y Ciencias, Departamento de Bioingenierías, Epigmenio González 500, San Pablo, Querétaro 76130, México

Carlos-Alonso Salas-Ramírez
Department of Chemical and Biochemical Engineering of the National Technology of Mexico/Technological Institute of Durango (TecNM/IT Durango)

Eraricar Salleh
Food and Biomaterial Engineering Research Group, School of Chemical and Energy Engineering, Faculty of Engineering, Universiti Teknologi Malaysia, 81310 Johor Bahru, Malaysia

Jorge Alberto Sánchez-Burgos
Teccnológico Nacional de México/Instituto Tecnológico de Tepic, Av. Tecnológico # 2595, Col. Lagos del Country, 63175 Tepic, México

Sonia Sáyago-Ayerdi
Teccnológico Nacional de México/Instituto Tecnológico de Tepic, Av. Tecnológico # 2595, Col. Lagos del Country, 63175 Tepic, México

E. Patricia Segura-Ceniceros
Nanobioscience Group, Chemistry School, Autonomous University of Coahuila, Blvd. V. Carranza e Ing. J. Cardenas V., 25280 Saltillo, Coahuila, Mexico

Crystel Aleyvick Sierra-Rivera
Laboratory of Immunology and Toxicology, Faculty of Chemistry, Autonomous University of Coahuila, Saltillo, Mexico

Sonia Yesenia Silva-Belmares
Research Group of Chemist Pharmacist Biologist, Chemistry School, Autonomous University of Coahuila, Blvd. V. Carranza e Ing. J. Cardenas V., 25280 Saltillo, Coahuila, Mexico

Octavio Torres-Hernández
Tecnologico de Monterrey, Escuela de Ingeniería y Ciencias, Campus Queretaro, Av. Epigmenio González 500, Fracc. San Pablo, CP 76130 Queretaro, Mexico

Angelica Valdez-Villarreal
Tecnológico Nacional de México/Instituto Tecnológico de Durango, Unidad de Posgrado, Investigación y Desarrollo Tecnológico (UPIDET), Blvd. Felipe Pescador 1830 Ote., Col. Nueva Vizcaya, 34080 Durango, Dgo., México

Alejandra I. Vargas-Segura
Faculty of Dentistry, Autonomous University of Coahuila, Saltillo 25280, Mexico

Cynthia Selene Vasquez-Ramos
Nanobioscience Group, Chemistry School, Autonomous University of Coahuila, Blvd. V. Carranza e Ing. J. Cardenas V., 25280 Saltillo, Coahuila, Mexico

Contributors

Miriam Vázquez-Segoviano
Tecnologico de Monterrey, Escuela de Ingeniería y Ciencias, Campus Queretaro,
Av. Epigmenio González 500, Fracc. San Pablo, CP 76130 Queretaro, Mexico

Dayang Norulfairuz Abang Zaidel
Food and Biomaterial Engineering Research Group, School of Chemical and Energy Engineering,
Faculty of Engineering, Universiti Teknologi Malaysia, 81310 Johor Bahru, Malaysia

Radik A. Zaynullin
Ufa State Petroleum Technological University, 1 Kosmonavtov Str., 450062 Ufa, Bashkortostan, Russia

Alejandro Zugasti-Cruz
Laboratory of Immunology and Toxicology, Faculty of Chemistry, Autonomous University of Coahuila,
Saltillo, Mexico

Abbreviations

α-MSH	α-melanocyte-stimulating hormone
AA	arachidonic acid
ACh	acetylcholine
AChE	acetylcholinesterase
AChEIs	acetylcholinesterase inhibitors
AD	Alzheimer's disease
AgRP	agouti-related neuropeptide expression
ALP	alkaline phosphatase
ALT	alanine aminotransferase enzyme
AMPK	AMP-activated protein kinase
ANOVA	analysis of variance
BSA	bovine serum albumin
BZDs	benzodiazepines
CAT	catalase
cGMP	cyclic guanosine monophosphate
CNS	central nervous system
DHQ	dihydroquercetin
DM	diabetes mellitus
DPPH	2,2-diphenyl-1-picrylhydrazyl
EGF	epidermal growth factor
EI	electron impact
EPM	elevated plus maze test
GABA	gamma-aminobutyric acid
GAS	group A *Streptococcus*
GBSS	granule-bound starch synthase
GGPP	geranylgeranyl pyrophosphate
GLUTs	glucose transporters
H/bp	herbal or botanical products
HHP	high hydrostatic pressure
HIPEF	high-intensity pulsed electric field
Hist	histamine
HPA	human pancreatic α-amylase
HPLC	high-pressure liquid chromatography
HPP	high-pressure processing

HUVECs	human umbilical vein endothelial cells
IAA	α-amylase inhibitors
IL	interleukin
iNOS	inducible nitric oxide synthase
IR	insulin receptor
JAR	Just About Right
LAM	Labeled Affective Magnitude
MAE	microwave-assisted extraction
MF	microfiltration
MHG	microwave hydrodiffusion and gravity
MW	molecular weight
NO	nitric oxide
PA	palmitic acid
PCA	principal component analysis
PEPCK	phosphoenolpyruvate carboxykinase
PG	propylene glycol
PI	isoelectric point
PI3K	phosphatidyl inositol 3-kinase
PKC	protein kinase C
PMA	phorbol myristate acetate
PME	pectin methyesterase
POD	peroxidase
POMC	proopiomelanocortin
PPO	polyphenol oxidase
PTD	prostaglandins
PTZ	pentylenetetrazole
QDA	qualitative descriptive analysis
RDS	rapidly digestible starch
RF	rheumatic fever
RHD	rheumatic heart disease
RO	reverse osmosis
ROS	reactive oxygen species
RS	resistant starch
RTD	ready-to-drink
SBE	starch-branching enzymes
SCI	subjective cognitive impairment
SDS	slowly digestible starch
SGLT	sodium-glucose type
SmF	submerged fermentation

SOD	superoxide dismutase
SREBP	sterile regulatory element-binding protein
SSF	solid-state fermentation
SSS	soluble starch synthase enzymes
TFPS	polysaccharide extracts of *T. fuciformis*
TNF-alpha	tumor necrosis factor alpha
UAE	ultrasonic-assisted extraction
UF	ultrafiltration
WHO	World Health Organization

Preface

This new volume provides potential applications and new developments in functional health foods and nutraceuticals. In addition to serving as a reference manual, it summarizes the current state of knowledge in key research areas and contains novel ideas for future research and development. Additionally, it provides an easy-to-read text suitable for teaching senior undergraduate and postgraduate students in the relevant areas.

This book addresses functional food product development and nutraceuticals from a number of perspectives:

- the process itself;
- health research that may provide opportunities;
- idea creation;
- regulation; and
- processes and ingredients.

It also features case studies that illustrate real-product development and commercialization histories. Ensuring that foods and nutraceuticals remain stable during the required shelf life is critical to their success in the market place, yet companies experience difficulties in this area.

This volume is a valuable reference for professionals involved in quality assurance and product development and researchers focusing on food and nutraceutical stability. It includes information on the chemical properties, dietary sources, intakes, efficacy, health effects, and safety of each bioactive compound, functional food, or nutraceutical. It provides instant access to comprehensive, cutting-edge data, making it possible for food scientists, nutritionists, and researchers to utilize this ever-growing wealth of information.

The book discusses the technology preparation, equipment involved, and changes that occur during processing in physical, chemical, microbiological, and organoleptic properties of the products. It is useful to students, teachers, researchers, food manufacturers, policy maker and to all those who are interested in better health and longevity.

PART I
Health-Promoting Properties in Functional Foods and Beverages

CHAPTER 1

Anxiolytic and Concentration-Stimulating Properties of *Valeriana* and *Passiflora*: Chemical Composition and Antioxidant Activity

DANIELA IBARRA-FLORES, GABRIELA PEREZ-SANCHEZ,
LIZ ELENA RODRIGUEZ-ROMO, ANAYELLI DEMENEGHI-RIVERO,
VÍCTOR OLVER-GARCÍA, PAOLA I. ANGULO-BEJARANO,
VICTOR MANUEL RODRIGUEZ-GARCÍA, and
ANABERTA CARDADOR-MARTINEZ[*]

Tecnologico de Monterrey, Escuela de Ingeniería y Ciencias, Departamento de Bioingenierías, Epigmenio González 500, San Pablo, Querétaro 76130, México

[*]*Corresponding author. E-mail: mcardador@tec.mx*

ABSTRACT

Nowadays, the population lives under constant stress conditions due to different factors that often conduct health issues such as anxiety, colitis, and migraines. The need to stay healthy, maintaining the lifestyle leads to the development of functional products using medicinal plants. *Passiflora incarnata* is well known for its anxiolytic properties and is recommended for nervous restlessness due to its active compounds such as flavonoids and alkaloids, among numerous other constituents. On the other hand, lemon balm (*Melissa officinalis*) is a medicinal plant that is used in aromatic beverages and as a natural tranquilizer and is commonly used for its antioxidant, antimicrobial, antistress, memory, and concentration improving properties. *M. officinalis* has shown to possess high levels of phenolics compounds and antioxidant properties and is a potent inhibitor of gamma-aminobutyric acid transaminase. Various combinations and numerous medicinal properties of its extract, oil, and leaves make it a multipurpose plant. *Passiflora* and *Melissa* phytochemicals were extracted and evaluated for antioxidant activity. Both plants showed high phenolic and flavonoid contents, as well as high antioxidant capacity evaluated by 2,2-diphenyl-1-picrylhydrazyl free-radical scavenging capacity.

1.1 INTRODUCTION

It is well recognized that food and beverages play a key role in human health, disease prevention, and treatment. This idea is not new: Hippocrates wrote "Let food be thy medicine and medicine be thy food" 2400 years ago and in Asian communities, the concept of functional food products and herbs were quite familiar (Valls et al., 2013). The demand for a "healthy" diet has increased in many parts of the world (Ozen et al., 2012), and with the diffusion of functional foods in the market, the line between pharma and nutrition has blurred out (Eussen et al., 2011).

Functional beverages represent an excellent delivering vehicle for bioactive compounds such as vitamins, probiotics, plant extracts, among others. This segment describes the fastest-growing segment worth $38.4 billion in 2007 (BBC Research, 2008) and is projected to reach USD 208.13 billion by 2024. This represents a huge opportunity since it is reported that an annual reduction of 20% in health-care costs is possible through the consumption of functional foods (Sun-Waterhouse, 2011). The diffusion of functional foods is due to several factors like busy lifestyles, insufficient exercise, and increased awareness of the link between diet and health, as a consequence of information by health authorities and media on nutrition (Corbo et al., 2014).

Herbal medicine is one of the principal components of traditional medicine in the most important cultures in human history. Medicinal plants are nowadays becoming popular due to the need to get rid of synthetic drugs. Thus, the biological activity of many medicinal plants has been studied scientifically, and new publications are added to scientific literature every single day. Natural products and their derivatives (including antibiotics) represent more than 50% of all drugs in clinical use in the world (Van Wyk and Wink, 2018). Pharmacies all over the world are experiencing a rapidly growing interest in healthy living, self-medication, and natural remedies (Corbo et al., 2014).

In Latin American, herbal medicine plays an important role as a curative element for the treatment of common illnesses such as respiratory and infectious diseases, which characterize the socioeconomic status of underdeveloped countries (Zolla, 1980). In Mexico, medicinal plants are regarded as part of a culture's traditional knowledge. Therefore, it is essential to research scientifically about the pharmacological properties of such plants.

Plant-based drugs are reported to have several effects on different biochemical pathways that may contribute to wellness and health. Also, it affects receptors in the body that result in different responses. Some medicinal plants traditionally used have no evidence of side effects, which makes it hard to dismiss medical claims about the safety and efficacy of the plant (Van Wyk and Wink, 2018). Conveniently, continuous research over the past decades has let us understand the scientific justification of these natural remedies. Herbal medicine represents a resource for the development of new drugs in the pharmaceutical industry and also offers a sustainable alternative to public health care in underdeveloped countries.

Plants that alter our minds have always been something amazing for humans, due to religious rituals, magic thoughts, and also because of mystic superstitions that people commonly use to do (Júnior et al., 2008). For many years the fact that plants have many great properties to treat mental diseases was ignored; nowadays, medicinal plants have a significant impact on our daily lives. Nevertheless, there is still a wrong belief that plants, coming from nature, are less toxic than synthetic drugs. Thus, plants could be a benefit in the treatment of mental diseases, such as anxiety.

Anxiety, defined as a resonant situation of emotional stimulation that contains the fear or worries feeling (Saki et al., 2014), is one of the most common mental disorders affecting people's health and it is associated with biochemical, cognitive, behavioral, and psychological changes. Data from the National Institute of Mental Health in the USA report that anxiety-related disorders would overwhelm every 7 of 10 individuals in the United States alone by 2020. Anxiety happens naturally, but some people experience it more than others; it has various mental and physical signs that can affect health and life expectancy, and the health care costs of a person with abnormal anxiety are 80% higher compared to an average person in the same age group (Dhawan et al., 2002).

Anxiolytic substances are used to treat anxiety, diminish tension, stress, and other central nervous systems (CNS) diseases. Benzodiazepine and its derivatives are the most prominent treatment for these problems, including insomnia. Azapirones and β-blockers are used in the treatment (Dhawan et al., 2002). However, the chemical character of these drugs often results in unfavorable risks and side effects such as dependence, abstinence syndrome, and amnesia (Lader and Morton, 1991). This reveals the need for potential anxiolytic drugs without these adverse reactions (Carlini, 2003), such as medicinal plants. Most herbal products are quite safe and produce less adverse reactions compared to conventional drugs used for the treatment of CNS deficiencies.

Anxiety pathophysiology still needs to be studied and established. However, it is well known that this includes neurobiology abnormalities of serotonergic, GABAergic, and glutamatergic transmission (Tyrer and Baldwin, 2006), and the involvement of these systems often reflect the efficacy of drugs in the treatment of anxiety.

There are many plants reported to have anxiolytic properties, but this chapter gives special attention to *Passiflora incarnata* and *Melissa officinalis* as principal compounds of a functional beverage with the capacity of decrease anxiety and stress levels, while maintaining mental concentration and memory functions.

Passionflower (*P. incarnata* L.) is one of 500 species that represent the genus *Passiflora* (Passifloraceae) but the only one with extensive traditional use and to achieve herbal drug status from the European Medicines Agency (EMA, 2008). It has been used for many years due to its sedative and anxiolytic properties, as well as anti-inflammatory and antioxidant capacity (Fellows and Smith, 1938). The effects on the CNS, including sedative, anticonvulsant, and anxiolytic properties, are reported in many scientific articles. However, the mechanism of action is not clear since identifying the major active ingredient in passionflower extract has been problematic. Some studies point out that *P. incarnata* has a pharmacological profile similar to benzodiazepines and, as these drugs, acts through gamma-aminobutyric acid (GABA) receptors (Jawna-Zboinska et al., 2016).

Lemon balm, *M. officinalis* L., is a medicinal plant that has long been used in different medical systems around the world. It is a potential treatment to a wide range of diseases, including cardiovascular and respiratory problems, but especially for anxiety and CNS, as well as a memory enhancer (Dastmalchi et al., 2008). Crude extracts and pure compounds isolated from this plant exhibit several pharmacological effects from which anxiolytic, antiviral, effects on mood, cognition, and memory have been shown in clinical trials. The main mechanisms proposed for the mentioned neurological effects are acetylcholinesterase inhibitory activity and stimulation of the acetylcholine and $GABA_A$ receptors (Shakeri et al., 2016).

On the side of neurodegenerative diseases, Alzheimer's disease (AD) is characterized by loss of learning ability with aging, due to the loss of basal forebrain cholinergic neurons (Mufson et al., 2008). These neurons are related to learning and memory processes (Mufson et al., 2003). The drugs that are currently available for AD are tacrine, donepezil, or galantamine, acting through inhibiting acetylcholinesterase increasing the level of acetylcholine in the brain. These drugs improve cognitive functions but cannot impede the progress of the disease (Birks, 2006). That is why new attempts are focusing on the progression of the disease from different pathways. Herbal extracts had been studying for this purpose as a solution to Alzheimer's disease.

The main goals of using medicinal plants for therapeutic purposes are to isolate bioactive compounds for direct use as drugs, to produce bioactive compounds to produce patentable entities of higher activity and lower toxicity, or to use the whole plant or part of it as a herbal remedy. In recent years, medicine has focused on the role of neuro-endocrinological abnormalities such as cytokine or steroidal alterations, abnormal circadian rhythm, and changes in GABAergic or glutamatergic transmission (Antonijevic, 2006).

1.2 P. incarnata

1.2.1 BACKGROUND OF PASSIFLORA GENUS

P. incarnata L. is a herbal medicine with a long tradition of medicinal use worldwide. The word *Passiflora* comes from the Latin word "Passio" due to Spanish "conquistadores" that described its flowers as symbols of the "passion of Christ" in 1529 (Kinghorn, 2001). The passionflower was known and used as a remedy by the Indians of South America, and also by the Brazilian healers. In 1569, the doctor Monardes discovered the passionflower in Peru. Forty years later, it was introduced in Europe as an ornamental plant, its branch and its floral structure fascinated botanists of the time. Besides *P. incarnata*, *Passiflora edulis* and *Passiflora laurifolia* fruits are also consumed (Wohlmuth et al., 2010). Archeological evidence indicates that native Americans developed human–plant mutualism at the Colombian period and *P. incarnata* was a common weed crop between anthropogenic habitats (Miroddi, Calapai, Navarra, Minciullo, and Gangemi, 2013).

P. edulis (Figure 1.1) was adequately described in 1818, from the material of purple fruits grown in England, and indicated as obtained from seeds received from Portugal, which were believed to be from Brazil (Bernacci et al., 2008). Nevertheless, reports on the species had already been written in Europe in 1648. Moreover, on the occasion of the effective publication of the genus *Passiflora*, mentions of references and illustrations that corresponded to *P. edulis* were made, but these references and illustrations were mixed with those of another species, *P. incarnata* L., to which it is related, and with which it presents great morphological similarities, although the latter is native from the United States (Bernacci et al., 2008).

FIGURE 1.1 *Passiflora edulis f. flavicarpa.* Dry tissue from leaves, flowers, and fruit of *P. edulis.* Courtesy of Vivian Aguilar

It was the Jesuits who gave the flower of passion its name in Latin. It is composed of "passio" for passion, "flos" of flower. Passionflower entered the world of phytotherapy in the second half of the past century, following its use in American homeopathy. During the First World War, passionflower was used as a soothing against what was called "fear of war."

Near to 400 species of the genus, *Passiflora* is known, and most of them have their origin in Tropical America. The majority are from the South of the United States, Central America, and South America. One hundred twenty of these species are native from Brazil. Although there is wide genetic variability represented by native biodiversity, in the commercial orchards of Brazil, a single species *P. edulis* predominates, with people's preference for fruits of the yellow-colored rind, the yellow Passion-fruit. The yellow Passion-fruit accounts for 95% of the passion fruit-cultivation area, owing to the quality of its fruits, vigor, yielding, juiciness, and the consumer's choice.

Botanists and plant lovers have long been monitoring the spread of passionflower throughout the world, although today it is found in many tropical and subtropical countries.

Passiflora genus has a broad range of different characteristics such as morphological and anatomical differences, and phylogenetic variability. *Passiflora* species are difficult to classify due to some of them vary widely in terms of morphology, and others closely resemble each other. Classifications of *Passiflora* are based on different characteristics such as morphological, ecological, and agronomic variations.

P. edulis is characterized by its production, fruit weight, its size, juiciness, and also because of the acidity of its juice; however, these variables do not represent precise quantitative traits at the taxonomic level (Devi Ramaiya et al., 2013). Moreover, existing inter- and intraspecies dissimilarity among the *Passiflora* species makes the link between morphological plasticity, genotypic diversity, and speciation understandable. Various interspecific hybrids have been produced in this genus (Devi Ramaiya et al., 2013); thus contributing to broad morphological variability. For example, *P. edulis* has been misidentified and designated by *P. incarnata*. Within the original description itself, the similarity of *P. edulis* with *P. incarnata* is indicated, but hairless seeds and vascular tissues could differentiate these, perennial woody habit, bracts evenly serrated and with glands on the tip of each tooth, the corona shorter than the calyx, and bright purple fruit in *P. edulis*, instead of villous (sic) seeds and pubescent main veins, annual herbaceous habit, bracts with some big glands, the corona longer than the calyx, and greenish-yellow fruit in *P. incarnata* (Bernacci et al., 2008). The name *P. incarnata* is associated with the species native from the United States because these species show little variation, including in the shape of its leaf blade and in the relative length of the filaments of the corona (only the size of the stalk sometimes reaches that of *P. edulis*, but it is usually longer and narrower), and also because the only material, identified by Linnaeus himself, corresponds to this species (Bernacci et al., 2008).

1.2.1 PASSIFLORA GENUS ETHNOPHARMACOLOGY

There is archeological evidence which indicates that the Native Americans commonly used *P. incarnata*, but at first, it was used as food. They did not have any other intention than using it for planting or cultivation. Years passed, and people started to discover the properties of these plants. Spanish conquerors learned about the sedative properties of it from Colombian people. Then, passionflower was introduced to the Europeans, who started to use it in their medicine so it started to become a popular traditional herbal remedy as well as a homeopathic remedy that was used to treat diseases such as anxiety, mental stress, and mild sleep disorder (Miroddi et al., 2013).

1.2.2 PHYTOCHEMICAL COMPOUNDS IN PASSIFLORA GENUS

The passion fruit is an exotic fruit popular for its beautiful flower and its aroma (Devi Ramaiya et al., 2013). Demand for the passion fruit is increasing due to its essential nutrients content, including micronutrients such as phenolic compounds and ascorbic acid content (Devi Ramaiya et al., 2013). Around America, especially in hot areas, *P. incarnata* is a well-known medicinal plant. This plant has been used for the treatment of anxiety and is recommended for insomnia (Soulimani et al., 1997).

1.2.3 ALKALOIDS

The alkaloids are one of the most important active principles that constitute the *Passiflora* (0.1%), it contains an indole alkaloid of tricyclic structure and in smaller concentration it has other alkaloids such as harman, harmalin, harmin, harmol, and harmalol (Soulimani et al., 1997).

An experiment was carried out in Cuba about the dynamics of monthly and daily accumulation of alkaloids in *P. incarnata,* and it was concluded that there are significant differences between the months for alkaloid yields. On the other hand, it is possible to observe the influence of the harvest hours and the months of the year, as well as the interaction of both factors, but without finding a trend that allows establishing practical recommendations.

1.2.4 GLYCOSIDES

There are some cyanogenic glycosides found in the *Passiflora* and they can be divided into four distinct structural types. Type I includes D-glucopyranosides of the enantiomeric cyanohydrins of 2-cyclopentenone, tetraphyllin A, and deidaclin. Type II includes compounds that could be regarded as a further structural elaboration of those belonging to type I. Thus; these glycosides contain an additional, rare sugar residue, a sulfate group, or additional oxygenation of the cyclopentene ring. Type III consists of linamarin, epilotaustralin, lotaustralin, and the corresponding gentiobioside. Finally, type IV consists of prunasin and its derivatives.

1.2.5 PHENOLIC COMPOUNDS

Phenolic compounds are secondary metabolites well known for their antioxidant potential due to the compounds that function in free radical scavenging activity, and their role in reducing the risk of cancer and cardiovascular disease (Devi Ramaiya et al., 2013).

Around 2.5% (Lutomski et al., 1981) of the plant compounds are flavonoids like vitexin, isovitexin, orientin, isoorientin, apigenin, and kaempferol (Table 1.1). The result of the total phenolic compounds has a strong correlation with antioxidant activity and other constituents such as flavonoid.

1.3 PASSIFLORA TOXICITY

Though herbal products are generally regarded as safe, the presence of cyanogenic constituents does not let to discard their toxicity. Given an adequate protein diet, human beings can detoxify cyanide satisfactorily (Jones, 1998). The US FDA (2000) have documented reports of erratic heartbeat, headache, hepatitis, diarrhea, weakness, intracerebral bleeding, and so forth due to the consumption of multi-ingredient medicinal products in which *Passiflora* species are present. However, all tested products were ingredient formulations, and the reported toxicities cannot be completely ascribed to Passionflower.

The toxicity of some species of *Passiflora* is reported due to several causes. For example, the unripe fruits of *P. adenopoda* have caused poisoning because of the presence of HCN, produced in the pericarp and aril, as well as in the primary stem, petiole, bracts, and stipules (Saenz and Nassar, 1972). On the other hand, *P. alata* is reported due to its content of etiologic agents of IgE-mediated asthma and rhinitis (Giavina-Bianchi et al., 1997), that is why this plant represents a persistent risk for the personnel who work with this species. In the case of *P. edulis*, previous reports have shown pancreatic toxicity in humans and animals (Maluf et al., 1991). By contrast, *Passiflora incarnata* is listed as a "safe herbal sedative" by the US FDA (HerbClip, 1996), any monograph mentions the toxicity of contraindication of this plant.

TABLE 1.1 Main Flavonoids Present in *Passiflora* Genus

Name	Molecular Formula	Chemical Structure
Isovitexin	$C_{21}H_{20}O_{10}$	
Apigenin	$C_5H_{10}O_5$	
Vitexin	$C_{21}H_{20}O_{10}$	
Orientin	$C_{21}H_{20}O_{11}$	
Chrysin	$C_{15}H_{10}O_4$	

Some individuals have experienced emesis after consumption of this plant, even at medicinal doses. It is used as an antispasmodic at moderate doses, but at excessive doses, it can produce spams and even paralysis in animals.

The advice is not to take *P. incarnata* with procarbazine antineoplastic drugs, to minimize CNS depression (Martin, 1978). Neuromuscular relaxing effects of *P. incarnata* have been enhanced by the use of aminoglycoside antibiotics (Dhawan et al., 2004). Phyto-constituents of several *Passiflora* species, called Vitexin, have demonstrated antithyroid activity in rats (Gaitan et al., 1995). In 2001, an American magazine US Pharmacist (2001) discouraged the use of *P. incarnata* in lactating mothers.

Thus, the presence of passiflorine and harmin alkaloids in *P. incarnata* makes it considered an environmental toxic grass (Environmental Toxicology Newsletter, 1983).

1.4 CLINICAL APPLICATIONS

Among the 500 species of the genus *Passiflora*, *P. incarnata* has shown most of the clinical applications throughout the world (Dhawan et al., 2004). It is an official plant drug in several Pharmacopoeias around the world, including the United States, France, and Germany.

P. incarnata has been evaluated for the treatment of different diseases that has been preclinically tested for hypertension and cancer, but it is most commonly used for treating anxiety and CNS disorders (Miroddi et al., 2013). Also, the plant extract has been subjected in combination with the extracts of other herbs to double-blind, placebo-controlled studies in patients with anxiety-related disorders (Schulz et al., 1997). Different kinds of preparations derived from *P. incarnata* are available, but dried extracts are the most valuable product.

Anxiolytic activity is the most researched pharmacological role of *P. incarnata*. According to Grundmann et al. (2008), passionflower and diazepam could share the same mechanism of action. Several pharmacological effects of this plant are mediated via GABA system modulation, including affinity to $GABA_A$ and $GABA_B$ receptors and effects on GABA uptake. The dysfunction of this system is implicated in neuropsychiatric conditions such as anxiety and depressive disorders (Miroddi et al., 2013). It is plausible that binding to the GABA-site of the $GABA_A$ receptor is one mechanism of action of this extract (Appel et al., 2011).

The anxiolytic capacity is evaluated by some experimental animal models such as the Elevated Plus Maze test (EPM), the staircase test, and the light/dark box choice test (Grundmann et al., 2009). These tests have evaluated hydro-alcoholic and aqueous extracts of aerial parts of *P. incarnata* on mice (Soulimani et al., 1997). Both extracts exhibited anxiolytic properties and the EPM test suggested that extracts obtained from roots and flowers have lower anxiolytic effects compared to leaf extract, recommended to maximize the anxiolytic action of this plant (Dhawan et al., 2004). Analgesic properties have been tried by tail-flick and hot-plate tests in rats, showing that *Passiflora incarnata* acts against pain (Speroni and Minghetti, 1988).

Also, variable flavonoid and GABA concentrations *Passiflora* extracts had demonstrated anticonvulsant properties against pentylenetetrazole (PTZ)-induced seizures (Elsas et al., 2010). Passionflower, according to preclinical evidence, can be used as a treatment for addictive behaviors caused by substances such as ethanol, nicotine, and cannabis (Capasso and Sorrentino, 2005).

There is a high-interest compound reported in *P. incarnata* called BZF (Dhawan et al., 2002) that has been tested for its ability to counter chronic drug use and substance dependence (Dhawan, 2003). Breivogel and Jamerson (2012) demonstrated that passionflower in aqueous extract could antagonize locomotor sensitization in rats with developed nicotine dependence and when administered morphine along with BZF chronically, the occurrence of morphine dependence was significantly attenuated (Dhawan et al., 2002). For nicotine, the administration of BZF showed benefits preventing the development of tolerance and dependence.

Also, the BZF compound exhibited a decrease in anxiety levels according to EPM in ethanol-dependent mice and showed less dependence when treated with ethanol–BZF combinations (Dhawan, 2003). Besides that, BZF can counter the effects on fertility and sperm count caused by nicotine consumption in male rats (Dhawan et al., 2002).

Antitussive and antiasthmatic actions were also demonstrated, first in a murine model of sulfur dioxide-induced cough (Dhawan et al., 2002) and dose-dependent effect against acetylcholine-induced bronchospasms in pigs.

The antitumoral activity has been exhibited due to the presence of chrysin and apigenin in passionflower, by inhibiting the growth of different kinds of cancer cells: breast carcinoma, human thyroid cancer, and human prostate tumors (Ingale and Hivrale, 2010).

For this chapter, anti-anxiety is the main clinical application to take into account since this kind of disorders are the most prevalent psychiatric disorders (Kessler et al., 2010) and induce further costs such as prescription medications, lower productivity, and suicide (Movafegh et al., 2008).

Benzodiazepines (BZDs) are one class of drugs used to treat anxiety disorders. However, this treatment often results in many complications such as sedation, tolerance, and dependence. *Passiflora* has been evaluated as a treatment for generalized anxiety disorder and its effectiveness in presurgery anxiety (Akhondzadeh et al., 2001).

1.4.1 ANTIOXIDANT EFFECTS

There are many molecules that are known due to their antiradical activity that inhibits oxidative damage, which provides them an antioxidant activity that prevents aging and neurodegenerative diseases (Zeraik et al., 2011). Some of these molecules are phenolic compounds, such as flavonoids. Antioxidant activity of these compounds depends on the structure, particularly on the position of the hydroxyl groups and the aromatic rings (Balasundram et al., 2006). Flavonoids are the most abundant components in many fruits and vegetables (Kim et al., 2015). Flavonoids are commonly found in the *Passiflora* genus (Zeraik et al., 2011). The presence of these molecules in the pulp of *P. edulis* is quite significant in comparison with other beverages that are famous because of their content of orange juice or sugarcane juice.

Most flavonoids are found as glycones. Depending on the position of the linkage between sugars and flavonoids, the biological function is different (Kim et al., 2015). Flavonoids normally accumulate in plants as *O*-glycosylated derivatives. Nevertheless, some species predominantly synthesize flavone-*C*-glycosides, which are stable to hydrolysis and are biologically active (Brazier-Hicks et al., 2009). One of the most famous flavonoid glycosides present in *P. edulis* is *C*-glycosyl flavonoids such as orientin, isoorientin, vitexin, and isovitexin (Xiao et al., 2016). There is limited information about the antioxidant activity of flavonoid *C*-glycosides, although studies proved that orientin reduces the amount of malondialdehyde, a highly reactive compound used as a marker for oxidative stress (Xiao et al., 2016). Different studies proved that in most cases *C*-glycosylflavonoids showed higher antioxidant potential than *O*-glycosylflavonoids.

1.4.2 ANTIDEPRESSION AND ANTI-ANXIETY

Several studies have been carried out to know more about *Passiflora* properties. In 2014, researchers from different universities of Iran conducted a study in which they investigated the antidepressant effect of the extract of *Passiflora* in forced swim test and tail suspension test in male mice to know more about the antidepression and anti-anxiety effects of this herb. The results of the study revealed that *Passiflora* has several compounds in its building, but among them, the β-carboline alkaloids such as harmaline, harmine, and harmalol have antidepression properties (Jafarpoor et al., 2014). In the tests carried out, passionflower decreased immobility time but increased swimming time, so it was concluded that this plant gives considerable responses to the previously mentioned effects.

1.4.3 INSOMNIA

Despite tremendous advancements in the field of medical therapeutics, the studies conducted at the National Institute of Mental Health, Bethesda, reveal the alarming increase in CNS-related disorders, the main two being anxiety and insomnia (Lopez and Murray, 1998).

Passiflora relieves nervous irritability resulting from a prolonged illness, menstrual disturbances, and mental overwork, which helps in reducing the problem. On the other hand, it calms the restlessness of typhoid fever. The sleep produced by the passionflower is rich in rapid eye movement sleep, which causes freshness on awakening. *Passiflora* leaf infusion is also a popular traditional European remedy as well as a homeopathic remedy for insomnia of infants and older people due to the vasculitis it causes to patients suffering it (Dhawan et al., 2004).

1.4.4 ANTI-INFLAMMATORY

Reactive oxygen species are implicated in many human diseases such as neurodegenerative disorders, cancer, cardiovascular diseases, atherosclerosis, cataracts, DNA damage, and inflammation. Many molecules, such as phenolic compounds, are known to prevent inflammatory conditions (Zeraik et al., 2011). As we mentioned before, the main glycones found in *Passiflora edulis* are the *C*-glycosides. The anti-inflammatory potential of these flavonoids has been poorly investigated. Nevertheless, Odontuya et al. (2005) proved that one of the flavonoids present in the species called isoorientin possesses the ability to inhibit thromboxane B2, which is a potent hypertensive agent and facilitates a homeostatic balance in the circulatory system (Xiao et al., 2016). On the other hand, studies in the literature pointing to the anti-inflammatory effects of flavonoids, for example, Ueda et al. (2004) demonstrated that substances of this class inhibited TNF in different inflammatory and allergy models.

The inflammatory response is well known to be characterized by increased vascular permeability, increased blood flow, as well as infiltration of neutrophils and macrophages. *Passiflora* genus has been studied concerning the anti-inflammatory activity. The leaf extracts of *P. edulis* is not only used in popular medicine in Brazil mainly as a sedative but also in the treatment of skin inflammation. (Zucolotto et al., 2009). Based on the results of Montanher et al. (2007), the aqueous extract from *P. edulis* was effective in inhibiting inflammatory parameters such as the reduction of leukocyte influx and decrease of neutrophils. Also, results shown in Zucolotto et al. (2009) study demonstrated that crude extract from leaves of *P. edulis* exhibited important anti-inflammatory activity in the inflammation model induced by carrageenan, a potent chemical that functions in stimulating the release of inflammatory and proinflammatory mediators. Typical signs of inflammation include edema, hyperalgesia, and erythema, which develop immediately following the treatment of carrageenan (Amdekar et al., 2012).

1.5 *M. officinalis*

1.5.1 BACKGROUND

Over the years, many common names have been associated with *M. officinalis,* including balm, English balm, garden balm, balm mint, common balm, melissa, sweet balm, heart's delight, and honey plant (Meyers, 2007). Melissa in Latin is the abbreviation of its original Greek name melissophyllum, which means mélitta "bee" and meli "honey," about the fact that these insects are attracted because of the nectar of it. In modern Latin, "officinalis" means "used in pharmacy." In ancient times, *Melissa* was grown to feed the bees. Its medicinal use dates back more than 2000 years. In the 10th century, Arab doctors used it to strengthen the heart and fight melancholy. From its Moorish introduction into Spain in the seventh century, its cultivation and use spread throughout Europe by the middle ages (Sevik and Guney, 2013). Lemon balm has been associated with the feminine, the moon, and water, and was considered a sacred herb in the temple of the Ancient Roman goddess Diana (William et al., 1990). According to magical folklore, the herb has powers of healing, success, and love, and can be made into healing incense and sachets or carried to help the bearer find love. Culpeper associated lemon balm with the planet Jupiter and the astrological constellation Cancer. Some herbalists believe lemon balm is also beneficial for the astrological signs Sagittarius and Aquarius (Wittig et al., 1995).

M. officinalis L. is a medicinal plant that has long been used in different ethnomedical systems, especially in the European Traditional Medicine and Traditional Iranian Medicine, for the treatment of several diseases. It is also widely used as a vegetable and to add flavor to dishes (Shakeri et al., 2016). This plant is cultivated in some parts of Iran because it is used in folk medicine for its digestive. Also, it is used as an analgesic and sedative. Due to its carminative properties, it has been used in various aromatic preparations together with other species; as a digestive stimulant, it presents antibacterial and sedative activity due to the essential oil, as well as antiviral activity of the polyphenolic fraction and indicated in the treatment of gastrointestinal spasms.

It is a perennial herb that can reach 80 cm to 1 m high and has opposite leaves. It is bushy, upright, serrated, margin, and hairy. Leaves are soft, hairy, and from 2 to 8 cm long or either heart-shaped. The leaf surface is coarse and deeply veined, and the leaf edge is scalloped or toothed (Ardalani et al., 2014). When crushing the herbaceous parts, these provided a characteristic lemon smell. Its flowers, which consist of small clusters of 4–12, are pedunculated, pentamers, grouped in verticillaster. Its ovate seeds are black or dark brown and have a hairy root system with many lateral roots that better adapt to different environmental conditions. Lemon balm is susceptible to frost, especially when there is no snow cover. New shoots start to appear early in spring, so they can also suffer from spring frosts. It blooms in summer (Seidler-Łożykowska et al., 2013).

This plant was introduced from Europe into North America as a culinary and medicinal herb. It is still cultivated in gardens, and sometimes it is grown commercially. Escaped plants are typically found in such habitats as thickets, fence rows, abandoned homesites, vacant lots, areas along roadsides, banks of ponds, floodplain areas along drainage canals, and waste areas. Fields with a history of disturbance are preferred. Lemon balm is especially likely to naturalize in urban and suburban areas, as this is where most cultivated plants occur.

1.5.2 PHYTOCHEMICAL COMPOUNDS IN M. officinalis

M. officinalis is widely used mainly for therapeutic purposes that give the plant significant value. Various combinations and different medicinal properties of its extract, oil, and leaves were studied many years ago. The different attributes uses are due to its different chemical compounds. Miraj et al. (2017) report that the main constituents of lemon balm were hydroxycinnamic acid derivatives such as rosmarinic acid, caffeic acid; tannins, flavonoids including apigenin and luteolin; monoterpene glycosides, sesquiterpenes, and volatile oils like citronella, citral a and b, geraniol, among others.

Thirty-three compounds were identified as 89.30% of the total oil leaf composition (Miraj et al., 2017); Table 1.2 summarizes the main compound category and the organ where it is present.

TABLE 1.2 Some Constituents of *Melissa officinalis* L.

Category	Name	Organ
Phenolic compounds	• Isogeraniol	
	• Verbenol	
	• Carane	
	• I R-α-Pinene	
	• Geraniol	
Nitrogen compounds	• Menthol	Leaf and oil
	• Eugenylglycoside	
	• Cubenole	
	• Cinerone	
	• β-Caryophyllene	
Terpenes	• Citronellal	
	• Citral	
	• Geranial	Leaf and oil
	• Neral	
	• Linalool	
	• Nerol	
	• 2-Pinen-4-one	

Adapted from Miraj, Rafieian, and Kiani (2017).

It is reported that the vegetal product of *M. officinalis* contains 0.64% flavonoids expressed in rutoside and 8.962% phenylpropane derivatives expressed as caffeic acid equivalents and six polyphenolic compounds were identified: caftaric acid, caffeic, *p*-cumaric acid, ferulic acid, luteolin, and apigenin (Hanganu et al., 2008).

In the case of lemon balm extract, the main compounds are hydroxycinnamic acid derivatives and flavonoids with rosmarinic acid, and caffeic acid (Miraj et al., 2017).

Apigenin has been revealed to have cytostatic and cytotoxic effects on various cancer cells, preventing hypertension, cardiac hypertrophy, autoimmune myocarditis, inhibits asthma, as well as other beneficial biological activities (Zhou et al., 2017). Also, it affects CNS by enhancing the blood–brain barrier and reducing 4/NF-kB mediated inflammation. Besides, in the corticosterone- and LPS-induced depressive animal model, this compound may attenuate the abnormal behavior that may be related to the inhibition of inflammatory cytokine production (Li et al., 2015).

Rosmarinic acid is antiviral and antioxidant; an enriched extract containing it is used against herpes viruses (Blaschek et al., 2013), while the essential oil is used as a digestive aid in pharmaceutical preparations. This chapter gives special attention to anxiolytic properties, found on alcohol extracts from *M. officinalis*.

Compared with the essential leaf oil, the infusion essential oil is markedly enriched in aldehydes (90%) with a citral level near 74%, but the lower quantity of citronellal (16%), which in essential oil represents 40% (Carnat et al., 1998).

1.5.3 TOXIC ASPECTS OF *M. officinalis* EXTRACTS

Traditional practices and use of medicinal plants give us an insight into the recommended doses. However, with plant-based products, it is often not clear what the optimal doses are to balance efficacy and safety (Ulbricht et al., 2005), since the composition and concentration vary in every single plant and the preparations also vary from batch to batch. This, added to the fact that active compounds of the plant are not always well identified, makes the standardization a complicated process and the clinical trials not so comparable.

Metal dispersion in food crops have been evaluated, suggesting that the grown soil may be contaminated by lead, as a consequence of environmental pollution and therefore the plant could contain trace amounts of this element in its physiology (Ulbricht et al., 2005).

Despite the insufficient available data on chronic toxicity, there are some contraindications reported in the literature. Due to the potential for thyroid hormone inhibition, it has to be used cautiously in patients with thyroid problems, and it is reported that lemon balm increases intraocular pressure, so the cautious use in patients with glaucoma is thoroughly recommended (Ulbricht et al., 2005).

Experimental studies on laboratory animals have shown that the hydroalcoholic extract of *M. officinalis* has a calming effect on the CNS (Namjoo et al., 2013). However, lemon balm interacts with some drugs such as alcohol, barbiturates, and selective serotonin reuptake inhibitors. In the case of alcohol, a combination with lemon balm increases its sedative effects (Kennedy et al., 2002). The hypnotic effects of barbiturates tend to increase with lemon balm according to animal assays (Soulimani et al., 1991). Also, based on preclinical studies, when combined, lemon balm with other sedatives results in additive effects (Gyllenhaal et al., 2000). Another interaction studied was the interference of extracts of lemon balm in thyroid hormone replacement therapy; this study suggested that some constituents of *M. officinalis* may block the binding of the stimulating thyroid hormone to its receptors by acting on the hormone and the receptor (Santini et al., 2003). Finally, talking about interactions with nicotine and scopolamine, lemon balm may displace drugs bound to the nicotinic and muscarinic receptors as clinical trials reported previously (Kennedy et al., 2003).

According to an experimental toxicity study of lemon balm hydroalcoholic extract on liver and kidney tissues of female BALB/C mice that evaluated enzyme changes and pathology to show side effects of the consumption of this plant. In this research, the activity of alanine aminotransferase enzyme (ALT) in group receiving 1350 mg/kg lemon balm showed a significant decrease compared to control group and injection of 450 and 1350 mg/kg doses of lemon balm hydro-alcoholic extract decreased significantly alkaline phosphatase (ALP) compared to control group (Namjoo et al., 2013). It seems that the decrease in the activity of ALT and ALP enzymes in treatment groups is due to the presence of apigenin compound that prevents the activity of the ATP precursors and aminotransferases. It can be concluded that the consumption of lemon balm extract dose-dependently causes toxicity in liver tissue.

1.5.4 CLINICAL ASPECTS OF *M. officinalis*

In many countries, herbal medicine is part of the culture. Many plants possess essential natural compounds that fulfill the benefits to cure diseases. *M. officinalis* historically has been described as a multibenefit plant. Some of its traditional uses are tranquilizing, hypotensive, antibacterial/antiviral/antifungal, and antioxidant activity. Also, it is effective in the treatment of depression, hysteria, headaches, psychosis, and insomnia.

High levels of free radicals in living systems can oxidize biomolecules leading to tissue damage. Oxidative stress is an imbalance of free radicals and antioxidants inside the body. It is implicated in degenerative processes, diseases, and even mutagenesis due to the inducing of protein and DNA damage, and instability of mitochondrial and nuclear genomes.

An antioxidant is defined as a compound that delays or inhibits oxidation. The deficiency of these molecules leads to oxidative stress. According to many studies, phenolic compounds, especially flavonoids, are characterized by being the major phytochemicals as antioxidants. The action as free-radical scavengers of phenolic compounds comes from their reducing properties as an electron-donating agent. *M. officinalis* contains flavonoids, phenolic compounds, glycosides, and tannins. Total phenolic compounds represent 11.8% of dry leaves, and flavonoids compound 0.5%.

The increasing consumption of medicinal plant products by different kinds of patients has led to worry about their safety and efficacy (Bishop and Lewith, 2010). Sometimes, the classification of these products is as medicinal products and occasionally as food supplements, depending on the regulation of the related (Miroddi et al., 2013).

1.6 GENETIC, PHYTOCHEMICAL, AND BIOLOGICAL CHARACTERIZATION OF *P. incarnata* AND *M. officinalis*

1.6.1 METHODOLOGY

1.6.1.1 DNA EXTRACTION

DNA extraction from *Passiflora incarnata* was performed by an improved CTAB method without liquid nitrogen, adapted from Doyle and Doyle (1987). This protocol was used to obtain pure DNA from leaves samples that frequently contain multiple contaminants, such as polyphenols. In every trial, 20 mg of the dry powdered plant was used.

The quality of the DNA samples was determined using 1% agarose gel electrophoresis. The DNA quantity and purity were further confirmed in a Nanodrop® spectrophotometer. The spectral quality of DNA was measured by the A_{260}/A_{280} ratio.

For the DNA extraction of *M. officinalis*, CTAB extraction protocol from Saghai-Maroof et al. (1984) with minor modifications on concentration and amount of the components was performed. *M. officinalis*, also known as lemon balm, contains high levels of polyphenols and polysaccharides, which represents a challenge for the isolation of high-quality DNA for many DNA extraction protocols. Polysaccharides make DNA viscous, glue-like, and non-amplifiable in the PCR by inhibiting *Taq* DNA polymerase activity and interfering in the accuracy and activity of restriction enzymes (Porebski et al., 1997). The protocol selected uses PVP and ammonium acetate to overcome the high levels of polyphenols and polysaccharides, as well as β-mercaptoethanol 0.2% and RNase.

1.6.1.2 BARCODING AMPLIFICATION

Gradient PCR was performed to determine the specific annealing temperature for the barcoding regions *mat*K, *rbc*L, and the intergenic region of *psb*A-*trn*H in *P. incarnata* and *M. officinalis*. The PCR reaction was performed in triplicate for reproducibility using an T100™ thermocycler (BioRad) with expected ranges from 100 to 800 bp.

1.6.2 QUANTIFICATION OF PHENOLIC COMPOUNDS

Total phenolic compounds were determined by using the Folin–Ciocalteu method adapted from Oomah et al. (2005). On a 96-well microtitration flat-bottom plate, 20 µL of the extracts and 150 µL of water HPLC were added, then it was oxidized with 50 µL of 2 N Folin–Ciocalteu reagent. After 5 min the reaction was neutralized with 50 µL sodium carbonate solution (20 %). After incubation for 2 h at 25 °C, the absorbance was reading on a spectrophotometer X-Mark BioRad at 760 nm. Quantification of TPC was performed using a calibration curve prepared with the gallic acid standard in a range of 50–300 µg/L. The analysis was performed in triplicate, and the results were expressed as mg of gallic acid/g of extract.

1.6.3 FLAVONOIDS QUANTIFICATION

The spectrophotometric assay for the quantitative determination of flavonoid content described by Oomah et al. (2005) was performed, mixing 50 µL of the methanolic extract, 180 µL of methanol, and 20 µL of a solution of 10 g/L 2-aminoethyl-diphenyl borate in each well of 96-well microtitration flat-bottom plate. After 15 min, the absorbance was monitored at 404 nm with a spectrophotometer X-Mark BioRad. The absorption of the extract was compared with a rutin standard at concentrations ranging from 0 to 50 μg/mL. Flavonoid content was expressed as milligrams of rutin equivalent per gram of dry sample. Samples were prepared in triplicate.

1.6.4 FREE-RADICAL DPPH SCAVENGING CAPACITY

2,2-Diphenyl-1-picrylhydrazyl (DPPH) is a free radical used to assess antioxidant activity. The reduction of DPPH by an antioxidant or by a radical species results in a loss of absorbance at 515 nm (Xie and Schaich, 2014). Antioxidant activity by the DPPH method was adapted for use with microplates as described by Oomah et al. (2005). First, a solution of DPPH (150 μM) was prepared in 80% (v/v) aqueous methanol, then 0.02 μL of the extract was added to a 96-well flat-bottom plate containing 0.2 μL of DPPH solution. The plate was covered and left in the dark at room temperature. After 90 min, the plate was read in a spectrophotometer X-Mark BioRad at 520 nm. Data are expressed as a percentage of DPPH discoloration. Samples were prepared in triplicate.

1.6.5 GAS CHROMATOGRAPHY–MASS SPECTROMETRY (GC–MS) ANALYSIS

For the identification of different compounds present in *P. incarnata* y *M. officinalis* extracts, a methanolic extract was used. The methanolic extract was analyzed by GC–MS by an Agilent Technologies 7890A GC System coupled to an Agilent Technologies 5975C Triple Quadrupole mass spectrometer equipped with a capillary silica DB-17ht column (30 m × 250 μm × 0.15 μm). The splitless mode was conducted. The carrier gas was helium (1.29 mL/min). The temperature program was set as follows: 50 °C hold for 1 min, raised at 30 °C/min to 80 °C, then increased to 110 °C at 10 °C/min. and finally reached 270 °C at 6 °C/min for 40 min.

1.7 RESULTS AND DISCUSSION

1.7.1 DNA EXTRACTION AND BARCODING AMPLIFICATION

The modifications in both *M. officinalis* and *P. incarnata* protocols gave better results in the extractions. In the case of *M. officinalis*, optimization of

CTAB DNA extraction buffer such as β-mercaptoethanol helped to remove tannins and polyphenols, and denature proteins. Also, additional RNAse helped to remove contaminations of RNA, which appreciated the first gels as smeared bands.

Figure 1.2 shows resulting rbcL region amplicons for *M. officinalis* around 1000 bp in which it is observed that while the temperature rises, the band is more precise, which means that it is more specific. Ideal temperatures for *M. officinalis* are from 51 to 53 °C. In the case of *P. incarnata*, no amplicons were obtained. According to the National Center for Biotechnology Information, U.S., (2019) to rbcL, the expected amplicons should be around 561 bp. In the case of *P. incarnata*, the size of the amplicons should be about 346 bp (National Center for Biotechnology Information, U.S., 2019).

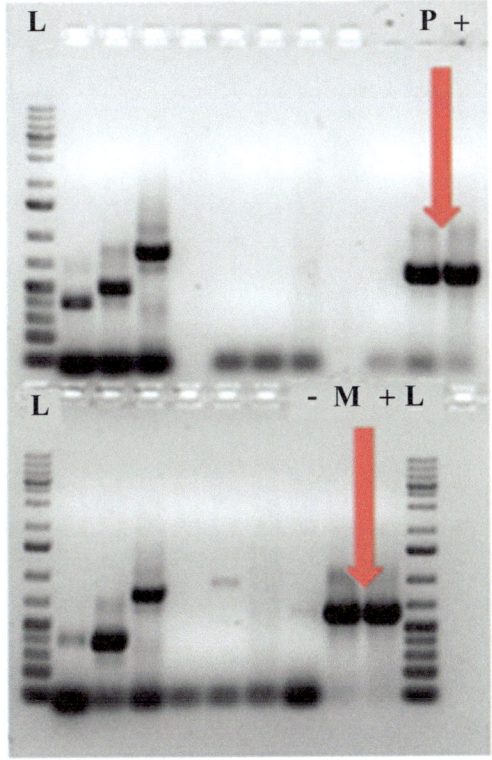

FIGURE 1.2 Amplicons of *Passiflora incarnata* and *Melissa officinalis* for rbcL region loaded in 2% agarose TAE 1X gel run at 80 V for 45 min at 59 °C for *P. incarnata* and for 56 °C *M. officinalis*. L: GeneRuler 1 kb DNA Ladder, P: Sample 1 from *P. incarnata*, M: Sample 1 from *M. officinalis*, +: GMO positive control from GMO Investigator Kit, and −: Negative control (DNAse free water)

As shown in Figure 1.3, amplicons of *P. incarnata* and *M. officinalis* for psbA-trnH intergenic regions were obtained. The amplicon of *Passiflora* (500 bp) is bigger than its negative control (400 bp), which means that the sample of *Passiflora* could have been contaminated by mishandling.

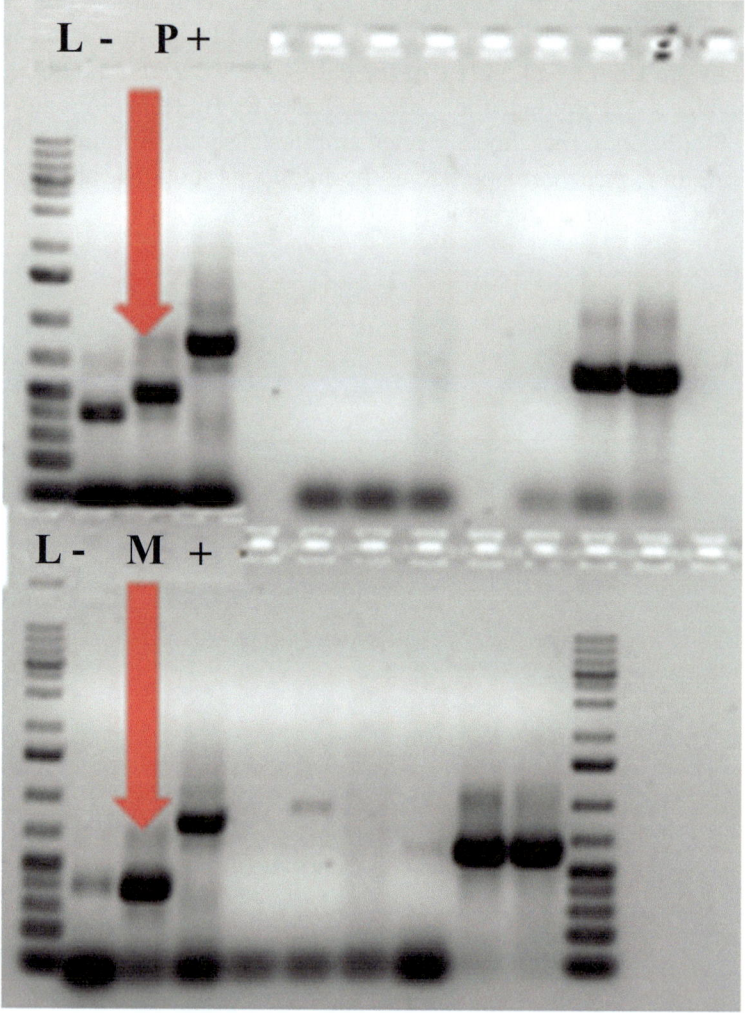

FIGURE 1.3 Amplicons of *Passiflora incarnata* and *Melissa officinalis* for psbA-trnH intergenic region loaded in 2% agarose TAE 1X gel run at 80 V for 45 min at 53 °C for *P. incarnata* and 51 °C for *M. officinalis* was. L: GeneRuler 1 kb DNA Ladder, P: Sample 1 from *P. incarnata*, M: Sample 1 from *M. officinalis*, +: GMO positive control from GMO Investigator Kit, and −: Negative control (DNAse free water).

On the other hand, the amplicons are bigger than expected according to the literature. Given that, it could not be assumed that the sample is *P. incarnata*. The expected size of *M. officinalis* should be around 410 bp (National Center for Biotechnology Information, U.S., 2019). The resulting band appears to be around 400 bp. Since the band is around the expected size, the sample seems to be *M. officinalis*.

1.7.2 QUANTIFICATION OF PHENOLIC COMPOUNDS AND FLAVONOIDS

The total phenolic content resulted in higher in *M. officinalis* extract. This content was in the range reported for other authors (18.17–64.17 mg gallic acid equivalent/g; Mabroukiet al., 2018; Petkova et al., 2017).

Passiflora phenolic content was 21.17 mg gallic acid equivalent/g dry weight (Table 1.3). This result was five times lower than obtained by Guimarães et al. (2019). Phenolic compounds are widely distributed in plants and have garnered attention due to their antimutagenic, antitumor, and antioxidant properties, which contribute to human health. The variation in TPC values may be attributed to the plant origins of the extractable compounds and the efficacy of the solvents used to recover the polyphenols from the plant materials (Devi Ramaiya et al., 2013).

On the other hand, flavonoid content in *Passiflora* was sevenfold than that of *Melissa* (Table 1.3). Hossain et al. (2009) reported a flavonoid content of 3.8 mg/g of dry lemon balm. In this study, the flavonoid content in *Melissa* was three times higher and similar to that (26 ± 3 mg QE/g dw) reported by Franco et al. (2018).

Guimarães et al. (2019) reported a flavonoid content of 150 µg/g dw in Passiflora leaves, almost twice the reported in this study. As previously mentioned, differences in the content of phenolics and flavonoids could be attributed to the origin of samples and management.

1.7.3 DPPH SCAVENGING CAPACITY

The capacity to scavenge DPPH free radical was similar for *Passiflora* and *Melissa* extracts despite the different phenolic content (Table 1.3), suggesting that the composition of phenolics could be different between both extracts.

Antioxidant compounds can act on free radicals by scavenging mechanisms, which may be attributed to its hydrogen and/or electron donating; thus, they might prevent reactive radical species from reaching biomolecules such as lipoproteins, polyunsaturated fatty acids, DNA, amino acids, proteins, and food systems (Franco et al., 2018).

TABLE 1.3 Content of Phenolics and Flavonoids in *Melissa officinalis* and *Passiflora incarnata*

Plant	Phenolics[a]	Flavonoids[b]	DPPH[c]
M. officinalis	40.03	13.67	85.08
P. incarnata	21.17	79.86	88.36

[a] *Total phenolic compounds concentration expressed as mg of gallic acid/g of sample.*
[b] *Flavonoid content expressed as µg of rutin equivalents/g of sample.* [c] *Scavenging capacity expressed as percent discoloration.*

1.7.4 CHEMICAL COMPOSITION OF MELISSA AND PASSIFLORA EXTRACTS

Methanol fraction of *Melissa* was analyzed by GC–MS to determine its chemical composition. The GC–MS analysis showed that both extracts contained a variety of compounds (Figure 1.4 and Table 1.4). By comparing the MS spectral data with those of standards and MS library, 63 compounds were identified for *M. officinalis* and 69 compounds for *P. incarnata* (Figure 1.5 and Table 1.5).

There were many compounds identified using the GC–MS analysis. For *Melissa*, the highest peak was 9,12,15-octadecatrien-1-ol,(Z,Z,Z), which is commonly known as linolenyl alcohol, it is a long-chain fatty primary alcohol that is octadecanol and has a role as an antibacterial agent damaging bacterial cell membranes (Rizzo, 2014).

2-Methoxy-4-vinylphenol was the second-highest peak, and is a naturally occurring phenolic compound used as a flavoring agent. This compound belongs to methoxyphenols, organic compounds that contain a methoxy group attached to the benzene ring of a phenol. It has been found in buckwheat, apple, peanut, and curry (Jeong et al., 2011). It is reported to induce the arrest of abnormal cell cycle progression by blocking the hyper-phosphorylation of retinoblastoma protein in chemical carcinogen-treated NIH 3T3 cells (Jeong and Jeong, 2010). Also, it exerts potent anti-inflammatory effects by inhibiting LPS-induced NO, PGE2 iNOS, and COX-2 in RAW264.7 cells (Jeong et al., 2011).

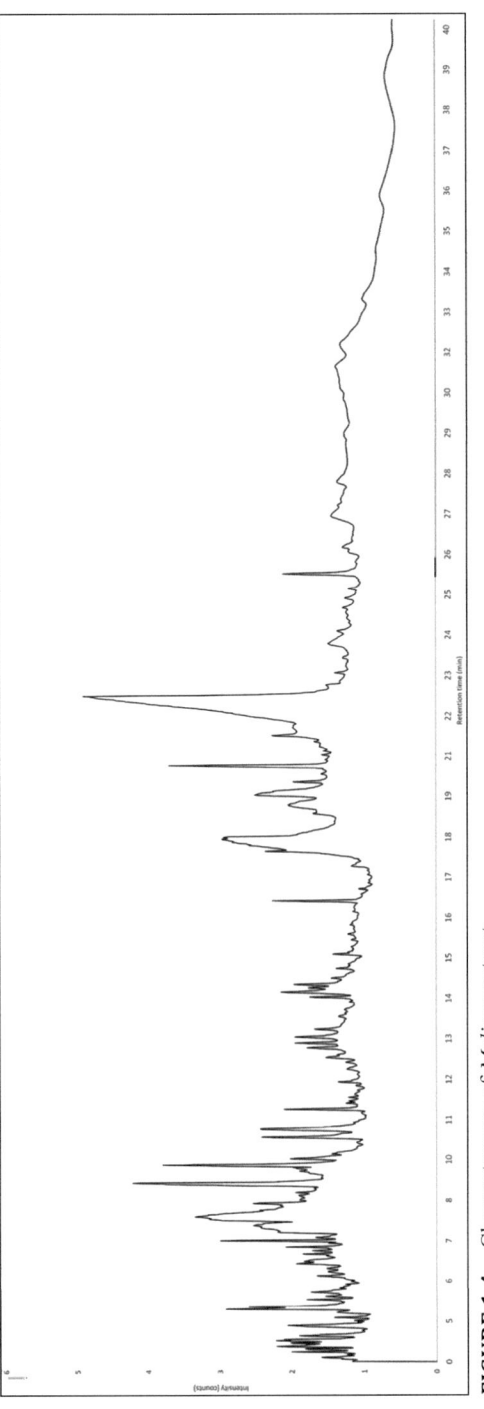

FIGURE 1.4 Chromatogram of *Melissa* extract.

TABLE 1.4 Compounds Identified by GC-MS in *Melissa* Extracts

Peak Number[a]	Compound Name	Molecular Formula	Molecular Weight (g/mol)	Match (%)
1	cis-linaloloxide	$C_{10}H_{18}O_2$	170	52
2	Phenol	C_6H_5OH	94	95
3	3-Piperdino-1,2.propanediol	$C_8H_{17}NO_2$	159	64
4	Benzyl alcohol	C_7H_8O	108	97
5	Benzofuran, 4,5,6,7.tetrahydro-3,6.dimethyl	$C_{10}H_{14}O$	150	81
6	Phenol,2-methoxy	$C_7H_8O^2$	124	93
7	4N-Mehylcytosine	$C_5H_7N_3O$	125	49
8	2(R), 3(S)-1,2,3,4-Butanetretol	$C_4H_{10}O_4$	122	59
9	1,2-Ethanediol,1-(2-phenyl-1,3,2-dioxaborolan-4-yl)-,[S-(R*,R*)]	$C_{10}H_{13}BO_4$	208	53
10	Benzene, 1-methoxy-4-(1-propenyl)-	$C_{10}H_{12}O$	148	89
11	Glycerin	$C_3H_8O_3$	92	50
12	1,2,3,4-Butanetetrol, [S-(R*,R*)]	$C_4H_{10}O_4$	122	53
13	Xylitol	$C_5H_{12}O_5$	152	50
14	3-Cyclohexen-1-one, 2-isopropyl-5-methyl-	$C_{10}H_{16}O$	152	96
15	Benzofuran, 2,3-dihydro-	C_8H_8O	120	58
16	Phenol,4-(2-propenyl)-	$C_9H_{10}O$	134	96
17	Resorcinol	$C_6H_6O_2$	110	53
18	1,2-Benzenediol	$C_6H_6O_2$	110	64
19	2-Methoxy-4-vinylphenol	$C_9H_{10}O_2$	150	93
20	Neric acid	$C_{10}H_{16}O_2$	168	50
21	Phenol, 2-methoxy-3-(2-propenyl)-	$C_{10}H_{12}O_2$	164	98
22	Benzonitrile, 4-methyl	C_8H_7N	117	81
23	p-Methoxyheptanophenone	$C_{14}H_{20}O_2$	220	95
24	Phenol, 2,6-dimethoxy-	$C_8H_{10}O_3$	154	93
25	Benzonitrile, 2-fluoro-6-methoxy-	C_8H_6FNO	151	53
26	Allyl (2-methylphenyl) sulfide	$C_{10}H_{12}S$	164	50
27	4-Oxatricyclo[4,3,1,1(3,8)undecane	$C_{10}H_{16}O$	152	50
28	3-Butenoic acid, 4-phenyl-, methyl ester	$C_{11}H_{12}O_2$	176	50
29	6-Hydroxymethyl,5-methyl-cicyclo(3,1,0)hexan-2-one	$C_7H_9NO_2$	140	53

TABLE 1.4 *(Continued)*

Peak Number[a]	Compound Name	Molecular Formula	Molecular Weight (g/mol)	Match (%)
30	Pyrrolidin-2-one, 1,5-dihydro-3,4-dimethyl,5-(1-piperidyl)	$C_{11}H_{18}N_2O$	194	50
31	1,12-Tridecadiene	$C_{13}H_{24}$	180	53
32	4-Methyl-2,5-dimethoxybenzaldehyde	$C_{10}H_{12}O_3$	180	64
33	Megastigmatrienone	$C_{13}H_{18}O$	190	84
34	Nonanoic acid	$C_9H_{18}O_2$	158	53
35	Hexadecanoic acid, methyl ester	$C_{17}H_{34}O_2$	270	98
36	1,1-Dimethyl-1-silacyclo-2,4-hexadiene	$C_7H_{12}Si$	124	53
37	*n*-Hexadecanoid acid	$C_{16}H_{32}O_2$	256	99
38	Phytol	$C_{20}H_{40}O$	296	70
39	9,12,15-Octadecatrienoic acid, methyl ester	$C_{19}H_{32}O_2$	292	99
40	Estra-1,3,5(10)-trien-17beta-ol	$C_{18}H_{24}O$	256	56
41	d-Glycero-d-galacto-heptose	$C_7H_{14}O_7$	210	55
42	Methyl 6,O-(1,mehylpropyl)-beta-d-galactopyranoside	$C_{11}H_{22}O_6$	250	51
43	9,12,15-Octadecatrien-1-ol,(Z,Z,Z)	$C_{18}H_{32}O$	264	91
44	9,12,15-Octadecatrienoic acid, ethyl ester, (Z,Z,Z)	$C_{20}H_{34}O_2$	306	95
45	1,4,8-Dodecatriene, (E,E,E)-	$C_{12}H_{18}$	162	74
46	(7r,8s)-*cis*-anti-*cis*-7,8-Epoxytricyclo[7,3,0,0(2,6)dodecane	$C_{12}H_{18}O$	178	97
47	Tricyclo(8,6,0,0(2,9)hexadeca-3,15,diene, trans-2,9-anti-9,10-trans-1,10-	$C_{16}H_{24}$	216	53
48	Cis-8-methyl-exo.tricyclo[5,2,1,0(2,6)]decane	$C_{11}H_{18}$	150	91
49	(7r,8s)-*cis*-anti-*cis*-7,8-Epoxytricyclo[7,3,0,0(2,6)dodecane	$C_{12}H_{18}O$	178	56
50	(7r,8s)-*cis*-anti-*cis*-7,8-Epoxytricyclo[7,3,0,0(2,6)dodecane	$C_{12}H_{18}O$	178	90
51	2(1H)-Naphthalenone, octahydro-4a,7,7-trimethyl-,trans-	$C_{13}H_{22}O$	194	91
52	1H-Pyrido[4,3-b]indole,2,3,4,4a,5,9b-hexahydro-2,8-dimethyl-5-(4-nitrobenzoyl)-,(4ar,9bs)-rel-	$C_{20}H_{21}N_3O_3$	351	55

TABLE 1.4 (Continued)

Peak Number[a]	Compound Name	Molecular Formula	Molecular Weight (g/mol)	Match (%)
53	1,5,Benzothiazepin-4-(5H)-one, 3-(acetyloxy)-5-[2-(dimethylamino)ethyl]-2,3-dihydro-2-(4-methoxyphenyl)-, (2S-cis)-	$C_{22}H_{26}N_2O_4S$	414	64
54	Cyclooctene,3.ethenyl-	$C_{10}H_{16}$	136	92
55	Cyclopropaneoctanal, 2-octyl-	$C_{19}H_{36}O$	280	64
56	9-Octadecenal, (Z)-	$C_{18}H_{34}O$	266	50
57	2-Dodecen-1-y(-)succinic anhydride	$C_{16}H_{26}O_3$	266	90
58	2,6,10-Dodecatrien-1-ol, 3,7,11-trimethyl-	$C_{15}H_{26}O$	222	70
59	(7r,8s)-cis-anti-cis-7,8-Epoxytricyclo[7,3,0,0(2,6)dodecane	$C_{12}H_{18}O$	178	95
60	Nonanoic acid, 9-(3-hexenyldenecyclopropylidene), 2-hydroxy-1-(hydroxymethyl)ethyl ester, (Z,Z,Z)-	$C_{21}H_{42}O_4$	352	95
61	2-Allylphenol	$C_9H_{10}O$	134	76
62	4H-1-Benzopyran-4-one, 5-hydroxy-7-methoxy-2-(4-methoxyphenyl)-	$C_{17}H_{14}O_5$	298	95
63	Nitrophenide	$C_{12}H_8N_2O_4S_2$	308	62

Another compound was benzyl alcohol that also has some applications in medicine for its local anesthetic properties, antispasmodic action, and bacterial and fungal inhibitory properties against *S. aureus*, *E. coli*, *B. subtilis*, among others (Carter et al., 1958).

Resorcinol has been used for many years in human medicine as an antiseptic and in keratolytic topical medications (Cassano et al., 1999). It is used in pharmaceutical preparations for topical treatments such as anti-acne or dermatitis. Anti-thyroidal activity has been observed in vitro following resorcinol exposure, due to the inhibition of thyroid peroxidase enzymes, resulting in decreased thyroid hormone synthesis. In animals and humans, it is reported that resorcinol affects the CNS by increasing locomotor activity, but more studies are needed to conclude this property of resorcinol.

2(*R*),3(*S*)-1,2,3,4-Butanetretol, commonly known as erythritol, is also present in a retention time of 6.1 min. This compound is used as a bulk sweetener in the food industry and it is safe for diabetics since it does not affect glucose and insulin levels (den Hartog et al., 2010). It is also an appealing HO radical scavenger with good bioavailability.

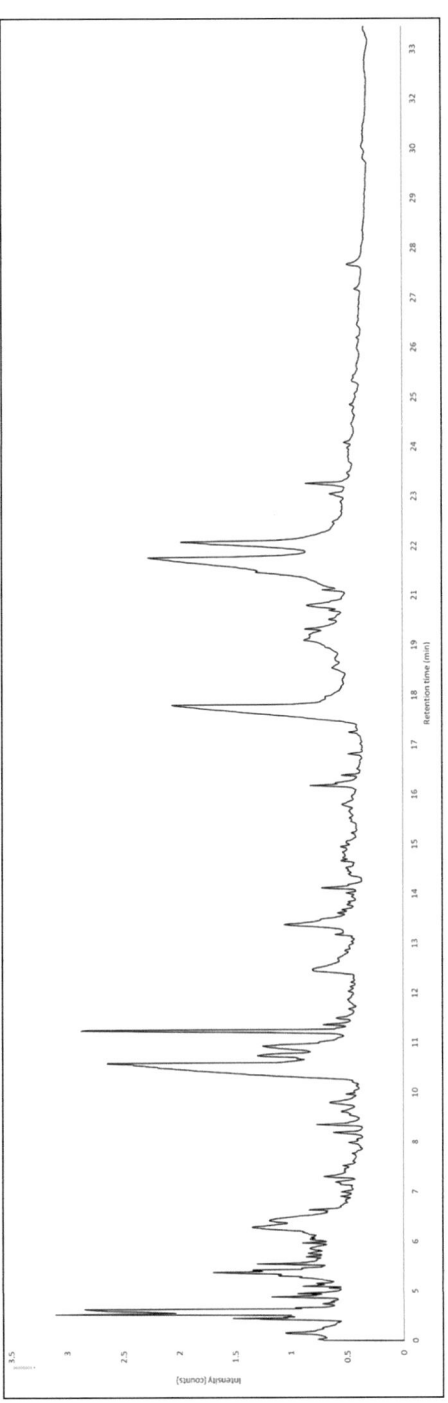

FIGURE 1.5 Chromatogram of *Passiflora* extract.

Table 1.5 shows the components found in *P. incarnata* identified in the GC–MS analysis.

TABLE 1.5 Compounds Identified in Passiflora Extract by GC-MS

Peak Number	Compound name	Molecular Formula	Molecular Weight (g/mol)	Match (%)
1	1-Pentanol,4-amino-	$C_5H_{13}NO$	103	50
2	Butyrolactone	$C_4H_6O_2$	86	64
3	Phenol	C_6H_5OH	94	87
4	2-Pyrrolidinemethanol, 1-methyl-	$C_6H_{13}NO$	115	59
5	2,4-Dimethyl-2-oxazoline-4-methanol	$C_6H_{11}NO_2$	129	78
6	Cycloserine	$C_3H_6N_2O_2$	102	37
7	Benzyl alcohol	C_7H_8O	108	60
8	Phenol,2-methoxy	$C_7H_8O_2$	124	91
9	Topotecan	$C_{23}H_{23}N_3O_5$	421	50
10	Glycerin	$C_3H_8O_3$	92	53
11	Xylitol	$C_5H_{12}O_5$	152	53
12	4H-Pyran-4-one, 2,3-dihydro-3,5-dihydroxy-6-methyl	$C_6H_8O_4$	144	93
13	d-Threitol	$C_4H_{10}O_4$	122	64
14	2-Methoxy-4-vinylphenol	$C_9H_{10}O_2$	150	93
15	Tetrahydropyran 12-tetradecyn-1-ol-ether	$C_5H_{10}O$	294	80
16	Octane,2,4,6-trimethyl-	$C_{11}H_{24}$	156	50
17	1,4-Cyclohexadiene	C_6H_8	80	50
18	1,3,5-Hexatriene	C_6H_8	80	50
19	Bicyclo(3,2,1)oct-6-en-3-one	$C_8H_{10}O$	122	70
20	3-Amino-2,6-dimethoxypyridine	$C_7H_{10}N_2O_2$	154	55
21	1,3-Cyclohexadiene	C_6H_8	80	50
22	9,12,15-Octadecatrienoic acid, methyl ester, (Z,Z,Z)	$C_{19}H_{32}O_2$	292	50
23	2,6-Dimethyl-3-aminobenzoquinone	$C_8H_9NO_2$	151	58
24	Benzenamine, N, N- diethyl	$C_8H_{11}N$	149	70
25	Phenol, 2-methoxy-4-(1-propenyl)-	$C_{20}H_{24}O_4$	164	95
26	Naphthalen-4a,8a-imine, octahydro-	$C_{10}H_{17}N$	151	58
27	Pidolic acid	$C_5H_7NO_3$	129	59
28	Perdeuterobenzene	C_6H_6	84	64

TABLE 1.5 *(Continued)*

Peak Number	Compound name	Molecular Formula	Molecular Weight (g/mol)	Match (%)
29	Pyridine-D5-	C_5D_5N	84	49
30	(1-(4-Amino-furazan-3-yl)-5-methyl-1H-(1,2,3)triazol-4-y)-piperidin-1-yl-methanone	$C_{11}H_{15}N_7O_2$	277	50
31	2(3H,4H)-Cyclopenta(b)furanone, 3a,6a-dihydro-	$C_7H_8O_2$	124	55
32	Cyclohexene, 1-methoxy	$C_7H_{12}O$	112	52
33	Borazine, 2,4,5-triethyl	$C_6H_{18}B_3N_3$	165	62
34	2,3,5,6-Tetrafluoroanisole	$C_7H_4F_4O$	180	70
35	N-(But-1-enyl)-pyrrolidin-2-one	$C_8H_{13}NO$	139	50
36	Tetradecanoic acid	$C_{14}H_{28}O_2$	228	84
37	Quinoline, 8-ethyl	$C_{11}H_{11}N$	157	64
38	Stevioside	$C_{38}H_{60}O_{18}$	804	38
39	Propylamine, N-[9-borabicyclo[3,3,1]non-9-yl]-	$C_{11}H_{22}BN$	179	76
40	Phenol, 2,6-dimethoxy-4-(2-propenyl)-	$C_{11}H_{14}O_3$	194	70
41	Hexadecanoic acid, methyl ester	$C_{17}H_{34}O_2$	270	96
42	3-Amino-3-(4-ethoxy-phenyl)-propionic acid	$C_{11}H_{15}NO_3$	209	86
43	Cyclohexanecarboxylic acid, 2-hydroxy-,ethyl ester	$C_9H_{16}O_3$	172	53
44	n-Hexadecanoic acid	$C_{16}H_{32}O_2$	256	99
45	13-Tetradecenal	$C_{14}H_{26}O$	210	55
46	Cholestan-3-ol	$C_{27}H_{48}O$	388	56
47	Phytol	$C_{20}H_{40}O$	296	46
48	2-Piperidinone, N-[4,bromo-n-butyl]-	$C_9H_{16}BrNO$	233	59
49	1-(2,4-Dichloro-phenyl)-4,4-dimethyl-2-oxa-apiro[5,5]undecane-3,5-dione	$C_{16}H_{19}BrO_4$	354	53
50	3,7,7-Trimethyl-8-(2-methyl-propenyl)-bicyclo[4,2,0]oct-2-ene	$C_{15}H_{24}$	204	50
51	9,12,15-Octadecatrienoic acid, methyl ester	$C_{19}H_{32}O_2$	292	55
52	Acetamide, N-(4-ethoxy-3-hydroxyphenyl)-	$C_{10}H_{13}NO_3$	195	59
53	Linoleic acid ethyl ester	$C_{20}H_{34}O_2$	308	83
54	2-(4-Hydroxybutyl)ciclohexanol	$C_{10}H_{20}O_2$	172	83

TABLE 1.5 *(Continued)*

Peak Number	Compound name	Molecular Formula	Molecular Weight (g/mol)	Match (%)
55	9,17-Octadecadienal, (Z)-	$C_{18}H_{32}O$	264	99
56	cis-7-Dodecen-1-yl acetate	$C_{14}H_{26}O_2$	226	98
57	9,12,15-Octadecatrien-ol, (Z,Z,Z)-	$C_{18}H_{32}O$	278	91
58	9,12-Octadecadienoic acid (Z,Z)-	$C_{18}H_{32}O_2$	280	95
59	cis-8-methyl-exo-tricyclo[5.2.1.0(2,6)] decane	$C_{11}H_{18}$	150	95
60	Imidazo[4,5-e][1,4]diazepin-8(1H)- one, 4,5,6,7-tetrahydro-4, 7-dimethyl- 5-thioxo-	$C_8H_{10}N_4OS$	210	70
61	5,10-Diethoxy-2,3,7,8-tetrahydro-1H,6H-dipyrrolo[1,2-a,1'2'-d]pyrazine	$C_{14}H_{22}N_2O_2$	250	83
62	Tridecanedial	$C_{13}H_{24}O_2$	212	89
63	2(1H)-Naphthalenone, octahydro-4a-methyl-7-1(1-methylethyl)	$C_{14}H_{24}O$	208	93
64	2,6-Nonadienoic acid,7-ethyl-9	$C_{18}H_{30}O_3$	294	64
65	(7-Methylbenzo(b)thien-3-yl)methanol	$C_{10}H_{10}OS$	178	64
66	Cis-8-methyl-exo-tricyclo[5.2.1.0(2,6)] decane	$C_{11}H_{18}$	150	90
67	2,4,7,14-Tetramethyl-4-vinyl-tricyclo[5.4.3.0(1,8)]tetradecan-6-ol	$C_{20}H_{34}O$	290	64
68	Phenol,2,2'-methylenebis [6-(1,1-dimethylethyl)]-4-ethyl	$C_{25}H_{36}O_2$	368	99
69	Stigmasta-5,22-dien-ol, acetate, (3.beta.)-	$C_{31}H_{50}O_2$	454	86

The highest peaks were 2-pyrrolidinemethanol, 1-methyl, 2,4-dimethyl-2-oxazoline-4-methanol, and 1,3,5-hexatriene, 2,6-dimethyl-3-amino-benzoquinone.

2,6-Dimethyl-3-amino-benzoquinone was the second-highest peak, its presence is reported in different plants such as coriander *(Coriandrum sativum*; Ravindran, 2017). Another compound found in *P. incarnata* extract was xylitol, reported to improve bone biochemical properties in rats after oral administration (Mattila et al., 1999).

There was also benzyl alcohol that has been used as a local anesthetic for brief superficial skin procedures (Wilson and Martin, 1999), being significantly less painful on injection than lidocaine with epinephrine. As in *M. officinalis*, 2-methoxy-4-vinylphenol was also present, in which anti-inflammatory properties are remarkable. Also, this compound had shown antimicrobial activity (Mastelic et al., 2010).

Cycloserine represents a broad-spectrum antibiotic with bacteriostatic and bactericidal properties, depending on the site and susceptibility of the organism, blocking the formation of peptidoglycans on bacteria walls. Cycloserine is a broad-spectrum antibiotic with only moderate anti-TB activity. It inhibits cell wall synthesis. The MIC of cycloserine in the Bactec 460-TB system is 25–75 µg/mL (Donald and McIlleron, 2009).

1.8 CONCLUSION

Lemon balm has shown to possess high phenolic content and antioxidant properties. Antioxidant activity of lemon balm has been shown, as evidenced by the reduction of DPPH free radical.

Passiflora also has high phenolic and flavonoid contents, which could be responsible for the antioxidant activity showed by the extracts of this plant.

The analysis of the chemical composition of both plants showed several compounds with different biological properties, suggesting that those compounds could contribute to health benefits.

Although more studies are needed, it could be expected that *Melissa* and *Passiflora* extracts can provide not only antioxidant capacity but also anxiolytic properties to nutraceutical formulations.

KEYWORDS

- **anxiolytic**
- **phenolics**
- **antioxidants**
- **Passiflora**
- **Melissa**

REFERENCES

Akhondzadeh, S., Kashani, L., Mobaseri, M., Hosseini, S. H., Nikzad, S., and Khani, M. (2001). Passionflower in the treatment of opiates withdrawal: a double-blind randomized controlled trial. *Journal of Clinical Pharmacy and Therapeutics, 26*(5), 369–373. doi: 10.1046/j.1365-2710.2001.00366.x

Amdekar, S., Roy, P., Singh, V., Kumar, A., Singh, R., and Sharma, P. (2012). Anti-inflammatory activity of lactobacillus on carrageenan-induced paw edema in male Wistar rats. *International Journal of Inflammation, 2012*. doi:10.1155/2012/752015

Antonijevic, I. A. (2006). Depressive disorders—is it time to endorse different pathophysiologies? *Psychoneuroendocrinology, 31*(1), 1–15. doi:10.1016/j.psyneuen.2005.04.004

Appel, K., Rose, T., Fiebich, B., Kammler, T., Hoffmann, C., and Weiss, G. (2011). Modulation of the gamma-aminobutyric acid (GABA) system by *Passiflora incarnata* L. *Phytotherapy Research, 25*(6), 838–843. doi:10.1002/ptr.3352

Ardalani, H., Eradatmand Asli, D., and Moradi, P. (2014). Physiological and morphological response of lemon balm (*Melissa officinalis* L.) to prime application of salicylic hydroxamic acid. *Electronic Journal of Biology, 10*(3), 93–97.

Balasundram, N., Sundram, K., and Samman, S. (2006). Phenolic compounds in plants and agri-industrial by-products: antioxidant activity, occurrence, and potential uses. *Food Chemistry, 99*(1), 191–203. doi:10.1016/j.foodchem.2005.07.042

BBCResearch. (2008). Global Nutraceutical Market Worth $176.7 Billion in 2013. Retrieved from https://www.bccresearch.com/pressroom/fod/global-nutraceutical-market-worth-$176.7-billion-2013

Bernacci, L. C., Soares-Scott, M. D., Junqueira, N. T. V., Passos, I. R. d. S., and Meletti, L. M. M. (2008). Passiflora edulis Sims: the correct taxonomic way to cite the yellow passion fruit (and of others colors). *Revista Brasileira de Fruticultura, 30*, 566–576. doi:10.1590/S0100-29452008000200053

Birks, J. (2006). Cholinesterase inhibitors for Alzheimer's disease. *Cochrane Database System Review* (1), CD005593. doi:10.1002/14651858.CD005593

Bishop, F. L., and Lewith, G. T. (2010). Who uses CAM? A narrative review of demographic characteristics and health factors associated with CAM use. *Evidence-Based Complementary and Alternative Medicine, 7*(1), 11–28. doi:10.1093/ecam/nen023

Blaschek, W., Hänsel, R., Keller, K., Reichling, J., Rimpler, H., and Schneider, G. (2013). *Hagers Handbuch der Pharmazeutischen Praxis: Folgeband 2: Drogen A K*: Springer-Verlag.

Brazier-Hicks, M., Evans, K. M., Gershater, M. C., Puschmann, H., Steel, P. G., and Edwards, R. (2009). The C-glycosylation of flavonoids in cereals. *Journal of Biological Chemistry, 284*(27), 17926–17934. doi:10.1074/jbc.M109.009258

Breivogel, C., and Jamerson, B. (2012). Passion flower extract antagonizes the expression of nicotine locomotor sensitization in rats. *Pharmaceutical Biology, 50*(10), 1310–1316. doi: 10.3109/13880209.2012.674535

Capasso, A., and Sorrentino, L. (2005). Pharmacological studies on the sedative and hypnotic effect of *Kava kava* and *Passiflora* extracts combination. *Phytomedicine, 12*(1–2), 39–45. doi:10.1016/j.phymed.2004.03.006

Carlini, E. A. (2003). Plants and the central nervous system. *Pharmacology, Biochemistry and Behavior, 75*(3), 501–512. doi:10.1016/s0091-3057(03)00112-6

Carnat, A. P., Carnat, A., Fraisse, D., Lamaison, J. L., Heitz, A., Wylde, R., and Teulade, J. C. (1998). Violarvensin, a new flavone di-C-glycoside from *Viola arvensis*. *Journal of Natural Products, 61*(2), 272–274. doi:10.1021/np9701485

Carter, D. V., Charlton, P. T., Fenton, A. H., Housley, J. R., and Lessel, B. (1958). The preparation and the antibacterial and antifungal properties of some substituted benzyl alcohols. *Journal of Pharmacy and Pharmacology, 10*(Supp), 149–157T; discussion 157-149T. doi: 10.1111/j.2042-7158.1958.tb10394.x

Cassano, N., Alessandrini, G., Mastrolonardo, M., and Vena, G. A. (1999). Peeling agents: toxicological and allergological aspects. *European Academy of Dermatology and Venereology, 13*(1), 14–23. doi:10.1111/j.1468-3083.1999.tb00838.x

Corbo, M. R., Bevilacqua, A., Petruzzi, L., Casanova, F. P., and Sinigaglia, M. (2014). Functional beverages: the emerging side of functional foods: commercial trends, research, and health implications. *Comprehensive Reviews in Food Science and Food Safety, 13*(6), 1192–1206. doi:10.1111/1541-4337.12109

Dastmalchi, K., Dorman, H. D., Oinonen, P. P., Darwis, Y., Laakso, I., and Hiltunen, R. (2008). Chemical composition and in vitro antioxidative activity of a lemon balm (*Melissa officinalis* L.) extract. *LWT–Food Science and Technology, 41*(3), 391–400. doi:10.1016/j.lwt.2007.03.007

den Hartog, G. J., Boots, A. W., Adam-Perrot, A., Brouns, F., Verkooijen, I. W., Weseler, A. R., ... Bast, A. (2010). Erythritol is a sweet antioxidant. *Nutrition, 26*(4), 449–458. doi:10.1016/j.nut.2009.05.004

Devi Ramaiya, S., Bujang, J. S., Zakaria, M. H., King, W. S., and Shaffiq Sahrir, M. A. (2013). Sugars, ascorbic acid, total phenolic content and total antioxidant activity in passion fruit (*Passiflora*) cultivars. *Journal of the Science of Food and Agriculture, 93*(5), 1198–1205. doi:10.1002/jsfa.5876

Dhawan, K. (2003). Drug/substance reversal effects of a novel tri-substituted benzoflavone moiety (BZF) isolated from *Passiflora incarnata* Linn.—a brief perspective. *Addiction Biology,* (4), 379–386. doi:10.1080/13556210310001646385

Dhawan, K., Dhawan, S., and Sharma, A. (2004). *Passiflora*: a review update. *Journal of Ethnopharmacology, 94*(1), 1–23. doi:10.1016/j.jep.2004.02.023

Dhawan, K., Kumar, S., and Sharma, A. (2002). Comparative anxiolytic activity profile of various preparations of *Passiflora incarnata* linneaus: a comment on medicinal plants' standardization. *Journal of Alternative and Complementary Medicine, 8*(3), 283–291. doi:10.1089/10755530260127970

Donald, P. R., and McIlleron, H. (2009). Chapter 59 - Antituberculosis drugs. In H. S. Schaaf, A. I. Zumla, J. M. Grange, M. C. Raviglione, W. W. Yew, J. R. Starke, M. Pai, and P. R. Donald (Eds.), *Tuberculosis* (pp. 608–617). Edinburgh: W.B. Saunders.

Doyle, J. J., and Doyle, J. L. (1987). *A Rapid DNA Isolation Procedure for Small Quantities of Fresh Leaf Tissue*. Retrieved from https://worldveg.tind.io/record/33886/

Elsas, S. M., Rossi, D. J., Raber, J., White, G., Seeley, C. A., Gregory, W. L., ... Soumyanath, A. (2010). *Passiflora incarnata* L. (Passionflower) extracts elicit GABA currents in hippocampal neurons in vitro, and show anxiogenic and anticonvulsant effects in vivo, varying with extraction method. *Phytomedicine, 17*(12), 940–949. doi:10.1016/j.phymed.2010.03.002

Eussen, S. R., Verhagen, H., Klungel, O. H., Garssen, J., van Loveren, H., van Kranen, H. J., and Rompelberg, C. J. (2011). Functional foods and dietary supplements: products at the interface between pharma and nutrition. *European Journal of Pharmacology, 668*(Suppl 1), S2–S9. doi:10.1016/j.ejphar.2011.07.008

Fellows, E. J., and Smith, C. S. (1938). The chemistry of *Passiflora incarnata*. *Journal of the American Pharmaceutical Association, 27*(7), 565–573. doi:10.1002/jps.3080270706

Franco, J. M., Pugine, S. M. P., Scatoline, A. M., and de Melo, M. P. (2018). Antioxidant capacity of *Melissa Officinalis* L. on biological systems. *Eclética Química Journal, 43*(3), 19–29.

Gaitan, E., Cooksey, R. C., Legan, J., and Lindsay, R. H. (1995). Antithyroid effects in vivo and in vitro of vitexin: a *C*-glucosylflavone in millet. *The Journal of Clinical Endocrinology and Metabolism, 80*(4), 1144–1147. doi:10.1210/jcem.80.4.7714083

Giavina-Bianchi, P. F., Jr., Castro, F. F., Machado, M. L., and Duarte, A. J. (1997). Occupational respiratory allergic disease induced by *Passiflora alata* and *Rhamnus purshiana*. *Annals of Allergy, Asthma and Immunology, 79*(5), 449–454. doi:10.1016/S1081-1206(10)63042-6

Grundmann, O., Wahling, C., Staiger, C., and Butterweck, V. (2009). Anxiolytic effects of a passion flower (*Passiflora incarnata* L.) extract in the elevated plus maze in mice. *Die Pharmazie, 64*(1), 63–64. doi:10.1691/ph.2009.8753

Grundmann, O., Wang, J., McGregor, G. P., and Butterweck, V. (2008). Anxiolytic activity of a phytochemically characterized *Passiflora incarnata* extract is mediated via the GABAergic system. *Planta Medica, 74*(15), 1769–1773. doi:10.1055/s-0028-1088322

Guimarães, S. F., Lima, I. M., and Modolo, L. V. (2019). Phenolic content and antioxidant activity of parts of *Passiflora edulis* as a function of plant developmental stage. *Acta Botanica Brasilica*. doi:10.1590/0102-33062019abb0148

Gyllenhaal, C., Merritt, S. L., Peterson, S. D., Block, K. I., and Gochenour, T. (2000). Efficacy and safety of herbal stimulants and sedatives in sleep disorders. *Sleep Med Rev, 4*(3), 229–251. doi:10.1053/smrv.1999.0093

Hanganu, D., Vlase, L., Filip, L., Sand, C., Mirel, S., and Indrei, L. L. (2008). The study of some polyphenolic compounds from *Melissa officinalis* L. (Lamiaceae). *Revista medico-chirurgicala a Societatii de Medici si Naturalisti din Iasi, 112*(2), 525–529.

Hossain, M. A., Kim, S., Kim, K. H., Lee, S.-J., and Lee, H. (2009). Flavonoid compounds are enriched in lemon balm (*Melissa officinalis*) leaves by a high level of sucrose and confer increased antioxidant activity. *HortScience, 44*(7), 1907–1913. doi:10.21273/HORTSCI.44.7.1907

Ingale, A., and Hivrale, A. (2010). Pharmacological studies of *Passiflora* sp. and their bioactive compounds. *African Journal of Plant Science, 4*(10), 417–426.

Jafarpoor, N., Abbasi-Maleki, S., Asadi-Samani, M., and Khayatnouri, M. H. (2014). Evaluation of antidepressant-like effect of hydroalcoholic extract of *Passiflora incarnata* in animal models of depression in male mice. *Journal of HerbMed Pharmacology, 3*(1), 41–45. Retrieved from http://herbmedpharmacol.com/Abstract/JHP_20150527123854

Jawna-Zboinska, K., Blecharz-Klin, K., Joniec-Maciejak, I., Wawer, A., Pyrzanowska, J., Piechal, A., . . . Widy-Tyszkiewicz, E. (2016). *Passiflora incarnata* L. Improves spatial memory, reduces stress, and affects neurotransmission in rats. *Phytotherapy Research, 30*(5), 781–789. doi:10.1002/ptr.5578

Jeong, J. B., Hong, S. C., Jeong, H. J., and Koo, J. S. (2011). Anti-inflammatory effect of 2-methoxy-4-vinylphenol via the suppression of NF-kappaB and MAPK activation, and acetylation of histone H3. *Archives of Pharmacal Research, 34*(12), 2109–2116. doi:10.1007/s12272-011-1214-9

Jeong, J. B., and Jeong, H. J. (2010). 2-Methoxy-4-vinylphenol can induce cell cycle arrest by blocking the hyper-phosphorylation of retinoblastoma protein in benzo[a]pyrene-treated NIH3T3 cells. *Biochemical and Biophysical Research Communications, 400*(4), 752–757. doi:10.1016/j.bbrc.2010.08.142

Jones, D. A. (1998). Why are so many food plants cyanogenic? *Phytochemistry, 47*(2), 155–162. doi:10.1016/s0031-9422(97)00425-1

Júnior, L. J. Q., Almeida, J., Lima, J. T., Nunes, X. P., Siqueira, J. S., de Oliveira, L. E. G., . . . Barbosa-Filho, J. M. (2008). Plants with anticonvulsant properties: a review. *Revista Brasileira de Farmacognosia, 18*(Suppl 0).

Kennedy, D. O., Scholey, A. B., Tildesley, N. T., Perry, E. K., and Wesnes, K. A. (2002). Modulation of mood and cognitive performance following acute administration of *Melissa officinalis* (lemon balm). *Pharmacology, Biochemistry and Behavior, 72*(4), 953–964. doi:10.1016/s0091-3057(02)00777-3

Kennedy, D. O., Wake, G., Savelev, S., Tildesley, N. T., Perry, E. K., Wesnes, K. A., and Scholey, A. B. (2003). Modulation of mood and cognitive performance following acute administration of single doses of *Melissa officinalis* (Lemon balm) with human CNS nicotinic and muscarinic receptor-binding properties. *Neuropsychopharmacology, 28*(10), 1871–1881. doi:10.1038/sj.npp.1300230

Kessler, R. C., Ruscio, A. M., Shear, K., and Wittchen, H. U. (2010). Epidemiology of anxiety disorders. In S. M and S. T (Eds.), *Behavioral Neurobiology of Anxiety and Its Treatment. Current Topics in Behavioral Neurosciences. Vol. 2* (2011/02/11 ed., Vol. 2, pp. 21–35). Berlin, Heidelberg: Springer.

Kim, B. G., Yang, S. M., Kim, S. Y., Cha, M. N., and Ahn, J.-H. (2015). Biosynthesis and production of glycosylated flavonoids in *Escherichia coli*: current state and perspectives. *Applied Microbiology and Biotechnology, 99*(7), 2979–2988.

Kinghorn, A. D. (2001). Pharmacognosy in the 21st century. *Journal of Pharmacy and Pharmacology, 53*(2), 135–148. doi:10.1211/0022357011775334

Lader, M., and Morton, S. (1991). Benzodiazepine problems. *British Journal of Addiction, 86*(7), 823–828. doi:10.1111/j.1360-0443.1991.tb01831.x

Li, R., Zhao, D., Qu, R., Fu, Q., and Ma, S. (2015). The effects of apigenin on lipopolysaccharide-induced depressive-like behavior in mice. *Neuroscience Letters, 594*, 17–22. doi:10.1016/j.neulet.2015.03.040

Lopez, A. D., and Murray, C. C. J. L. (1998). The global burden of disease, 1990–2020. *Nature Medicine, 4*(11), 1241–1243. doi:10.1038/3218

Lutomski, J., Segiet, E., Szpunar, K., and Grisse, K. (1981). The importance of the passionflower in medicine. *Pharmazie in Unserer Zeit, 10*(2), 45–49.

Mabrouki, H., Duarte, C. M. M., and Akretche, D. E. (2018). Estimation of total phenolic contents and in vitro antioxidant and antimicrobial activities of various solvent extracts of *Melissa officinalis* L. *Arabian Journal for Science and Engineering, 43*(7), 3349–3357. doi:10.1007/s13369-017-3000-6

Maluf, E., Barros, H., Frochtengarten, M. L., Benti, R., and Leite, J. (1991). Assessment of the hypnotic/sedative effects and toxicity of *Passiflora edulis* aqueous extract in rodents and humans. *Phytotherapy Research, 5*(6), 262–266. doi:10.1002/ptr.2650050607

Martin, E. W. (1978). *Drug Interactions Index: 1978/79*: Lippincott Williams and Wilkins.

Mastelic, J., Blazevic, I., and Kosalec, I. (2010). Chemical composition and antimicrobial activity of volatiles from *Degenia velebitica*, a European stenoendemic plant of the *Brassicaceae* family. *Chemistry and Biodiversity, 7*(11), 2755–2765. doi:10.1002/cbdv.201000053

Mattila, P., Knuuttila, M., Kovanen, V., and Svanberg, M. (1999). Improved bone biomechanical properties in rats after oral xylitol administration. *Calcified Tissue International, 64*(4), 340–344. doi:10.1007/s002239900629

Meyers, M. (2007). Lemon balm: an herb society of America guide. *The Herb Society of America*, Kirtland, OH, USA.

Miraj, S., Rafieian, K., and Kiani, S. (2017). *Melissa officinalis* L: A review study with an antioxidant prospective. *Evidence-Based Complementary and Alternative Medicine, 22*(3), 385–394. doi:10.1177/2156587216663433

Miroddi, M., Calapai, G., Navarra, M., Minciullo, P. L., and Gangemi, S. (2013). *Passiflora incarnata* L.: ethnopharmacology, clinical application, safety and evaluation of clinical trials. *Journal of Ethnopharmacology, 150*(3), 791–804. doi:10.1016/j.jep.2013.09.047

Montanher, A. B., Zucolotto, S. M., Schenkel, E. P., and Fröde, T. S. (2007). Evidence of anti-inflammatory effects of *Passiflora edulis* in an inflammation model. *Journal of Ethnopharmacology, 109*(2), 281–288. doi:10.1016/j.jep.2006.07.031

Movafegh, A., Alizadeh, R., Hajimohamadi, F., Esfehani, F., and Nejatfar, M. (2008). Preoperative oral *Passiflora incarnata* reduces anxiety in ambulatory surgery patients: a double-blind, placebo-controlled study. *Anesthesia and Analgesia, 106*(6), 1728–1732. doi:10.1213/ane.0b013e318172c3f9

Mufson, E. J., Counts, S. E., Perez, S. E., and Ginsberg, S. D. (2008). Cholinergic system during the progression of Alzheimer's disease: therapeutic implications. *Expert Review of Neurotherapeutics, 8*(11), 1703–1718. doi:10.1586/14737175.8.11.1703

Mufson, E. J., Ginsberg, S. D., Ikonomovic, M. D., and DeKosky, S. T. (2003). Human cholinergic basal forebrain: chemoanatomy and neurologic dysfunction. *Journal of Chemical Neuroanatomy, 26*(4), 233–242. doi:10.1016/s0891-0618(03)00068-1

Namjoo, A., MirVakili, M., and Faghani, M. (2013). Biochemical, liver and renal toxicities of *Melissa officinals* hydroalcoholic extract on balb/C mice. *Journal of HerbMed Pharmacology, 2*(2).

Odontuya, G., Hoult, J., and Houghton, P. (2005). Structure-activity relationship for antiinflammatory effect of luteolin and its derived glycosides. *Phytotherapy Research: An International Journal Devoted to Pharmacological and Toxicological Evaluation of Natural Product Derivatives, 19*(9), 782–786.

Oomah, B. D., Cardador-Martínez, A., and Loarca-Piña, G. (2005). Phenolics and antioxidative activities in common beans (*Phaseolus vulgaris* L). *Journal of the Science of Food and Agriculture, 85*(6), 935–942. doi:10.1021/jf020296n

Ozen, A. E., Pons, A., and Tur, J. A. (2012). Worldwide consumption of functional foods: a systematic review. *Nutrition Reviews, 70*(8), 472–481. doi:10.1111/j.1753-4887.2012.00492.x

Petkova, N., Ivanov, I., Mihaylova, D., and Krastanov, A. (2017). Phenolic acids content and antioxidant capacity of commercially available *Melissa officinalis* L. teas in Bulgaria. *Bulgarian Chemical Communications, 49*, 69–74.

Porebski, S., Bailey, L. G., and Baum, B. R. (1997). Modification of a CTAB DNA extraction protocol for plants containing high polysaccharide and polyphenol components. *Plant Molecular Biology Reporter, 15*(1), 8–15. doi:10.1007/bf02772108

Ramaiya, S. D., Bujang, J. S., and Zakaria, M. H. (2014). Assessment of total phenolic, antioxidant, and antibacterial activities of *Passiflora* species. *The Scientific World Journal, 2014*, 10. doi:10.1155/2014/167309

Ramaiya, S. D., Bujang, J. S., and Zakaria, M. H. (2014). Genetic diversity in *Passiflora* species assessed by morphological and its sequence analysis. *The ScientificWorld Journal, 2014*, 598313. doi:10.1155/2014/598313

Ravindran, P. (2017). *The Encyclopedia of Herbs and Spices*: CABI.

Rizzo, W. B. (2014). Fatty aldehyde and fatty alcohol metabolism: review and importance for epidermal structure and function. *Biochimica et Biophysica Acta (BBA), 1841*(3), 377–389.

Saenz, J. A., and Nassar, M. (1972). Toxic effect of the fruit of Passiflora adenopoda D. C. on humans: phytochemical determination. *Revista de Biología Tropical, 20*(1), 137–140. Retrieved from https://www.ncbi.nlm.nih.gov/pubmed/4347223

Saghai-Maroof, M. A., Soliman, K. M., Jorgensen, R. A., and Allard, R. (1984). Ribosomal DNA spacer-length polymorphisms in barley: Mendelian inheritance, chromosomal location, and population dynamics. *Proceedings of the National Academy of Sciences, 81*(24), 8014–8018.

Saki, K., Bahmani, M., and Rafieian-Kopaei, M. (2014). The effect of most important medicinal plants on two importnt psychiatric disorders (anxiety and depression)—a review. *Asian Pacific Journal of Tropical Medicine, 7S1*, S34–42. doi:10.1016/S1995-7645(14)60201-7

Santini, F., Vitti, P., Ceccarini, G., Mammoli, C., Rosellini, V., Pelosini, C., . . . Pinchera, A. (2003). In vitro assay of thyroid disruptors affecting TSH-stimulated adenylate cyclase activity. *Journal of Endocrinological Investigation, 26*(10), 950–955. doi:10.1007/BF03348190

Schulz, V., Hübner, W. D., & Ploch, M. (1997). Clinical trials with Phyto- Psychopharmacological agents. *Phytomedicine*, 4(4), 379–387.

Seidler-Łożykowska, K., Bocianowski, J., and Król, D. (2013). The evaluation of the variability of morphological and chemical traits of the selected lemon balm (*Melissa officinalis* L.) genotypes. *Industrial Crops and Products, 49*, 515–520. doi:10.1016/j.indcrop.2013.05.027

Sevik, H., and Guney, K. (2013). Effects of IAA, IBA, NAA, and GA3 on rooting and morphological features of *Melissa officinalis* L. stem cuttings. *The Scientific World Journal*, 2013, 909507. doi:10.1155/2013/909507

Shakeri, A., Sahebkar, A., and Javadi, B. (2016). Melissa officinalis L.—A review of its traditional uses, phytochemistry and pharmacology. *Journal of Ethnopharmacology, 188*, 204–228. doi:10.1016/j.jep.2016.05.010

Soulimani, R., Fleurentin, J., Mortier, F., Misslin, R., Derrieu, G., and Pelt, J. M. (1991). Neurotropic action of the hydroalcoholic extract of *Melissa officinalis* in the mouse. *Planta Medica, 57*(2), 105–109. doi:10.1055/s-2006-960042

Soulimani, R., Younos, C., Jarmouni, S., Bousta, D., Misslin, R., and Mortier, F. (1997). Behavioural effects of *Passiflora incarnata* L. and its indole alkaloid and flavonoid derivatives and maltol in the mouse. *Journal of Ethnopharmacology, 57*(1), 11–20. doi:10.1016/s0378-8741(97)00042-1

Speroni, E., and Minghetti, A. (1988). Neuropharmacological activity of extracts from *Passiflora incarnata*. *Planta Medica, 54*(06), 488–491.

Sun-Waterhouse, D. (2011). The development of fruit-based functional foods targeting the health and wellness market: a review. *International Journal of Food Science and Technology, 46*(5), 899–920.

Tyrer, P., and Baldwin, D. (2006). Generalised anxiety disorder. *Lancet, 368*(9553), 2156–2166. doi:10.1016/S0140-6736(06)69865-6

Ueda, H., Yamazaki, C., and Yamazaki, M. (2004). A hydroxyl group of flavonoids affects oral anti-inflammatory activity and inhibition of systemic tumor necrosis factor-α production. *Bioscience, Biotechnology, and Biochemistry, 68*(1), 119–125.

Ulbricht, C., Brendler, T., Gruenwald, J., Kligler, B., Keifer, D., Abrams, T. R., . . . Collaboration, N. S. R. (2005). Lemon balm (*Melissa officinalis* L.): an evidence-based systematic review by the Natural Standard Research Collaboration. *Journal of Herbal Pharmacotherapy, 5*(4), 71–114. Retrieved from https://www.ncbi.nlm.nih.gov/pubmed/16635970

Valls, J., Pasamontes, N., Pantaleón, A., Vinaixa, S., Vaqué, M., Soler, A., . . . Gómez, X. (2013). Prospects of functional foods/nutraceuticals and markets. *Natural Products*, 2491–2525.

Van Wyk, B.-E., and Wink, M. (2018). *Medicinal Plants of the World*: CABI.

Wilson, L., and Martin, S. (1999). Benzyl alcohol as an alternative local anesthetic. *Annals of Emergency Medicine, 33*(5), 495–499.

Wohlmuth, H., Penman, K. G., Pearson, T., and Lehmann, R. P. (2010). Pharmacognosy and chemotypes of passionflower (*Passiflora incarnata* L.). *Biological and Pharmaceutical Bulletin, 33*(6), 1015–1018. doi:10.1248/bpb.33.1015

Xiao, J., Capanoglu, E., Jassbi, A. R., and Miron, A. (2016). Advance on the flavonoid C-glycosides and health benefits. *Critical Reviews in Food Science and Nutrition, 56*(sup1), S29-S45. doi:10.1080/10408398.2015.1067595.

Xie, J., and Schaich, K. (2014). Re-evaluation of the 2, 2-diphenyl-1-picrylhydrazyl free radical (DPPH) assay for antioxidant activity. *Journal of Agricultural and Food Chemistry, 62*(19), 4251–4260. doi:10.1021/jf500180u

Zeraik, M. L., Serteyn, D., Deby-Dupont, G., Wauters, J.-N., Tits, M., Yariwake, J. H., … Franck, T. (2011). Evaluation of the antioxidant activity of passion fruit (*Passiflora edulis* and *Passiflora alata*) extracts on stimulated neutrophils and myeloperoxidase activity assays. *Food Chemistry, 128*(2), 259–265. doi:10.1016/j.foodchem.2011.03.001

Zhou, X., Wang, F., Zhou, R., Song, X., and Xie, M. (2017). Apigenin: A current review on its beneficial biological activities. *Journal of Food Biochemistry, 41*(4), e12376. doi:10.1111/jfbc.12376

Zolla, C. (1980). Traditional medicine in Latin America, with particular reference to Mexico. *Journal of Ethnopharmacology, 2*(1), 37–51. doi:10.1016/0378-8741(80)90028-8

Zucolotto, S. M., Goulart, S., Montanher, A. B., Reginatto, F. H., Schenkel, E. P., and Fröde, T. S. (2009). Bioassay-guided isolation of anti-inflammatory C-glucosylflavones from *Passiflora edulis*. *Planta Medica, 75*(11), 1221–1226. doi:10.1055/s-0029-1185536

CHAPTER 2

Tremella fuciformis and Its Polysaccharides as an Alternative Therapy Against Oxidative Stress

RUBÉN F. GONZÁLEZ-LAREDO[*], ANGELICA VALDEZ-VILLARREAL,
NURIA ELIZABETH ROCHA-GUZMÁN,
MARTHA ROCÍO MORENO-JIMÉNEZ, and
JOSÉ ALBERTO GALLEGOS-INFANTE

Tecnológico Nacional de México/Instituto Tecnológico de Durango, Unidad de Posgrado, Investigación y Desarrollo Tecnológico (UPIDET), Blvd. Felipe Pescador 1830 Ote., Col. Nueva Vizcaya, 34080 Durango, Dgo., México

[*]Corresponding author. Email: rubenfgl@itdurango.edu.mx

ABSTRACT

Tremella fuciformis is an edible mushroom of highly nutritive value that for centuries has been attributed with many health benefits. It has been used as traditional medicine or prophylactic agent particularly in Chinese civilization. Based on scientific studies, it is revealed that polysaccharide extracts are the main active components of *T. fuciformis*; these components possess biological activities that interact with the human body promoting health thanks to their pharmacological and nutraceutical properties. Recently, it has been used in antitumor treatments to improve the immune function of the organism, and based on various reports, it is attributed to anticancer, anti-inflammatory, antioxidant, and neuroprotective effects. Also, it is currently considered as a potent prospect for its bioactive compounds to be used for the development of cosmetics with a focus on the prevention and treatment of aging symptoms, regulating the effects of oxidative stress on the skin, and promoting longevity. Polysaccharides from *T. fuciformis* (TFPS) have shown positive effects after regular consumption when they are considered as ingredients in supplements or phytopharmaceutical products for human health. In this chapter, it is displayed an updated state-of-the-art on the fungus based on the scientific literature reported lately, particularly on the bioactive TFPS, emphasizing their nutraceutical importance and potential applications as a therapeutic agent against oxidative stress and its consequences.

2.1 INTRODUCTION

Historically, mushrooms have been always consumed by mankind; since the ancient Greeks that believed they delivered strength to the fighting soldiers, to the Romans, who appreciated them as the *food of gods*. The Chinese culture for centuries has valued mushrooms as a healthy food *per se*, or as the closest thing to the mythological *elixir of life*. Mushrooms have been labeled as an attractive culinary ingredient and part of the universal and millennial culture based on their sensorial characteristics and properties. Today, they are prized as a popular and healthy food because of their low content of fat, sodium, carbohydrates, and calories, with no cholesterol (Chang and Wasser, 2012). Besides, they contain an important proportion of protein and fiber, as well nutrients as riboflavin, niacin, vitamin D, selenium, and potassium. The significance of mushrooms, alongside their extended record as a food source, is the value of their healing properties in traditional medicine. The beneficial effects on health and the treatment of some diseases have been already described (Valverde et al., 2015). The nutraceutical potential of many mushrooms is promising, alike the prevention or treatment factors against different diseases.

From the estimated one and half million fungi in the world that produce large enough fruiting bodies to be considered as mushrooms, >14,000 species have been already classified. The mushroom industry can be divided into three main categories: edible mushrooms, medicinal mushroom products, and wild mushrooms (Chang, 2005). The medicinal properties of mushrooms have been known for millennia and used for the benefit and health of humans in the passage of ancient civilizations from China, Europe, and Africa to Mesoamerica. The identity and chemical structure of polysaccharides from mushrooms and its relationship with biological activities for its experimental testing and clinical use as antitumor or immunostimulating polysaccharides are in progress (Wasser, 2011). Numerous bioactive polysaccharides or polysaccharide–protein complexes from medicinal mushrooms appear to enhance innate and cell-mediated immune responses and exhibit antitumor activities in vivo. Many mushroom polymers have been already reported with immunotherapeutic properties, inhibiting tumor cells. Although the stimulation and modulation of the host immune responses by these mushroom polysaccharides seem a key point, the mechanisms of their antitumor actions are still not completely understood. Therefore, for modern medicine therapies polysaccharides with antitumor and immunostimulating properties are very important. Few of the studied mushroom polysaccharides have passed phases I, II, and III in clinical trials and now are used extensively in Asia to treat some types of cancer. Currently, more than a hundred medicinal functions are labeled to medicinal mushrooms and fungi including antibacterial, antidiabetic, antifungal, antihypercholesterolemia, antioxidant, antiparasitic, antitumor, antiviral, cardiovascular, detoxification, hepatoprotective, immunomodulating, and radical scavenging effects, among others activities (Chang and Wasser, 2012).

Mushrooms occupy a large lucrative segment in the market thanks to its uses for the prevention and treatment of diseases. Their products are presented mainly as the powdered whole fruiting bodies alone, as their dried extracts or as an ingredient in cocktail preparations, which have been introduced as nutraceuticals in Western countries, although not always with the appropriate systematic scientific verification (Chang and Buswell, 2008). The demand in North America for medicinal mushrooms and derived products has shown annual growth rates of about 20%–40% depending on the species. The current worldwide market value of mushroom nutraceuticals and related dietary supplement products is estimated to be well above USD $14 billion per year.

Since ancient times, mainly in civilizations of antique China, it has been believed that *Tremella fuciformis* has brought many health benefits, when it has been used in formulation in different types of foods and used in traditional medicines. In the past decades and at present time, the polysaccharide extracts of *T. fuciformis* (TFPS) have received increasing attention thanks to the fact that they are attributed with different dietary properties and pharmacological activities as an alternative therapy against oxidative stress. These include antioxidant, antitumor, and immunostimulant principles, proving effectiveness in promoting immunity, reducing blood sugar, and displaying antithrombosis, antitumor, antihepatitis, antimutagenic, and antiradiation effects (Cho et al., 2006; Zhu et al., 2011, 2012). Also for its important human health actions, it has been widely recommended in health products as a humectant ingredient, anti-wrinkle agent (Wen et al., 2016), and antiaging factor (Wang et al., 2015).

Oxidative stress can cause damage to cell biomolecules such as DNA, proteins, and lipids. It may be induced by free radicals that occur in response to redox reactions, radiation interactions, or biochemical reactions; also by mitochondrial dysfunction, by a weakened antioxidant system, or by a combination of all these factors. Such stress is associated with the onset and development of multiple pathologies, such as atherosclerosis, diabetes mellitus, cancer, rheumatoid arthritis, myocardial infarction, chronic inflammation, cardiovascular diseases, stroke and septic shock, premature aging, and neurodegenerative diseases in humans (Uttara et al., 2009). Many studies have indicated that abnormal concentration of hydrogen peroxide and mitochondrial reactive oxygen species (ROS), which are produced in cells at normal physiological conditions during respiration and metabolism, may promote degenerative and age-related diseases in the human body. Therefore, exogenous bioactive antioxidants might modulate important pathways of mitochondrial ROS to delay damage or increase adaptation to stress prior to the inception of diseases related to oxidative stress (Sena and Chandel, 2012).

2.2 TREMELLA fuciformis

Tremella fuciformis, also commonly called "white jelly," "silver ear," or "snow fungus," is well known for its bioactive polysaccharides content. It is a type of edible and medicinal mushroom that has been used since ancient times, especially in East Asian countries and through different generations. It is an old favorite in Chinese medicine, offering mainly their dried and powdered fruiting bodies, although it is also consumed as a food and usually as a special dessert. *T. fuciformis* is a species of basidiomycetes fungus that belongs to the order of Tremellales, to the Tremellaceae family, and the Tremella genus (Kuo, 2008). *T. fuciformis* is a parasitic yeast that does not form an edible fruiting body without parasitizing another fungus. It grows as a slimy, mucous-like film until finds its favorite hosts (i.e., various species of Annulohypoxylon fungi, which were reclassified from Hypoxylon in 2005). The fungus then invades and develops the hostile mycelial growth to form the known fruiting body. Therefore, cultivators usually pair cultures of *T. fuciformis* with this species for commercial production.

Fruiting bodies of *T. fuciformis* mushrooms contain microscopic egg-shaped or almost ball-shaped basidia. In mass, they look similar to a chrysanthemum flower, ranging from a transparent to bleached or yellowish-brown color depending on the degree of maturity, age of growth or drying condition (Figure 2.1). Their consistency is gelatinous and displays a smooth, bright, and lobed surface. They are clustered by a number of flat flaky or corrugated leaflets, or clustered by brain-like limbs. Their size can reach up to 25 cm and their weight from some grams to a few hundred grams, having common diameters of 3.6 to 11.3 cm and a height of 2.3 to 3.4 cm (López et al., 2014).

FIGURE 2.1 Image of fruiting bodies from *Tremella fuciformis*
Source: Reprinted from López et al., 2014.

Although *T. fuciformis* is found mainly in subtropical regions, it has also been reported from tropical and temperate regions to even colder zones; however, it is best characterized as mesophilic in regard to temperature. In respect to moisture, *T. fuciformis* is quite adaptable because its mycelium is relatively resistant to drought and the fruiting body can dry and shrink, but in the presence of sufficient moisture, it can get back its characteristic gelatinous texture.

T. fuciformis uses to grow on the rotten trunks of hardwood trees such as oak and willow (i.e., wood log culture). Actually, it could be cultivated on trees of 58 species from 18 families, and among them, 28 species were considered as exceptional sources of bed logs for the cultivation of the fungus (Chen and Hou, 1977). As an example, the mango tree (*Mangifera indica* L.)

is the most popular for cultivation of the silver ear mushroom in Taiwan. Today, *T. fuciformis* is a commercially cultivated edible mushroom grown under controlled conditions (i.e., plastic bags culture), using as substrates mainly forestry and agricultural lignocellulosic residues such as sawdust, cottonseed husks, and wheat bran (Chang and Miles, 2004).

Many bioactive substances have been reported in *T. fuciformis*, including fatty acids, proteins, enzymes, polysaccharides, phenols, flavonoids, dietary fiber, and trace elements. The most important bioactive components that have been identified are the polysaccharides found in its fruiting body, spores, and mycelium, to which most of the health benefits are attributed. They contribute positively to human physiology, improving the immune function and providing antitumor, antioxidant, anti-aging, hypoglycemic, hypolipidemic, and neuroprotection, among some other activities.

2.2.1 POLYSACCHARIDES OF TREMELLA FUCIFORMIS

Different fractions of polysaccharides of *Tremella fuciformis* (TFPS) can be obtained in different experimental settings. Typical molecular weight of polysaccharides mixtures might vary from 580 to 3740 kDa and normally, isolated monosaccharides in TFPS include glucose, galactose, fucose, mannose, xylose, and glucuronic acid. They are integrated in a linear (1®3)-linked α-D-mannose backbone, which is highly branched with side chains of β-D-xylose, α-D-fucose, and β-D-glucuronic acid. The TFPS exhibit diverse physiological benefits, promoting healthy effects such as immunomodulation, antitumor, hypolipidemic, neuroprotection, antioxidant, anti-aging, hypoglycemic, among others (Yang et al., 2019).

For many years, the molecular weight (MW) of a molecule was distinguished as a key parameter to dictate its biological properties. In the case of polysaccharides from mushrooms, high MW glucans are commonly believed to exhibit higher bioactivity, but they have first to bond receptors or proteins in order to trigger modulation events. Therefore, apparently higher MW polysaccharides would exhibit more connections to potential receptors or proteins, besides having more repeating units, providing more variability and increased probabilities for interaction. However, there are some exceptions as several antitumor mushroom polysaccharides, such as (1→3)-α-glucuronoxylomannans that are not so dependent on MW. Their hydrolyzed fractions, containing glucuronoxylomannans of lower MW from 53 to 1000 kDa also present such activity.

The high structural variability of TFPS is implied by three major factors. It is established by the location of monosaccharide residues, the position and chirality of glycosidic linkages, and the sequence of the monosaccharide moieties (Meng et al., 2016). To solve these factors, modern analytical techniques have to be used. Purified polysaccharides can be analyzed using high-performance liquid chromatography to determine their molecular weight. Subsequently, it may follow a spectral analysis with ultraviolet spectroscopy, infrared spectroscopy, gas chromatography, mass spectrometry, and nuclear magnetic resonance that can be used to determine the polysaccharide structure.

The initial primary structures of extracted and purified TFPS was reported by Yui et al. (1995) as α-D-mannose in the backbone chain, and β-D-xylose, β-D-gluconic, acid and β-D-xylobiose linked to the C-2 of the mannose main chain. The backbone displayed a symmetrical left-handed triple helix formed by six mannose residues, which besides the three side-chain groups form a repeating unit alongside the 2.42 nm central axis. Lately, a neuroprotective TFPS fraction of 2033 kDa molecular weight was isolated by Jin et al. (2016) with a backbone composed mainly of (1→2)- and (1→4)-linked-mannose and (1→3)-linked-glucans. In most cases, more than the primary repeating unit structure of the TFPS, it is published only the monosaccharide composition and the molecular weight.

Particularly, acidic heteroglycans (410, 250, 34, and 20 kDa) from the body of *T. fuciformis* were able to induce human monocytes to produce interleukin-1 (IL-1), IL-6, and tumor necrosis factor (TNF) in vitro. They were composed of a 3-linked mannose backbone and side chains containing glucosyl, mannosyl, fucosyl, xylosyl, and glucuronic acid residues, having 1.9%–2.9% of acetyl groups. The products from Smith degradation, lithium degradation, and deacetylation also induced good secretion of IL-1 as the original polysaccharides, indicating that xylosyl and glucuronic acid residues, as well as acetyl groups were not key promoters of the cytokine-stimulating activity (Gao et al., 1997).

More recently, a novel fungus of *T. fuciformis* was isolated and used to produce bioactive polysaccharides by optimized laboratory-scale fermentation (initial pH, optimal temperature, and broth volume were 6.0, 26 °C, and 80 mL in a 500 mL flask, respectively; Ge et al., 2020). The maximum yield was 9.05 g/L, about 59% over the basic one (5.69 ± 0.02 g/L). Purified polysaccharides in DEAE columns showed a molecular weight of 1140 kDa, consisting of mannose, glucuronic acid, glucose, galactose, xylose, and rhamnose units (at 3.5:1.2:2:1.6:1.4:3 molar ratios). Compound structure was determined by methylation derivation, infrared spectroscopy analysis, and nuclear magnetic resonance experiments. Major glycosidic linkages were identified as 1,4-xylp, 1,4-manp, 1-xylp, 1-manp, 1,4-glcp, and 1,3,4-galp.

Additionally the polysaccharides exhibited antioxidant activity according to assays of scavenging ROS and hydroxyl radicals.

Generally, chemical modification of TFPS may be consider (i.e., acetylation, alkylation, carboxymethylation, phosphorylation, selenization, and sulfation) to change their physicochemical and biological properties (Wu et al., 2019). To do so, the major reactive groups in the polysaccharide structure (i.e., hydroxyl, carboxyl, and amino groups) are used to chemically introduce the new functional groups. Among the modification reactions, sulfation is maybe the most common chemical modification to enhance water solubility and pharmacological activities. When modifying chemically the TFPS, they not only keep their original biological activities, but improve the cellular immunity. As an example, novel grafting has also been performed adding catechin to TFPS, producing a derivative with higher bioactivity than the original polysaccharide (Liu et al., 2016). In performing these chemical modifications, it may be possible to control the final structure of the polysaccharides and also control their specific physiological functions.

2.2.2 EXTRACTION OF POLYSACCHARIDES FROM TREMELLA fuciformis

There are different methods recommended for the extraction of bioactive components from this fungus, which include several conventional methods using hot water extraction; also, it may be wise to add alkali to improve the extraction, or giving some assistance with sonication or microwave irradiation to provide a more efficient extraction process (Table 2.1). Although simple, using hot water technology is the major conventional extraction method recommended for polysaccharides (Yan et al., 2011); however, it is usually related with higher temperatures, longer extraction times, and lower extraction yields, implying some hydrolysis or degradation. Recently, ultrasonic-assisted extraction (UAE) has been widely used for the extraction of important phytochemicals due to the higher mass transfer rates among phases under an intense low-frequency micro stirring. Combining moderate traditional hot water extraction with ultrasound, minimal structural changes and degradation may be provoked in polysaccharides (Zhou and Ma, 2006). Therefore, UAE may be a recommended technology for the efficient extraction of TFPS. In summary, at lower energy and water consumption, UAE offers simplified manipulation and higher process reproducibility at the shorter times (Vilkhu et al., 2008).

The method and variables to choice for the extraction of key components are of utmost importance, since this determines the quality, performance, and final chemical structure of the final product. Polysaccharides usually play the role of an important constituent of food. Research studies have shown that the polysaccharides present in foods as in mushrooms have important nutritional and biological functions (Zhang et al., 2010). Traditionally, TFPS are extracted with hot water, optimizing the analysis by the use of response surface methodology. For example, extracts at the highest yields (3%) with significant antioxidant and antitumor activities have been obtained under appropriate conditions of temperature (100 °C), liquid-to-solid ratio (5:1), and extraction time (4.5 h; Chen et al., 2010).

Polysaccharide extraction usually follows the hot water reflux method (Zhang et al., 2010, 2014). Therefore, for *T. fuciformis,* the extraction method commonly used since a long time ago is the agitation of the previously dried and pulverized fruiting bodies in hot water (Chen et al., 2010). However, this common method normally requires longer extraction times and higher extraction temperatures. To accelerate the extraction procedure and avoid reducing the bioactivity of the polysaccharides at an excessively high extraction temperature (Liu and Miao, 2010), ultrasound extraction has been suggested as an assistance to conventional polysaccharide extraction (Chen et al., 2010; Zhang et al., 2010; Yin et al., 2014).

Alternatively, microwave-assisted extraction (MAE) has also been used to extract TFPS with water. Again, experimental designs have considered the optimization of polysaccharide yields (~65%), considering the appropriate level of extraction variables such as microwave power (750 W), extraction time (60 s), and their liquid-to-solid ratio (20:1; Chen et al., 2012). Likewise, MAE has received attention as a relatively novel extracting method that uses microwave irradiation as an energy source to enhance the extraction of bioactives from natural resources. MAE has some advantages over traditional procedures, such as promoting quicker and more efficient extractions at top quality level and keeping operating costs relatively low.

Different factors are related to the extraction performance, including temperature, liquid-to-solid ratio, extraction time, extraction medium, power or ultrasonic amplitude, and microwave irradiation (Table 2.1). After choosing the optimal method to perform the extraction, the supernatant goes through filtration and centrifugation to remove residues. The resultant is then precipitated with anhydrous ethanol at three volumes for 24 h at 4 °C. After that time, the precipitate is collected by centrifugation and lyophilized or dried at low temperature to obtain the crude polysaccharide. For the separation and purification of polysaccharides by ion exchange and molecular weight, the use of methods such as column chromatography is very useful (Li et al., 2018; Wu et al., 2019).

TABLE 2.1 Reported Procedures for the Extraction of Polysaccharides from *Tremella fuciformis*

Method	Solvent	Conditions	Yield (%)	References
Hot water extraction	Distilled water	100 °C, 4.5 h ratio liquid/solid 5 (v/v)	3.08	Chen, 2010
Microwave-assisted extraction	Distilled water	Microwave output power 750 w extraction time 1 min liquid/solid ratio (20)	65.07 ± 0.99	Chen et al., 2012
Ultrasound-assisted extraction	Distilled water	50 °C, 30 min ratio liquid/solid (34 mL/g)	8.26	Yan et al., 2011
Ultrasound-assisted extraction	Polyethylene glycol-based (PEG)	90 °C, 60 min ratio solid/solvent (0.133 g/mL)	16.2 ± 0.8 13.8 ± 0.7 (using just water)	Zhang and Wang, 2016
Ultrasound-assisted extraction	Distilled water	85 °C, 2 h sonication intensity (6 W/cm^2) ratio liquid/solid (46 mL/g)	8.95	Zou and Hou, 2017
Alkaline extraction	NaOH, 0.7 M	80 °C, 3 h ratio liquid/solid (90 mL/g)	10.49	Wu et al., 2012
Alkaline extraction	NaOH, 0.76 M	3.56 h ratio liquid/solid 81.89	na	Wu et al., 2007

2.3 APPROVED MEDICATIONS

It is known that many polysaccharides from mushrooms, particularly of the (1→3)-β-D-glucan type, show immunomodulatory activity and may present therapeutic effects, such as anti-inflammatory and antitumor activities (Ma et al., 2007). In the case of *Tremella* species, their polysaccharides are mainly heteroglycans sharing a (1→3)-α-D-mannan backbone with some side chains bearing xylose and glucuronic acid moieties (Chen, 2010). Polysaccharides from fruiting bodies of *T. fuciformis* are able to induce the production in vitro of TNF-α and IL-1 and IL-6 by human monocytes (Gao et al., 1998). Additionally, the acidic heteropolysaccharides (i.e., D-glucurono-D-xylo-D-mannans) are able to reduce the level of cholesterol in the blood serum along their immunomodulatory capabilities (Cheung, 1996).

The enteric-coated capsules based on tremella polysaccharide was approved by the Chinese Food and Drug Administration in 2002 for treating cancer patients with leukopenia induced by chemotherapy and radiotherapy. This drug was approved according to the SFDAN National Drug standard WS-10001- (HD-0881) 2002. Their original names are Yin Er Bao Tang Chang Rong Jiaonang or Yin Er Bao Tang Jiao Nang. It is also used as an adjuvant drug for treating chronic persistent hepatitis and chronic active hepatitis (Zhang et al., 2008). *T. fuciformis* is a long-time preferred mushroom in China, especially for medicinal purposes, and its dried fruiting bodies are commonly prepared and presented in most pharmacies specializing in Chinese medicines, recommended as a tonic for weakness and to treat aging effects. There are nine pharmaceutical companies in China, which have obtained certification for production and credit numbers of certified drugs. The main components (more than 60%) in the product are polysaccharides.

There is a great number of publications related to biochemical and clinical studies identified by searches through bibliographical databases; they inform mainly on the biological and pharmacological activities of *Tremella fuciformis* polysaccharides (Yang et al., 2019). Within the classification of their pharmaceutical activities, we can define, immunomodulation, antitumor, antioxidant and anti-aging, hypoglycemic, hypolipidemic, and hypocholesterolemic activities, as well as neuroprotection effects, among some others.

2.4 IMMUNOMODULATION

At this time, there are different immunomodulatory agents reported in some foods or medicinal plants that contain active ingredients with therapeutic potential in the treatment of some diseases. These bioactive compounds have a pharmacological effect that acts as key promoters of the immune response (López-Luengo, 2008). Compounds that are called immunomodulators act by modulating the immune response through stimulation or suppression processes. This immunomodulatory protection exerts functions in some viral and bacterial infections and as an adjuvant in treatments used in cancer such as radiation, chemotherapy, or some other medications (Martínez-Manrique, 2006). These bioactive polysaccharides play an important role in improving immune activities, which helps the activation of macrophages, T lymphocytes, B lymphocytes, and the regulation of non-specific immunity, humoral immunity, and cellular immunity.

Using fractionation with ammonium sulfate and ion-exchange chromatographic techniques, Hung et al. (2014) isolated a new homodimeric protein (24 kDa, containing no glycan residues) from dried fruiting bodies of *T. fuciformis* and tested their ability to stimulate primary murine macrophages. This immunological activation protein can be an important bioactive compound for the development of healthy foods and nutraceutical products. The mushroom was able to stimulate the production of TNFα, IL-1β, IL-1ra, and IL-12, besides to CD86/MHC class II expression, m1NA expression of chemokines of type M1, and NF-nuclear κ accumulation B in murine peritoneal macrophage cells. In addition, *T. fuciformis* failed to stimulate TLR4-neutralized and TLR4-knockout macrophages, suggesting that TLR4 is a required receptor for the dry fungus signaling in macrophages. In summary, the results point out that the *T. fuciformis* improves proinflammatory gene expression and activates mouse macrophages, promoting M1-type polarization through a TLR4-Nf-kB signaling pathway and potentially stimulating the additional adaptive immune response T helper 1 (Th1; Hung et al., 2014).

TFPS inhibit lipopolysaccharide-induced oxidative stress and inflammation by inhibiting miR-155 expression and NF-κB activation in macrophages significantly. The miR-155 is one of the key small RNAs that regulate NF-κB and inflammation in macrophages, which was significantly downregulated. TFPS also decreased cytokine and ROS levels and attenuated cell inflammation after the treatment with lipopolysaccharide . The latter suggests that TFPS may be a potent reagent for inhibiting the development of inflammation (Ruan et al., 2018).

2.5 ANTITUMOR

Cancer is a pathology that is proliferated by a cell group that multiplies autonomously without any control, and can invade locally or involve other tissues. When the administration of certain polysaccharides is combined with radiotherapy or chemotherapy, it can promote the immune function of patients, which in turn prevents symptoms of adverse effects. That is why they could be a powerful therapeutic drug for the treatment of cancer patients. Several studies have shown that polysaccharides protect hematopoietic function, which helps to increase the amount of erythrocytes, hemoglobin, and leukocytes in the peripheral blood, also promoting DNA synthesis in mice that suffered radiation injuries, chemotherapy, or side effects from some other drugs. As an example, Xu et al. (2007) have synthesized three sulfated TFPS (with variable MW and degree of sulfation) and tested their anti-tumor effect in vivo. The best inhibitory rates to lymphoma at 12 mg/kg dose were around 60% and 70%, which are significantly different respect to the control groups. Therefore, sulfated TFPS might be innovative compounds with potential antitumor effect.

In another report (Li et al., 2008), the combined use of TFPS and 5-fluorouracil (i.e., 5-FU, a common medical drug for the treatment of a diversity of cancers) was tested in vivo. The antitumor and immune regulatory effects of the mixture were evaluated in mice implanted with tumor cells (i.e., sarcoma 180 and hepatocarcinoma 22 cells). It was observed a more potent antitumor effect than the TFPS or 5-FU treatments alone. At the experimented doses (5, 10, 20 mg/kg), TFPS could improve the depleted immune organ index and the carbon granule clearance index induced by 5-FU. Hence, the TFPS and 5-FU combination showed a promising additive anti-tumor activity with the plus from the polysaccharide that it could enhance the down-regulatory function induced by the 5-FU. Similarly, Ma et al. (2007) studied the antitumor effects of TFPS alone and in combination with chemotherapeutic agents. In this case, TFPS (at 25, 50, 100 mg/kg) combined with cyclophosphamide (5, 10, 20 mg/kg) were tested in S180, H22, and Lewis lung carcinoma tumor models in mice. Significant inhibition of tumor growth was observed (58.2%–76.0%). Additionally, the treatment did not impact the weight of mice, balancing the body weight gain due to the use of cyclophosphamide. Thus, TFPS show noticeable antitumor activity and detoxifying effects when used along with cyclophosphamide.

2.6 HYPOGLYCEMIC, HYPOLIPIDEMIC, AND HYPOCHOLESTEROLEMIC

T. fuciformis has very high dietary fiber content, which provides a potential hypocholesterolemic effect similar to other high-fiber foods from the human diet. The effect of the edible fungus in serum and liver lipids, fecal neutral steroids, and bile acid excretion in rats has been positively observed (Cheung, 1996). At the intended dose (5% of mushroom dry powder), it was found a 19% reduction of cholesterol, 31% of serum low-density lipoprotein, while the fecal excretion levels of neutral fecal steroids and bile acids increased to 51% and 36%, respectively. Hence, *T fuciformis* was an effective hypocholesterolemic supplement in vivo. The possible mechanism to explain the increase in fecal excretion of bile acids, it that dietary fiber and some other components may interact with bile acids and reduce their absorption in the enterohepatic circulation, which upturns its expression, increasing the liver conversion of cholesterol to bile acids (Cheung, 1996).

Oral administration of TFPS exhibited a significant dose-dependent hypoglycemic activity in normal mice according to Kiho et al. (1994), showing also a significant activity in streptozotocin-induced diabetic mice by intraperitoneal administration. Likewise, the hypoglycemic activity of exopolysaccharides produced by submerged mycelial cultures of *T. fuciformis* in mice has been explored. The outcome suggested that exopolysaccharides showed significant hypoglycemic effects and improved the insulin sensitivity by regulating peroxisome proliferators-activated receptor γ (PPARγ)-mediated lipid metabolism (Cho et al., 2007). These results highlight particularly *T. fuciformis,* among other mushrooms species, with higher potential as a functional food or as a key oral ingredient, impacting in the hypoglycemic nutraceuticals market for the management of diabetes mellitus (Perera and Li, 2011). Therefore further research is needed to identify their active compounds and mechanisms of action in the prevention and therapeutics of diabetes.

Complementary, the inhibitory effects of TFPS on 3T3-L1 adipocyte differentiation has been estimated by the reduction of PPARγ translation (Jeong et al., 2008). Particularly, the TFPS precipitated with 80% ethanol showed the highest inhibitory activity. Additionally, they inhibited significantly the triglyceride accumulation, Oil Red-O staining, and mRNA expression of PPARγ, C/EBPα, and leptin in a dose-dependent manner. Therefore, *T. fuciformis* mushroom, presenting inhibitory activity on the differentiation of 3T3-L1 adipocytes, might be recommended additionally as a potential anti-obesity supplement.

2.7 NEUROPROTECTION

In people of middle and advanced ages, the quality of life has been affected by neurodegenerative disorders such as cognitive impairment, memory problems, brain aging, and Alzheimer's dementia. There are increasing indications that polysaccharides from medicinal mushrooms show neuroplastic effects. *T. fuciformis* has been reported to be a medicinal mushroom with neuroprotective action or a source of nutraceuticals to prevent such neurodegenerative conditions. For instance, *T. fuciformis* hot water extracts were tested and significantly increased the neurite proliferation induced by nerve growth factor and ameliorated the neurotoxic effect of β-amyloid peptide on PC12h cells (Park et al., 2007). They were also able to stimulate peripheral nerve regeneration by enhancing the expression of the neurotrophic gene in the nerve injury model (Hsu et al., 2013). The neurotrophic effects were observed with significantly reduced toxicity when cells were treated with *T. fuciformis* extracts before the treatment with β-amyloid peptide. This study confirmed that *T. fuciformis* extracts could be used as a potential precautionary agent in neurodegenerative conditions such as Alzheimer's disease (Park et al., 2007).

T. fuciformis is also a good candidate for future studies that may end in its clinical application. Research examining the effects of mushroom activation on additional behavioral tasks will help elucidate whether binding protein signaling can improve different types of memory. For example, an in vivo treatment with *T. fuciformis* reduced trimethylamine-induced learning and memory deficits in the Morris water maze, and showed a protective effect on the decrease of cholinergic neurons and cAMP responsive element-binding protein (CREB; Park et al., 2012). The experiment also improved the glucose activity of the hippocampus and the growth of PC12h cell neurites.

Lately, Ban et al. (2018) has shown that the safe and tolerable oral administration of silver ear fungus to volunteers with subjective cognitive impairment (SCI) has improved their short-term memory, reducing complaints about their SCI condition. SCI is identified when a person recognizes a self-perceived cognitive problem without any diagnosable or apparent reason. The individuals in the treated group showed greater improvements in the total scores on the subjective memory complaint questionnaire and also significantly greater improvements in short-term memory and executive functions compared with those in the placebo group. Although, the study does not report a dose-dependent response, apparently, there are no reports of significant health risks related to the consumption of the mushroom. A daily administration of 1200 mg of *T. fuciformis* would be a safe and easily accessible intervention for people with SCI condition.

In the therapeutic actions of *T. fuciformis* supplementation, structural changes of the brain were additionally observed by neuroimaging, which for the case was a novel application. The brain regions exhibited an increased volume of gray matter after supplementation with the mushroom, relating the change of gray matter in the dorsolateral prefrontal cortex with the improvement of executive functions. This in vivo evidence can support the favorable neurotrophic effect of *T. fuciformis*.

A typical preparation of *T. fuciformis* would start with properly grounding the dried fruit bodies into a fine powder. Then, it would be extracted under pressure (121 °C, 20 min), with fifty volumes of acidified water (citric acid, 10%). The hot water extract is then adjusted with potassium carbonate (5%) to a suitable pH (4.5–5) and incubated (55 °C) overnight with MP cellulase

(10% w/v). Afterward, the extract is filtered and concentrated (8 °Brix). Finally, the mushroom concentrate is encapsulated by spray drying with dextrin (20% w/w) prior to packaging and eventual consumption (Park et al., 2012).

Complementary, the protective effect of a purified TFPS fraction against glutamate-induced cytotoxicity in differentiated DPC12 cells was reported by Jin el al. (2016). The aqueous extract of *T. fuciformis* was purified through DEAE-52 cellulose anion exchange and Sepharose G-100 columns. An active fraction improved the cell viability and suppressed the reactive oxygen species accumulation, the lactose dehydrogenase release, and caspase-3 activity, and inhibited the irregular mitochondrial changes by glutamate. Besides, it enhanced the level of B-cell lymphoma 2 (Bcl-2), and suppressed Bax expression and the release of cytochrome C in glutamate-treated cells. The active fraction treatment was able to reverse the increased activity of caspases-3, -8, and -9, due to the glutamate exposition. Additionally, the presence of a caspase-3 inhibitor (i.e., Ac-DEVD-CHO) noticeably boosted the effectiveness of the bioactive fraction, improving the viability of glutamate-exposed DPC12 cells. These results support the purified TFPS as an experimental neuroprotective agent against glutamate-induced DPC12 cell damage, primarily through the caspase-dependent mitochondrial pathway.

2.8 ANTIOXIDANT

Some foods as *T. fuciformis* may contain antioxidant ingredients that help to regulate the effects of reactive oxygen species when interacting abnormally in humans cells. Free radicals are molecules that have one or more unpaired electrons that can trigger propagation chain reactions with stable substrates causing potential biological damage. Therefore, many studies have shown that polysaccharides could improve the rate of elimination or stabilization of free radicals, concluding that they are able to show antioxidant capacity (Liu et al., 2005; Yan et al., 2006). Typical methods used to determine antioxidant activity are described by Li et al. (2014) when analyzing partly purified methanol extracts from edible white jelly mushroom (*T. fuciformis*). They applied the assays: ABTS+ radical scavenging activity, 1,1-diphenyl-2-picrylhydrazyl radical scavenging activity, and the inhibitory activity of human low-density lipoprotein oxidation. Among the fractions tested, the solubles in chloroform presented the strongest antioxidant activity, the highest total phenolic content (66.31 μg CAE/mg), and flavonoids content (5.12 μg QE/mg). Furthermore, the chloroform subfraction revealed anti-inflammatory activity through the inhibition of nitric oxide production and inducible nitric oxide synthase expression in RAW 264.7 cells. Main bioactive phenolic acids identified in the mushroom extract were 4-hydroxybenzoic acid (323 mg/kg d.w.), gentisic acid (174 mg/kg d.w.), and 4-coumaric acid (30 mg/kg d.w.).

The free-radical degradation and antioxidant activity of polysaccharides from *T. fuciformis* were also investigated by Zhang et al. (2014). They proposed the combination of Fe^{2+}, ascorbic acid, and H_2O_2 as degradation reagents to generate lower molecular weight products. The resultant oxidative-reduction degradation did not change the principal polysaccharide structures under the tested conditions. The antioxidant activity in vitro of five degraded polysaccharide fractions was evaluated. The degraded samples with the lower molecular weight presented the higher antioxidant activities. This may invite to hypothesize that the free-radical degradation enhances the antioxidant activity by decreasing the molecular weight of polysaccharides. In other words, the molecular weight significantly affected the antioxidant response. The lower molecular weight samples contain freer hydroxyl groups, which may have in turn an important effect on the antioxidant capacity. The functional groups in degraded samples have more possibilities to get in contact with the free radicals because of the better water solubility and more availability of surface and access areas.

Recently, two TFPS (MW 45,461 and 25,981 kDa) from two major cultivation regions of China were characterized by Li et al. (2018). They comprised protein, high uronic acids, and similar monosaccharides, although in different ratios, and only one showing a triple-helix conformation. Both polysaccharides exhibited high water and oil holding capacities, and retained antioxidant (i.e., scavenging of hydroxyl free radicals by the Fenton reaction) and immunomodulatory capacities after simulated gastrointestinal digestion in vitro. Thus, TFPS with their prebiotic capability can be used as a potential antioxidant, functional ingredient and immunomodulatory agent in the food and nutraceuticals industry.

2.9 ANTI-AGING AND SKIN WHITENING EFFECT

For years some benefits of skin preservation have been attributed to the consumption of different types of mushrooms and edible fungi, due to its great nutritional value and active compounds. At present time, interest has aroused in using organic and natural cosmetics for skincare. Representative chemical constituents in most edible fungi are polyphenols, terpenoids, selenium, polysaccharides, vitamins, and volatile organic compounds. These compounds show excellent anti-wrinkle and whitening skin, and moisturizing effects, which makes them valuable candidates for antioxidant action and anti-aging in cosmetic products (Wu et al., 2016). *T. fuciformis* extract is an unusual ingredient in integral skincare products that are suggested in the treatment of neurodermatitis and sclerodermatitis (De Baets and Vandamme, 2001).

Hot water TFPS obtained with no added chemical reactives, have shown moisturizing activity and inhibition of melanin formation as an innovative effect that contributes to the lightening of the skin and dark spots removal (Yan et al., 2006). In addition, they have shown excellent moisturizing effects and improved retention capacity of moisture when compared with hyaluronic acid (Liu et al., 2012). However, the effects and mechanisms of action of *T. fuciformis* extract in skin whitening or anti-wrinkle efficacy are still in progress.

Additionally, the optimized extraction of *T. fuciformis* fruiting bodies along two other plants yielded a bacteriostatic extract with a similar moisturizing effect as 1% sodium hyaluronate products, and free radical activity over 80% at the same control conditions. These investigations are delivering a theoretical basis for the application of combined herbal extracts in cosmetics (Zhang et al., 2013).

The protective effects of TFPS have been analyzed by Shen et al. (2017) on a proposed model that considered the induced injury of human skin fibroblasts by hydrogen peroxide. Their results showed that TFPS exhibited protective properties, notably inhibiting ROS production, reducing the skin oxidative stress induced by hydrogen peroxide, and by positively regulating the expression of SIRT1. SIRT1 is a member of the sirtuin protein family that converts nicotinamide adenine dinocleotide-dependent deacetylases. This enzyme deacetylates proteins that contribute to cell regulation; therefore, it has a protective function in the development of many age-related diseases. Thus, it is suggested that TFPS can interact physiologically as a potential agent for skin diseases from oxidative stress and aging.

Aqueous *T. fuciformis* extracts (80 °C) containing polyphenol (12 mg/g) and flavonoid (8.5 mg/g) reduced melanin contents and tyrosinase activity in α-MSH-stimulated melanocytes cell model. The expression factors of melanin formation were significantly down-regulated. Likewise, extracts promoted the synthesis of procollagen and reduced the mRNA expression of matrix metalloproteinase 1 (MMP-1) in the human dermal fibroblast cells. These data showed that extracts induce repression of cellular melanogenesis and protect against skin wrinkles caused by UVB-stimulated damage. Therefore, *T. fuciformis* extracts might be a cosmeceutical candidate for skin anti-wrinkle and whitening effects (Lee et al., 2016).

TFPS can promote immune functions in the human body, but also with anti-aging and antioxidant effects, so it may become important as a moisturizing agent especially in the cosmetics industry (e.g., products such as nourishing creams, replenishment masks, oral liquid for beauty, slimming capsules, etc.). TFPS are particularly competitive because their cost is <10% of the hyaluronic acid, which is a common moisturizing additive in cosmetics (Liu and He, 2012). Therefore, TFPS could be obtained as a high MW alkaline sample (NaOH, 1 M) to enhance extraction and solubility. Then a carboxymethylation may be performed for improving their biological properties, evaluating them after different degrees of substitution (Wang et al., 2015). The resulting chemical composition and Fourier transform infrared analysis showed the successful TFPS modification, proving that the carboxymethylation could effectively enhance their biological properties. The carboxymethylated TFPS showed significant moisture absorption and retention properties because of their carboxyl group from the carboxymethyl and uronic acid units. This ability may be reinforced by the chemical structure, the stereoscopic network, and three-dimensional network with special peripheral acidic groups that modulate also their rheological properties.

2.10 CONCLUSIONS AND RESEARCH PROSPECTS

Consumption of the fungus of *Tremella fuciformis* has been reported with health benefits since ancient civilizations. It has been used as curative medicine thanks to its pharmacological effects shown against various pathologies. Although the benefits and science of TFPS have been studied in recent years, today they represent a growing boom as a powerful prospect as a therapeutic product in alternative medicine, as an ingredient in the food industry, and as a source of pharmaceutical, nutraceutical, and cosmeceutical agents. The current studies have focused more on the structural properties of TFPS (i.e., mainly polysaccharides of the glucuronoxylomannans type), on their molecular weight, their monosaccharide composition, and side-chain position and how these are related with their biological functions. However, further scientific research is still required to provide more information on the mechanisms of action of the complex TFPS oriented to the multiple physiopathologies in which they are attributed with a positive effect for the prevention and treatment of diseases related to oxidative stress.

KEYWORDS

- **bioactives**
- **cosmeceutical**
- **extracts**
- **immunomodulation**
- **mushrooms**
- **nutraceutical**

REFERENCES

Ban, S., Lee, S.L., Jeong, H.S., Lim, S.M., Park, S., Hong, Y.S., Kim, J.E. (2018). Efficacy and safety of *Tremella fuciformis* in individuals with subjective cognitive impairment: A randomized controlled trial. *J. Med. Food,* 21(4), 400–407.

Chang, S.T., Wasser, S.P. (2012). The role of culinary-medicinal mushrooms on human welfare with pyramid model for human health. *Int. J. Med. Mushrooms,* 14, 95–134.

Chang, S.T., Buswell, J.A. (2008). Safety, quality control, and regulation aspects relating to mushroom nutraceuticals. *Proc. 6th Intl Conf. Mushroom Biology and Mushroom Products.* GAMU Gmbh, Krefeld, Germany, pp. 188–195.

Chang, S-T. (2005). The world mushroom industry: Trends and technological development. *Int. J. Med Mushrooms,* 8(4), 297–314.

Chang S.T., Miles P.G. (2004). *Mushrooms, Cultivation, Nutritional Value, Medicinal Effect, and Environmental Impact,* 2nd ed., CRC Press, Boca Raton, FL, USA, p. 451.

Chen, Y., Zhao, L., Liu, B., Zuo, S. (2012). Application of response surface methodology to optimize microwave-assisted extraction of polysaccharide from Tremella. *Physics Procedia,* 24, 429–433.

Chen, B. (2010). Optimization of extraction of *Tremella fuciformis* polysaccharides and its antioxidant and antitumour activities in vitro. *Carbohyd. Polym.* 81(2), 420–424.

Chen, P.C., Hou, H.H. (1977). Studies on the host range of jelly fungus *Tremella fuciformis* Berk. *Taiwan Mushrooms,* 1(2), 65–70.

Cheung, P.C.K. (1996). The hypercholesterolemic effect of two edible mushrooms: Auricularia auricula (tree-ear) and *Tremella fuciformis* (white jelly-leaf) in hypercholesterolemic rats. *Nutr. Res.* 16, 1721–1725.

Cho, E.J., Hwang, H.J., Kim, S.W., Oh, J.Y., Baek, Y.M., Choi, J.W, Bae, S.H., Jung, J.W. (2007). Hypoglycemic effects of exopolysaccharides produced by mycelial cultures of two different mushrooms *Tremella fuciformis* and *Phellinus baumii* in ob/ob mice. *Appl. Microbiol. Biotechnol.* 75(6), 1257–1265.

Cho, E.J., Oh, J.Y., Chang, H.Y., Yun, J.W. (2006). Production of exopolysaccharides by submerged mycelial culture of a mushroom *Tremella fuciformis*. *J. Biotechnol.* 127(1), 129–140.

De Baets, S., Vandamme, E.J. (2001). Extracellular Tremella polysaccharides: Structure, properties and applications. *Biotech. Lett.* 23, 1361–1366.

Gao, Q., Berntzen, G., Jiang, R., Killie, M. K., Seljejid, R. (1998). Conjugates of Tremella polysaccharides with microbeads and their TNF-stimulating activity. *Planta Med.* 64, 551–554.

Gao, Q., Killie, M.K., Chen, H., Jiang, R., Seljelid, R. (1997). Characterization and cytokine-stimulating activities of acidic heteroglycans from *Tremella fuciformis*. *Planta Med.* 63, 457–460.

Ge, X., Huang, W., Xu, X., Lei, P., Sun, D., Xu, H., Li, S. (2020). Production, structure, and bioactivity of polysaccharide isolated from *Tremella fuciformis* XY. *Int. J. Biol Macromol.* https://doi.org/10.1016/j.ijbiomac.2020.01.021

Hsu, S.H., Chan, S.H., Weng, C.T., Yang, S.H., Jiang, C.F. (2013). Long-term regeneration and functional recovery of a 15 mm critical nerve gap bridged by *Tremella fuciformis* polysaccharide-immobilized polylactide conduits. *Evid. Based Comp. Alternat. Med.* 2013, 959261.

Hung, C. L., Chang, A. J., Kuo, X. K., Sheu, F. (2014). Molecular cloning and function characterization of a new macrophage-activating protein from *Tremella fuciformis*. *J. Agric. Food Chem.* 62(7), 1526–1535.

Jeong, H.J., Yoon, S.J., Pyun, Y.R. (2008). Polysaccharides from edible mushroom Hinmogi (*Tremella fuciformis*) inhibit differentiation of 3T3-L1 adipocytes by reducing mRNA expression of PPARγ, C/EBPα, and Leptin. *Food Sci. Biotechnol.* 17(2), 267–273.

Jin, Y.X., Hu, X.Y., Zhang, Y. Liu, T.J. (2016). Studies on the purification of polysaccharides separated from *Tremella fuciformis* and their neuroprotective effect. *Mol. Med. Rep.* 13, 3985–3992.

Kiho, T., Tsujimura, Y., Sakushima, M., Usui, S., Ukai, S., Zasshi, Y. (1994). Polysaccharides in fungi. XXXIII. Hypoglycemic activity of an acidic polysaccharide (AC) from *Tremella fuciformis*. *J. Pharm. Soc. Jpn.* 114 (5), 308–315.

Kuo, M. (2008). *Tremella fuciformis*. Retrieved from the MushroomExpert.Com Web site: http://www.mushroomexpert.com/tremella_fuciformis.html

Lee, K.H., Park, H.S., Yoon, I.J., Shin, Y.B., Baik, Y.C., Kooh, D.H., Kim, S.K., Jung, H.K., Sim, M.O., Cho, H.W., Jung, W.S., Kim, M.S. (2016). Whitening and anti-wrinkle effects of *Tremella fuciformis* extracts. *Kor. J. Med. Crop Sci.* 24(1), 38–46.

Li, P., Jiang, Z., Sun, T., Wang, C., Chen, Y., Yang, Z., Du, B., Liu, C. (2018). Comparison of structural, antioxidant, and immuno-stimulating activities of polysaccharides from *Tremella fuciformis* in two different regions of China. *Int. J. Food Sci. Technol.* 53(8), 1942–1953.

Li, H., Lee, H.S., Kim, S.H., Moon, B., Lee, C. (2014). Antioxidant and anti-inflammatory activities of methanol extracts of *Tremella fuciformis* and its major phenolic acids. *J. Food Sci.* 79(4), 460–468.

Li, Y.C., Ma, E.L., Wang, X.L., Wang, M.W. (2008). Anti-tumor effect of Tremella polysaccharide and 5-fluorouracil combination in implanted with sarcoma 180 and hepatocarcinoma 22. *Chin. J. Hosp. Pharm.* 28(3), 209–211.

Liu, J., Meng, C., Yan, Y.H., Shan, Y.N., Kan, J., Jin, C.H. (2016). Structure, physical property and antioxidant activity of catechin grafted *Tremella fuciformis* polysaccharide. *Int. J. Biol. Macromol.* 82, 719–724.

Liu, H., He, L. (2012). Comparison of the moisture retention capacity of Tremella polysaccharides and hyaluronic acid. *J. Anhui Agric. Sci.* 40, 13093–13094.

Liu, J., Miao, X.Q. (2010). Optimization of polysaccharides extraction from Lotus Bee pollen by response surface methodology. *Food Sci.* 31 (14), 101–105.

Liu P.X., Gao X.R., Xu W.Q., Zhou Z.W., Shen X. (2005). Antioxidation activities of polysaccharides extracted from *Tremella fuciformis* Berk. *Chin. J. Biochem. Pharm.* 26(3), 169–170.

López-Luengo, M.T. (2008). Plantas medicinales con actividad inmunomoduladora. *Revision Offarm.* 27(11), 58–61.

López, A., García, J., González, A. (2014). Tremellales: Tremellaceae. *Tremella fuciformis. Funga Veracruzana.* 140, 1–4.

Ma, E.L., Li, Y.C., Wu, J., Jia, F.M., Wang, N., Shen, F.X. (2007). The anti-tumor effect of Tremella polysaccharide. *J. Shenyang Pharm. Univ.* 24(7), 426–428.

Martínez-Manrique, C.E. (2006). Modulación de la respuesta inmune: tendencias actuales. *Rev. Cub. Invest. Biomed.* 25(4), 1–8.

Meng, X., Liang, H., Luo, L. (2016). Antitumor polysaccharides from mushrooms: a review on the structural characteristics, antitumormechanisms and immunomodulating activities. *Carbohyd. Res.* 424, 30–41.

Park, K.J., Lee, S.Y., Kim, H.S., Yamazaki, M., Chiba, K., Ha, H.C. (2007). The neuroprotective and neurotrophic effects of *Tremella fuciformis* in PC12h cells. *Mycobiology,* 35, 11–15.

Park, H.J., Shim, H.S., Ahn, Y.H., Kim, K.S., Park, K J., Choi, W.K., Ha, H.C., kang, J.I., Kim, T.S., Yeo, I.H., Kim, J.S., Shim, I. (2012). *Tremella fuciformis* enhances the neurite outgrowth of PC12 cells and restores trimethyltin-induced impairment of memory in rats via activation of CREB transcription and cholinergic systems. *Behav. Brain Res.* 229(1), 82–90.

Perera, P.K., Li, Y. (2011). Mushrooms as a functional food mediator in preventing and ameliorating diabetes, Func. *Foods Health Dis.* 4, 161–171.

Ruan, Y., Li, H., Pu, L., Shen, T., Jin, Z. (2018). *Tremella fuciformis* polysaccharides attenuate oxidative stress and inflammation in macrophages through miR-155. *Anal. Cell. Pathol.* 2018, 5762371, 1–10.

Sena, L.A., Chandel, N.S. (2012). Physiological roles of mitochondrial reactive oxygen species. *Mol. Cell,* 48(2), 158-167.

Shen, T., Duan, C., Chen, B., Li, M., Ruan, Y., Xu, D., Shi, D., Yu, D., Li, J., Wang, C. (2017). *Tremella fuciformis* polysaccharide suppresses hydrogen peroxide-triggered injury of human skin fibroblasts via upregulation of SIRT1. *Mol Med Rep.* 16(2), 1340–1346.

Uttara, B., Singh, A.V., Zamboni, P., Mahajan, R.T. (2009). Oxidative stress and neurodegenerative diseases: A review of upstream and downstream antioxidant therapeutic options. *Curr. Neuropharmacol.* 7(1), 65–74.

Valverde, M. E., Hernández-Pérez, T., Paredes-López, O. (2015). Edible mushrooms: Improving human health and promoting quality life. *Int. J. Microbiol.* Article ID 376387, 1–14.

Vilkhu, K., Mawson, R., Simons, L., Bates, D. (2008). Applications and opportunities for ultrasound assisted extraction in the food industry—A review. *Innov. Food Sci. Emerg. Technol.* 9, 161–169.

Wang, R. (2015). Historical note: Scientific exploration of the snow fungus (*Tremella fuciformis* Berk.). *Indian J Hist Sci.* 50(2), 340–344.

Wang, X., Zhang, Z., Zhao, M. (2015). Carboxymethylation of polysaccharides from *Tremella fuciformis* for antioxidant and moisture-preserving activities. *Int. J. Biol. Macromol.* 72, 526–530.

Wasser, S.P. (2011). Current findings, future trends, and unsolved problems in studies of medicinal mushrooms. *Appl. Microbiol. Biotechnol.* 89(5), 1323–1332.

Wen, L., Gao, Q., Ma, C.W., Ge, Y., You, L., Liu, R.H., Fu, X., Liu, D. (2016). Effect of polysaccharides from *Tremella fuciformis* on UV-induced photoaging. *J. Funct. Foods,* 20, 400–410.

Wu, Y.J., Wei, Z.X., Zhang, F.M., Linhardt, R.J., Sun, P.L., Zhang, A.Q. (2019). Structure, bioactivities and applications of the polysaccharides from *Tremella fuciformis* mushroom: A review. *Int J Biol Macromol.* 121, 1005–1010.

Wu, Y., Choi, M.H., Li, J., Yang, H., Shin, H.J. (2016). Mushroom cosmetics: The present and future. *Cosmetics*, 3(3), 1–13.

Wu, Q., Dai, Y.G., Zou, X.F., Chen, L.N., Zhang, L.H., Gao, C.C., Chen, X. (2012). Isolation of alkali soluble polysaccharide from *Tremella fuciformis* and its antioxidant activity in vitro. *IEEE Int. Symp. IT in Med. Educ.* 711–714.

Wu, Q., Zheng, C., Ning, Z., Liu, B., Dong, H. (2007). Study on extraction and antioxidant effects of crude alkai-soluble Tremella polysaccharide. *Food Sci.* 28(6), 153–155.

Xu, W.Q., Gao, W.Y., Wang, Y.Y., Li, M.J. (2007). Studies on synthesis of sulfated *Tremella fuciformis* polysaccharides and its inhibition effect on tumor. *Chin. Pharm. J.* 42(8), 630–632.

Yan, Y.L., Yu, C.H., Chen, J., Li, X.X., Wang, W., Li, S.Q. (2011). Ultrasonic-assisted extraction optimized by response surface methodology, chemical composition and antioxidant activity of polysaccharides from *Tremella mesenterica*. *Carbohyd. Polym.* 83(1), 217–224.

Yan, J., Guo, X.Q., Wu, X.Y., Xu, G.Y., Gou, X.J. (2006). Research on the ability to scavenge free radical of Tremella polysaccharide. *J. Chengdu Univ. Nat. Sci.* 25(1), 35–38.

Yang, D., Liu, Y., Zhang, L. (2019). Tremella polysaccharide: The molecular mechanisms of its drug action. *Prog. Mol. Biol. Translat. Sci.* 163, 383–421.

Yang, S.H., Liu, H.I., Tsai, S.J. (2006). *Edible Tremella Polysaccharide for Skin Care*. U.S. Patent US20060222608, 5 October 2006.

Yin, X., You, Q., Su, X., (2014). A comparison study on extraction of polysaccharides from Tricholoma matsutake by response surface methodology. *Carbohyd. Polym.* 102, 419–422.

Yui, T., Ogawa, K., Kakuta, M., Misaki, A. (1995). Chain conformation of a glucuronoxylomannan isolated from fruit body of *Tremella fuciformis* Berk. *J. Carbohyd. Chem.* 14, 255–263.

Zhang, L., Wang, M. (2016). Polyethylene glycol-based ultrasound-assisted extraction and ultrafiltration separation of polysaccharides from *Tremella fuciformis* (snow fungus). *Food Bioprod. Proc.* 100, 464–468.

Zhang, Z., Wang, X., Zhao, M., Qi, H. (2014). Free-radical degradation by $Fe^{2+}/Vc/H_2O_2$ and antioxidant activity of polysaccharide from *Tremella fuciformis*. *Carbohyd. Polym.* 112, 578–582.

Zhang, K., Meng, X.Y., Sun, Y., Guo, P.Y. (2013). Preparation of Tremella, Speranskiae tuberculatae and *Eriocaulon buergerianum* extracts and their performance in cosmetics. *Deterg. Cosmet.* 36, 28–32.

Zhang, J., Gao, Y., Zhou, X., Hu, L., Xie, T. (2010). Chemical characterisation of polysaccharides from Lilium davidii. *Nat. Prod. Res.* 24, 357–369.

Zhang, L., Li, H.S., Bai, H.S. (2008). Comparison of the determination of polysaccharide with 3,5- dinitrosalicylic acid and phenol sulfate in Tremella polysaccharide enteric-coated capsules. *Tianjin Pharm.* 20(3), 14–17.

Zhou, C., Ma, H. (2006). Ultrasonic degradation of polysaccharide from a red algae (Porphyra yezoensis). *J. Agric. Food Chem.* 54, 2223–2228.

Zhu, H., Tian, B., Liu, W., Zhang, S., Cao, C., Zhang, Y., Zou, W. (2012). A three-stage culture process for improved exopolysaccharide production by *Tremella fuciformis*. *Biores. Technol.* 116, 526–528.

Zhu, H., Cao, C., Zhang, S., Zhang, Y., Zou, W. (2011). pH-control modes in a 5-L stirred-tank bioreactor for cell biomass and exopolysaccharide production by *Tremella fuciformis* spore. *Bioresour. Technol.* 102, 9175–9178.

Zou, Y., Hou, X. (2017). Extraction optimization, composition analysis, and antioxidation evaluation of polysaccharides from White Jelly Mushroom, *Tremella fuciformis* (Tremellomycetes). *Int. J. Med. Mushrooms*, 19(12), 1113–1121.

CHAPTER 3

Natural Extracts and Compounds as Inhibitors of Amylase for Diabetes Treatment and Prevention

CYNTHIA SELENE VASQUEZ-RAMOS,[1]
MELANY GUADALUPE GARCIA-MORENO,[1] DANIEL GARCÍA-GARCÍA,[1]
GLORIA ALICIA MARTÍNEZ-MEDINA,[1]
SUJEY ABIGAIL NIÑO-HERRERA,[1] HUGO LUNA-GARCÍA,[1]
SANDRA PACIÓS-MICHELENA,[1] ANNA ILYINA,[1]
E. PATRICIA SEGURA-CENICEROS,[1] MÓNICA L. CHÁVEZ- GONZÁLEZ,[1]
MAYELA GOVEA-SALAS,[1] JOSÉ L. MARTÍNEZ-HERNÁNDEZ,[1]
S. YESENIA SILVA-BELMARES,[2] RADIK A. ZAYNULLIN,[3]
RAIKHANA V. KUNAKOVA,[3] RODOLFO RAMOS-GONZÁLEZ,[4] and
ROBERTO ARREDONDO-VALDÉS[1*]

[1]*Nanobioscience Group, Chemistry School, Autonomous University of Coahuila, Blvd. V. Carranza e Ing. J. Cardenas V., 25280 Saltillo, Coahuila, Mexico*

[2]*Research Group of Chemist Pharmacist Biologist, Chemistry School, Autonomous University of Coahuila, Blvd. V. Carranza e Ing. J. Cardenas V., 25280 Saltillo, Coahuila, Mexico*

[3]*Ufa State Petroleum Technological University, 1 Kosmonavtov Str., 450062 Ufa, Bashkortostan, Russia*

[4]*CONACYT—Autonomous University of Coahuila, 25280 Saltillo, Coahuila, México Saltillo, COAH, México. Blvd. V. Carranza e Ing. J. Cardenas V., Saltillo, Coahuila, Mexico, CP 25280*

*Corresponding author. E-mail: r-arredondo@uadec.edu.mx

ABSTRACT

Digestive enzymes with hydrolytic activity, such as α-amylase, ensure hydrolysis of starch as a high-molecular-weight food component. The regulation of the enzymatic activity of amylase makes it possible to control the level of hydrolysis and absorption of simple sugars from starch. This control is essential for the treatment of metabolic diseases like diabetes and obesity. The release of amylase inhibitors from plant material is of great practical importance for the development of functional foods intended for the treatment and prevention of diabetes. The present contribution describes the advances in the use of plants as a source of compounds with inhibitory activity of the enzyme α-amylase, including the own experiences in this area of research.

3.1 INTRODUCTION

World Health Organization (WHO) recognized *Diabetes mellitus* (DM), along with cardiovascular diseases, cancer, and chronic respiratory diseases, is one of the most common among noninfectious diseases. In 2019 there were 464 million persons aged between 20 and 79 years suffering from diabetes. Despite the significant successes in the clinical and experimental treatment of diabetes mellitus (DM) achieved in the last 20 years, the incidence of diabetes is only increasing, and, according to WHO predictions, this trend will continue (WHO, 2019). DM is a group of metabolic diseases, which is recognized by hyperglycemia, when there is an increase in blood glucose at higher levels than usual, as a result of a deficiency in insulin secretion (type 1) or an altered tissue sensitivity to insulin (type 2) (Arredondo-Valdés, 2013; WHO, 2019).

Type 2 diabetes is diagnosed in approximately 90% of all patients. To decrease blood glucose level, the most common methods for the type 2 diabetes treatment include regulation of the sensitivity of tissues to insulin, an increase of insulin activity in specific tissues (targets), and inhibition of enzymes, which break down carbohydrates (Arredondo-Valdés, 2013; Uitdehaag et al. 1999). The first two methods include a wide range of treatments with the use of various synthetic drugs, which are characterized by undesirable side effects. At the same time, the treatment of type 2 diabetes with specific inhibitors of α-amylase or α-glucosidase can reduce blood glucose without significant metabolic complications. The α-amylase and α-glucosidases are the enzymes that hydrolyze starch, oligo- and disaccharides, or glycogen. These have a crucial role in the absorption of glucose into the blood. The inhibition of these enzymes considered one of the promising strategies for the treatment of hyperglycemia. The inhibitors include compounds that compete with dietary carbohydrates for the binding sites of gastrointestinal enzymes. An example of such an inhibitor is acarbose (Figure 3.1) (Maximov et al., 2016).

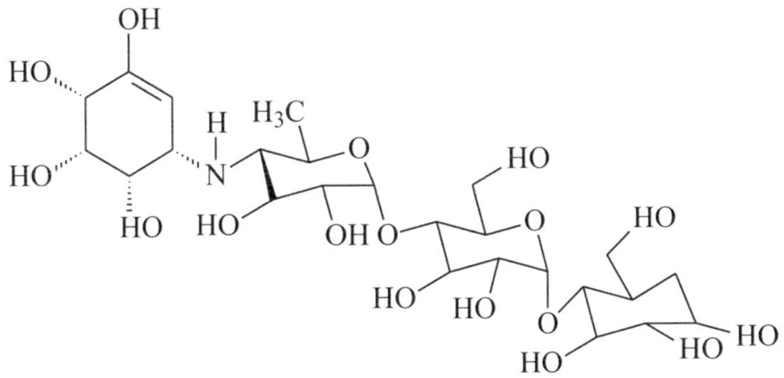

FIGURE 3.1 Chemical structure of acarbose.

In the treatment of diabetes, α-amylase and α-galactosidase inhibitors are used to reduce the amount of glucose in the blood of patients. In medical practice, it is often possible with the help of antienzyme preparations (the same inhibitors) to stabilize the state of patients with a shocking symptom, normalize some biochemical parameters, and remove patients from toxemia. However, this is only a small part of examples of the use of α-amylase inhibitors (IAA) (Arredondo-Valdés et al., 2013).

α-Amylase or α-1,4-glucan-4-glucanohydrolase is a hydrolytic enzyme that breaks down glycogen and starch to oligosaccharides of various lengths with α-configuration, such as maltose, maltotriose, and branched oligosaccharides with 6–8 glucose residues, containing α-1,4 and α-1,6 bonds. Other amylolytic enzymes are involved in starch degradation, but the contribution of α-amylase is to initiate this process. Human α-amylase is an enzyme consisting of 512 amino acid residues, with a molecular weight of 57.5 kDa, and divided into three domains (Uitdehaag et al. 1999; Marc et al., 2002). Slightly acidic properties characterize α-amylase, the isoelectric point (PI) ranges from pH 4.2–5.7. It is a calcium-dependent enzyme and usually contains from 1 to 4 calcium atoms. During acid or thermal inactivation of α-amylase, Ca^{2+} ions have a stabilizing effect. In terms of thermal stability, α-amylases differ depending on their origin: fungal α-amylase more rapidly degraded at elevated temperature, malt α-amylase more thermostable, and bacterial α-amylase is the most heat-resistant enzyme. The active catalytic center of the α-amylase family, frequently, is represented by the amino acid residues of aspartic (Asp) and glutamic (Glu) acids, which act as Ca^{2+} ligands. During substrate transformation (starch) in the active center of the enzyme, sequences of amino acid residues of arginine, histidine, and tyrosine are commonly involved (Kunakova et al., 2016). Pancreatic and salivary amylases have shown to have a high degree of homology of the amino acid sequence (97% of the entire sequence and the 92% of catalytic domains are similar). Amylase is a protein with a tertiary structure that can bind to a substrate and cause hydrolysis of α-1,4 glycosidic bonds. The active center of α-amylase is located between the terminal carboxyl groups and the B-domain. Ca^{2+} atoms are localized between the A and B domains and can act as stabilizers of the three-dimensional structure of the enzyme and, at the same time, as allosteric activators (Nakamura et al., 1993).

There are five subspecies of α-amylase, the genes of which are localized in the long arm of the first chromosome (1q21): AMY1A, AMY1B, and AMY1C—encoding salivary amylase, and AMY2A and AMY2B—encoding pancreatic amylases, respectively (Kunakova et al., 2016). In humans, the number of copies of these genes varies from 2 to 15 per somatic cell genome, and, accordingly, the production of enzymes by tissues is proportional to the number of copies. Interpopulation differences were established in the average number of copies of genes, presumably arising under the influence of positive selection during adaptation to food containing starch (Perry et al., 2007), which seen on the example of genetic characteristics of indigenous peoples of the Far North.

It is known that the traditional diet among the northern peoples (Sami, Nenets, Eskimos, and others) is mainly limited to meat and animal fat; their food contains a minimum of vital carbohydrates. It turned out that carbohydrate metabolism in these people is different from the metabolism of residents of non-Arctic regions. In such a way that is not required to metabolize large amounts of exogenous carbohydrates, the synthesis of the necessary monosaccharides stimulated from amino acids that come with dietary proteins. For this reason, the production of digestive enzymes that break down poly- and disaccharides, including α-amylase, is significantly reduced. On the other hand, the diets of residents of non-Arctic regions mainly include disaccharides (sucrose, lactose) and, consequently, the metabolism has developed an enzymatic system production for rapid assimilation of complex sugars mediated by hydrolysis to monosaccharides for rapid and effective absorption through the membrane of the intestinal wall (Kunakova et al., 2016).

This example demonstrates that excessive intake of readily digestible and available carbohydrates is not essential for the survival of human beings. Therefore the widespread introduction of carbohydrate breakdown enzyme inhibitors in the daily diet is not critical for human metabolism.

The research focused on the characterization of various inhibitors for enzymes involved in sugar metabolism creates the basis for the tasks to introduce these substances as nutraceuticals in the food and medicine sectors, including the practice of diabetes treatment. The present review focuses on plants capable of inhibiting the α-amylase, as well as the identification of some natural compounds useful for the elaboration of nutraceuticals for the treatment of diabetes.

3.2 STARCH

Starch is the primary source of energy storage in vegetables and found in large quantities in different varieties of plants, such as cereal grains, which contain between 60% and 75% by dry weight starch, tubers, leguminous seeds, and some immature fruits, and its concentration varies with their maturity status (Hernández-Medina et al., 2008). Starch is one of the primary energy sources in the human diet. The release of glucose from starchy foods and its relevance to obesity, diabetes, and other metabolic disorders have resulted in much interest in the dietary intake of starch necessary to maintain a state of good health of an individual.

Structurally it is formed by two glucose polymers: amylose and amylopectin. These molecules are organized in concentric rings to originate the granular structure. The distribution of the amylose within the concentric rings differs between the center and the periphery of the granule since it only occupies the available places left by the amylopectin after being synthesized (Agama-Acevedo et al., 2013). A linear polymer unit of glucose linked by α (1–4) bonds is amylose, in which some α (1–6) bonds may be present. This molecule is not soluble in water, but it can form hydrated micelles for its ability to bind neighboring molecules by hydrogen bonds and can generate a helical structure. (Hernández-Medina et al., 2008). While amylopectin is a branched polymer of glucose units linked in 94%–96% by α (1–4) bonds and 4%–6% with α (16) bonds, as demonstrated in Figure 3.2—these branches located approximately every 15–25 glucose units. Amylopectin is partially soluble in hot water (Guan and Hanna, 2004).

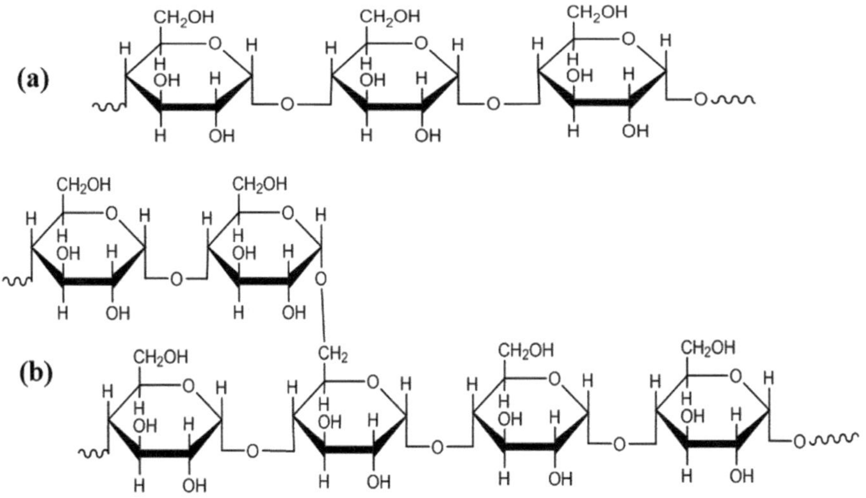

FIGURE 3.2 (a) Structure of amylose and (b) structure of amylopectin.

One of the main properties of native starch is its semicrystallinity, where amylopectin is the dominant component of crystallization in most starches. The crystalline portion is composed of double helix structures formed by hydrogen bonds between the hydroxyl groups in the linear chains of the amylopectin molecule and by external chains of amylopectin linked with portions of amylose. (Meneses et al., 2007). The biosynthesis of amylopectin involves the participation of soluble starch synthase enzymes (SSS), which bind glucose molecules through α 1–4 bonds, producing linear chains with different degrees of polymerization. Starch-branching enzymes (SBE) bind glucan chains via α 1–6 bonds, creating a branched molecule. Once amylopectin formed, granule-bound starch synthase (GBSS) begins to bind glucose to synthesize only the amylose chains. Each of these enzymes present isoforms as GBSS I and II; SSS I, II, III, and IV; and SBE I and II, which in turn can be of type "a" or "b," which differ in the mechanism of action during the synthesis of starch, generating structures that differ in the degrees of polymerization of the linear chains, in the density and the length of the branches; these isoforms may or may not be associated with the starch granule (Agama-Acevedo et al., 2013).

The digestion of starch, both amylose and amylopectin, is mainly mediated by amylases, dextrinases, and disaccharidases that act by hydrolyzing starch to glucose monomers that are absorbed directly through the intestinal mucosa. One of the factors that affect the digestibility of starch and its physiological response, it is attributed to the size of the granule, molecular structure, and the amylose/amylopectin ratio, in addition to the botanical origin that determines the organization and crystalline morphology (Villarroel et al., 2018).

There are two types of amylases: α and β. β-amylase is maltose, it cuts the starch chains in two glucose units, starting at the nonreducing end. The activity stops when it finds a 1–6 link. Thus β-amylase completely degrades amylose to maltose, while the product of its action on amylopectin is maltose and borderline dextrins. On the other hand, α-amylase degrades starch in a much more messy way: it attacks bonds 1–4 in random areas of the chain, even on both sides of links 1–6. Depending on the contact time, a set of varied, partly branched oligosaccharides, called dextrins, of varying molecular weight is obtained as the product of this reaction. If the reaction continues, the straight chains end up becoming maltose and maltotriose.

Resistant starch is defined as a starch sum and degradation product of all nonabsorbed starches in the intestine of healthy individuals. The resistance to digestion starch mainly attributed to the particular physical structure, determined in part by a higher amount of amylose concerning amylopectin, which makes it possible to constitute a more compact structure that is less susceptible to enzymatic hydrolysis. Studies in humans slide the potential positive effects of resistant starch in improving the health of the microbiota, reducing fat accumulation, improving insulin sensitivity, glycemic regulation, and lipid metabolism control (Fuentes-Zaragoza et al., 2010). The attributes of resistant starch explain much of these benefits as soluble fiber, its ability to generate short-chain fatty acids, and to form viscous solutions in the intestinal lumen. Resistant starch consumption has associated with systemic physiological effects, including glycemic metabolism control.

3.3 CHARACTERISTICS OF α-AMYLASE

α-Amylases (α-1,4-glucan-4-glucanohydrolases; E.C.3.2.1.1) are endoglycosidases that hydrolyze (1–4) glucosides linkages, play an essential role in the digestion of starch and glycogen, and are found in several organisms (Mahmood, 2016). Classification of enzymes degrading starch and related poly- and oligosaccharides can be made by the following divisions in their behavior: (1) endo-versus exo-mode of attack; (2) inversion versus retention of the anomeric configuration of the substrate; (3) preference for poly-, oligo-, or disaccharides; (4) a-(1–4), a-(1–6), or dual bond-type specificity; and (5) hydrolytic versus glucosyl-transfer activity; by the combination of these five categories, one can cover the diversity of reactions catalyzed by the starch-degrading and related enzymes (Søgaard et al. 1993).

These enzymes have been classified into glycosyl hydrolase family 13 (1–3) from their primary structures, a family that also includes α-glucosidases and cyclodextrin glucanotransferases (Numao et al. 2004; Schwarz et al. 2007). Several assays demonstrate that tertiary structures reveal that the catalytic domain of family 13 enzymes, especially the active site, is conserved.

In humans, the significant tissues that produce amylase are the pancreas and salivary glands. The ease of detection of salivary and pancreatic amylases in serum contributed to extensive clinical literature, relating the levels of amylase isozymes to various disease states of human salivary α-amylase (Michelle de Sales et al., 2012). By X-ray crystallographic methods, it recognized that these enzymes found to be composed of three structural domains capable of binding to the substrate and, by the action of highly specific catalytic groups, promote the breakage of the glycoside links (Ochiai et al., 2014). The human pancreatic α-amylase (HPA) domain A (residues 1–99 and 169–404) contains $(\beta/\alpha)8$-barrel motif; domain B (residues 100–168) consists of an open-loop that contains several helices and β-strands; and domain C (residues 405–496) is composed of 10 β-strands of which the eighth forms a Greek key motif (Mahmood, 2016). There are five α-amylase genes grouped in chromosome 1, with location 1q21, in humans. Three genes code for salivary amylase, AMY1A, AMY1B, and AMY1C, and the other two genes code for pancreatic amylase, AMY2A and AMY2B (Søgaard et al. 1993).

By other way humans salivary α-amylase (HAS) consists of 496 amino acids and is found in two forms in human saliva: a glycosylated isoform (apparent molecular weight 62 kDa) and a nonglycosylated form of 56 kDa; It binds a chloride ion near the active site in domain A and is coordinated by the side chains of Arg^{195}, Asn^{298}, and Arg^{337} and a calcium ion that is coordinated by His^{201} from domain A and Asn^{100}, Arg^{158}, and Asp^{167} from domain B (Mahmood, 2016; Brayer et al., 1995). The C domain presents a β sheet linked to the A domain by a simple polypeptide chain and seems like to be an independent domain with unknown function. The active site (substrate-binding) of the α-amylase is located in a long cleft located among the carboxyl end of mutually A and B domains. The calcium (Ca^{2+}) situated in the middle of A and B domains acts in stabilizing the three-dimensional configuration and as an allosteric activator. The substrate-binding site comprehends five subsites (−3 − 2 −1 +1 +2) (Michelle de Sales et al., 2012). α-Amylase catalyzes the hydrolysis of starch via a double displacement mechanism involving the formation and hydrolysis of a covalent β-glycosyl enzyme intermediate by using active site carboxylic acids for it. The remains, in particular, Asp^{197}, Glu^{233}, and Asp^{300} were described to function as catalytic residues. Probably, Asp^{197} acts as the nucleophile that works on the substrate at the anomeric sugar center, establishing a covalently bound reaction intermediate. In this step, the reducing end of the substrate is cleaved off the sugar structure. In the second step, a water molecule attacks the anomeric center to break the covalent bond between Asp^{197} and the substrate, attaching a hydroxyl group to the anomeric center. In the two steps, Glu^{233} and Asp^{300} either individually or collectively action as acid/base catalysts (Brayer et al. 1995).

In fact, the active site of human α-amylase consists of several major binding subsites identified through kinetic studies (Michelle de Sales et al., 2012). Recently several studies of the mechanisms on the rate and extent of starch digestion by α-amylase are revised in the light of current widely used organizations for (1) the fractions of rapidly digestible starch (RDS), slowly digestible starch (SDS), and resistant starch (RS), centered on in vitro digestibility, and (2) the types of RS: RS 1, RS 2, RS 3, and RS 4, centered on physical and chemical form. Based on methodological advances and new mechanistic insights, it is proposed that both classification systems should be modified. Several kinetic analyses of digestion profiles provide a robust set of parameters that should replace the classification of starch as a combination of RDS, SDS, and RS from a single enzyme digestion experiment. This should consist of the determination of the minimum number of kinetic procedures needed to describe the full digestion profile, organized with the proportion of starch implicated in each procedure, and the kinetic properties of the procedure. The current organization of RS types: RS 1, RS 2, RS 3, and RS 4, can be replaced by one which identifies the essential kinetic nature of RS, and that there are two essential steps for resistance based on (1) rate-determining access/binding of the enzyme to the substrate and (2) rate-determining conversion of substrate to a product once bound (Dhital et al., 2017).

3.4 THE EFFECT OF TEMPERATURE, pH, AND SALT ON α-AMYLASE

This enzyme can hydrolyze the internal α-1,4 glycosidic linkages to glucose, dextrin, maltotriose, and maltose while retaining the α-anomeric configuration in the products. α-Amylase is produced by many species, such as animals, plants, and microorganisms (Zhang et al., 2017). The α-amylase enzymes produced by an organism can adapt to function at the organism's growth conditions (Hiteshi and Gupta, 2014). Generally, the temperature, pH, and salt concentration conditions in which α-amylase enzymes are going to be used in areas like biotechnology, medicine, and industrial applications are rather extreme. Also, the activity of the α-amylase enzyme inhibitors presents dependence on temperature, pH, and presence of some particular ions.

Regularly, enzymes are most temperature sensitive in dilute solution and without substrate (Bisswanger, 2014). Accurately, the temperature has been reported to affect the activity of the α-amylase enzyme inhibitors. Moreover, the thermostability of this kind of enzymes is an essential property due to its potential biotechnological and industrial applications (Hiteshi and Gupta, 2014). The enzymes generally become denatured at a temperature of around 50–60 °C, resulting in a total loss of enzyme activity (Danson et al. 1996). However, the α-amylase enzyme is a representative of thermostable enzymes. Thermostable α-amylase enzymes have been isolated from a wide variety of sources, like animals, plants, and microorganisms, where the enzyme plays an essential role in the metabolism of carbohydrates (Prakash and Jaiswal, 2010).

Bertoft et al. (1984) studied the effect of different conditions on the stability of barley α-amylase enzyme. The authors reported that the enzyme was stable from 30 to 60 °C. Nevertheless, at higher temperatures, the enzyme loses its activity according to time. On the other hand, Awasthi et al. (2018) reported the activities and stability of α-amylase produced from bacteria (*Brevibacillus borstelensis* and *Bacillus licheniformis*). The bacterial strains presented a maximum enzymatic activity at 60 °C. Mainly, the α-amylase enzyme isolated showed a wide range of thermal stability, from 50 to 80 °C. Asgher et al. (2007) produced the α-amylase enzyme from *Bacillus subtilis*. The crude enzyme revealed that optimum activity was reached at 70 °C. The enzyme was stable for 1 h at 60 and 70 °C, whereas at 80 and 90 °C, 12% and 48% of the activity were lost, respectively. On the other hand, Kılıç Apar and Özbek (2004) reported the effects of heat on the activity of α-amylase from *Bacillus* sp., *Aspergillus oryzae*, and *B. licheniformis*. The study revealed that temperature and processing time are involved in the inactivation of the α-amylase enzyme, and the inactivation of the α-amylase enzymes studied at different temperatures was different and specific to the enzymes.

On the other hand, α-amylase enzyme catalytic activity is limited at low pH by protonation of the nucleophile (Asp) and at high pH by deprotonation of the hydrogen donor (Glu), in consequence, the pH–activity profile of the enzyme is determined by the pK_a values of these two active site groups, permitting the enzyme to catalyze at a wide range of pH (Nielsen et al., 2001). Asgher et al. (2007) produced α-amylase from *B. subtilis* and revealed that optimum enzymatic activity was achieved at pH 8. Although Awasthi et al. (2018) defined that the optimum pH for α-amylase enzymatic activity ranges from 6 to 8. Nevertheless, α-amylase enzymes from animals, precisely mammalian sources, possess an optimum pH activity near the physiological pH of 7.5 (Bisswanger, 2014). Yadav and Prakash (2011) studied the mechanisms of denaturation of α-amylase under acidic conditions. The study shows that the enzyme presents an optimum activity at pH range from 4.5 to 7, and a decrement in the activity was found at lower values.

Otherwise, α-amylase enzymes characterized for their requirement of calcium ions to maintain structural integrity, and in consequence, removal of the ions leads to a decrease in the thermal stability and the enzymatic activity (Nielsen et al., 2003). Adda et al. (2014) studied the effect of salt stress on α-amylase activity. The authors found that the effect of salinity on α-amylase activity depends on the concentration of NaCl, the enzymatic activity was increased at lower stress salinity and concluded that an increase in NaCl stress affects the α-amylase enzyme activity negatively. As a counterpart, Jabbour et al. (2013) reported a study of a salt-tolerant α-amylase enzyme, which was isolated from a pilot-plant biogas reactor. The α-amylase activity was tested in the presence of NaCl and $CaCl_2$ salts, and the optimal activity of the enzyme reached in the presence of 5% of NaCl, and 1 mM in the case of $CaCl_2$.

3.5 α-AMYLASE PRODUCTION

Enzymes are a group of protein molecules with the capacity of catalyzing diverse class of reactions that generally produced with the same aim in living organisms; however is frequent promote their use for supporting the generation of diverse products and services for human benefits, this as result of their extensive advantages as an environmental and cleaner process, a generation of particular reactions, safer products and the use of milder conditions representing a less energy-consuming procedure (Jegannathan et al. 2013); at worldwide level, the enzyme market dominated by hydrolases, that catalyze the breakdown of different polymeric substances as protein, starches, lipids or fibers into simplex molecules, with proteases, cellulose, and amylases as heading produced enzymes (Arbige et al., 2019), where α-amylase represent about 65% of enzyme commercially available (Simair et al., 2017).

The α-amylase (EC 3.2.1.1) is a protein that catalyzes the randomly endo-hydrolysis of a glyosidic linkage type D-α 1-4 in polysaccharides, generating glucose, maltose, and maltotriose units (Panneerselvam et al., 2015). This is a ubiquitous enzyme produced by plants, animals, and microbes (Monteiro de Souza et al., 2010). It is known as a biological catalyst and possesses a wide range of application in various fields, inside food production in the bakery for enhancing some physical properties in doughs or syrups production, as a detergent ingredient and improving biofuel generation as previous step promoting fermentable sugar release, also in textile, paper industries, and medical applications as part of controlled drug delivery systems, or as part of therapy in digestive disorders (Sharma and Satyanarayana, 2013; Azzopardi et al., 2016; Mobini-Dehkordi and Afzal-Javan 2012). This enzyme is produced from different organisms, but nevertheless for industrial scale the microbial source is the most viable when is compared with plant or animal sources, due to their biochemical versatility such as wide range of pH or temperature of action, stability, and their presence in plenty microbial strains, owing to their advantages over the plant and animal amylases and achieving a name of valuable biotechnological tools for enzyme production including α-amylase, as submerged and solid fermentation (Hansen et al., 2015; and Hashemi et al., 2013). These two biotechnological processes could apply to different types of microorganisms, also possess a different array of advantages making possible a selection of one of them depending on the procedures challenges and the expected results.

The submerged fermentation (SmF) is one of the most used procedures in enzyme production. This technique is characterized by use of liquid media (liquid nutrient broth), resulting in a highly controlled process (Rani et al., 2010). In this process it is possible to regulate a set of parameters like growth media composition, pH, temperature, aeration, or agitation that directly impacts in microorganism metabolism and metabolite yields (Hasen et al., 2015). On the other hand the solid-state fermentation (SSF) constitutes a biotechnological process with the key difference in moisture implied in process using just the moisture necessary to support organisms growth making a spoilage resistant media and low energy requirement due to the unnecessary sterilization process, this technology possesses an special compatibility with filamentous fungi being very close at their natural growth media promoting a high productivity yields and allowing the use a different agroindustrial waste as support/substrate (Soccol et al., 2017). Both techniques are explored using different strategies as focusing in engineering aspects as Ghobadi (Ghobadi et al., 2017), who studied the effect of two different agitation geometries in SmF using *Aspergillus oryzae* for α-amylase production, also different statistical approaches were used for enhancing the same enzyme production as Prajapati did (Prajapati et al., 2015), who used a screening methodology and a posterior surface methodology, and other authors also made efforts for an extensive study of medium components including different agrowastes and control parameters to promote a better α-amylase generation (Dar et al., 2015). Further studies about α-amylase as realized by Hashemi et al. (2013) compared not only productivity in the process but also analyzed the enzyme performance during SmF and SSF, another aspect that is exploited in these biotechnological techniques is the fact that microorganism could be genetically modified for further use in enzyme production as Wang (Wang et al., 2019) that use a novel gene encoding α-amylase production, from highly efficient fungus Rhizomucor miheni for their expression in Pichia pastoris; likewise, experts try to take advantage of complex metabolic tools in microorganisms and the heterogenicity in agro-industrial wastes for simultaneous production of different enzymes class as Qreshi (Qureshi et al., 2016) that study the co-production of protease and α-amylase in different lignocellulosic materials for a Bacillus sp. strain during SSF. The progressive intensification in enzyme including α-amylase, usage as bio-catalysis strategy in industrial and other sections make the field of study a huge opportunity for Scientific's with plenty of questions to solve, holding attention and efforts in microbial sources due to their multiple benefits and plasticity that can offer.

3.5.1 BACTERIAL AMYLASE

The α-amylase applications acquire a broad range of conditions, as previously mentioned microorganisms, contains a full group of metabolic tools allowing their survival in different conditions making the microbial source advantageous for industrial and human employment, each industrial target process possess unique characteristics; syrup production needs acidic α-amylase while detergent industry demands alkali environment active amylase; also some procedures need enzymes which are functional and stable at high temperature or do not depend on ions (Bhanja et al., 2016).

One of the most exploited bacterial genera through the history for industrial α-amylase production is *Bacillus* sp. (Erick et al., 2000), nevertheless actually horizons have been expanded toward the implementation and hunt of novel organism as extremophiles for α-amylases with unique characters as Wang (Wang et al., 2019) that study and purify an α-amylase from *Pseudoalteromonas* sp. a bacterium from Antarctic sea, or Allala (Allala et al., 2019) who study the potential of an extracellular thermal- and alkali-stable α-amylases from bacteria *Tepidimonas fonticaldi* isolated from thermal springs; however, this extremophiles organism not always possess a reasonable productivity rates and is necessary employs genetic engineering for their expression as heterologous proteins (Sindhu et al., 2017), also actually the scientist are looking for conjugate novel and aforementioned classical approaches as Ahmed et al. (2017) who study the production of α-amylase by different *Bacillus* species using SmF and different agrowastes, but exploring mono and cocultures for enhance the enzyme production, some authors report that coculture may impact in microorganisms metabolite production (Wo et al., 2019).

3.5.2 FUNGAL AMYLASE

Fungi represent a particular group of organisms in which one could find filamentous fungi and yeast. In nature the fungi kingdom is distributed in a wide range of different environments, and this group is characterized by playing an essential role in nutrient recycling and in biochemical reactions involved, attributed mainly to their broad enzyme collection; in industry the fungal enzymes possess advantages because these are highly resistant to adverse conditions attributed to glycosylation process (Seppälä et al., 2017, and Østergaard et al., 2011). The most predominant fungal α-amylase genera are *Penicillium* sp. and *Aspergillus* sp. (Kathiresan et al., 2006), nevertheless it is essential to find alternative sources, as Tallapragada (Tallapragada et al., 2017) who studied the production and the potential application in detergent industry of an α-amylase produced by *Monascus sanguineus* under SSF employing beetroot peels, other authors also found and purified another α-amylase from *Aspergillus oryzae* produced by SSF with the particular character of being alcohol tolerant (Bhanja et al., 2015); however other environments are explored as marine habitat (Homaei, et al., 2016), likewise other efforts are applied as immobilization techniques for *Aspergillus awamori* enzymes (Karam et al., 2017). Furthermore, the same approach of employ molecular tools is attempted, as Li (Li et al, 2018) promoted changes in the amino acid sequence of α-amylase produced by *Rhizopus oryzae* to enhance their activity parameters like pH and temperature stability. Additionally, Wang (Wang et al., 2018) studied that α-amylase gene family from *Aspergillus niger* encodes eight putative α-amylases and these were expressed in *P. pastoris* with different biochemical characteristics, report novel amylases with industrial potential, also.

3.6 A-AMYLASE SALIVA ISOENZYME

3.6.1 SALIVA

Saliva is a complex biological fluid produced and secreted by salivary glands (parotid, submandibular, sublingual glands, and minor salivary glands). That has represented an essential role in maintaining the health of the oral cavity and gastrointestinal tract by aiding in lubrication, mastication, protecting against teeth demineralization, and preventing dental wear, buffering action to maintain oral pH. Saliva not only maintains oral hygiene due to its antimicrobial and antiviral activity but also promotes the healing of wounds in several ways (Peytro and Breslin, 2016; Tiwari, 2011; Mandel et al., 2010). Besides these activities, saliva also plays an essential role in the digestion and metabolism of food; nevertheless, the exact mechanism of digestion is unclear.

Saliva contains a lot of different substances such as lipids, carbohydrates, electrolytes, hormones, water, and a considerable amount of proteins, peptides, and enzymes. Examples of these are lipases, peptidases, lysozymes, and hydrolases (Peytro and Breslin, 2016; Loo et al., 2010). These molecules play different roles in different functions to maintain the health of the oral cavity and gastrointestinal tract mentioned in previous lines. Mainly, a very important hydrolase that is in saliva and contributes primarily to digestion is the α-amylase. This enzyme is responsible for starch digestion due to its action of breakdown of α-1,4-glycosidic linkages and thus converts starch to dextrin and subsequently into smaller saccharides (Mandel et al. 2010).

3.6.2 SALIVARY α-AMYLASE

As it is mentioned in previous lines, α-amylase is an enzyme that catalyzes the hydrolysis of the internal α-1,4-glycosidic linkages in starch, obtaining low-molecular-weight products such as glucose, maltose, and maltotriose units (de Souza et al., 2010). This enzyme can be found in different organisms, in the human being the enzyme is found in the pancreas and oral cavity, particularly in saliva (Butterworth et al., 2011); both types of α-amylase, pancreatic and salivary, share about 94%–97% homology (Alpers, 2003). Although both enzymes are similar, they are encoded by different genes: AMY1 and AMY2 (Bonnefond et al., 2017). Salivary and pancreatic amylases act together, but in different stages, to carry out digestion. The primary digestion of dietary starch in humans accomplished by salivary α-amylase; this amylolytic digestion starts during mastication in the oral cavity and continues within the stomach. Then the mixture passes into the small intestine, where pancreatic amylase completes starch hydrolysis.

α-Amylase was first described in saliva in 1831 by Erhard Leuchs when he discovered that starch broke down when it was mixed with saliva. A few years later Payen and Persoz carried out similar experiments and isolated an enzyme which broke down starch (Butterworth et al., 2011; Zakowski and Bruns, 1985).

Since its discovery, salivary α-amylase has been widely studied due to its essential role in starch digestion and also, according to some studies it could be used in clinical practice not only as a biomarker of stress and other neurological illness (Vineetha et al., 2014) but also as a possible diabetes indicator (Mortazavi et al., 2014). According to different investigations, salivary α-amylase can be used as a target for drug design in attempts to treat diseases that affect a big part of world population, such as DM, obesity, hyperlipidemia, and caries. For these diseases, the control of carbohydrate digestion and monosaccharide absorption could be helpful to avoid further complications. At least potentially, the control of carbohydrate digestion and monosaccharide absorption can be brought about employing enzyme inhibitors and in this particular aspect, IAA are especially promising (Kim and Nho, 2004).

α-Amylase is one of the most abundant enzymes in the oral cavity accounting for 40%–50% of salivary protein since it plays an essential role in starch digestion, which is a common carbohydrate in the human diet (Dhital et al., 2015). This enzyme is produced by serous cells of the parotid gland but also by sublingual, submaxillary, and minor glands (Nikitkova et al., 2013).

On average, the concentration of α-amylase in human saliva ranges from 0.04 to 0.4 mg/mL (Jacobsen et al. 1972). Nevertheless, the quantity and enzymatic activity of salivary amylase, however, show significant variation among individuals. This is due to a few environmental factors, including stress levels and circadian rhythms. According to different studies, the concentration of salivary amylase can increase during food consumption and high levels of stress. Additionally, there is evidence that salivary amylase expression upregulated by a diet high in starch. Genetically, salivary amylase levels are influenced by individual copy number variation (CNVs) of the AMY1 gene on chromosome 1p21, which codes for salivary amylase (Mandel et al., 2010).

3.6.3 ISOENZYMES

There is not much clear information about salivary α-amylase isoenzymes. Some authors report that this enzyme consists of two families of isoenzymes: Form A and Form B. Form A has a higher molecular weight than Form B which consists of approximately between 57,000 and 60,000 Da, on the other hand, Form B consists a molecular weight between 54,000 and 57,000 Da (Zakowski and Bruns, 1985). The difference between Forms A and B is an attribute to the presence of glycosylation on Form A, and thus the absence of this in Form B (Nater et al., 2009). These isoenzymes have been studied for many years through the use of different types of electrophoresis or chromatography, other authors have reported more than two isoenzymes of salivary α-amylase, for example, Eckersall and Beeley (1981) mentioned in their study that the enzyme possessed not only five mayor isoenzymes but also a few minors. These isoenzymes had different PIs and different phenotypes.

3.7 PHYTOCONSTITUENTS WITH α-AMYLASE INHIBITORY ACTIVITY

3.7.1 SYNTHETIC CONSTITUENTS WITH α-AMYLASE INHIBITORY ACTIVITY

The natural extracts that obtained from plants with high antioxidant agents content with α-amylase enzyme inhibitory activity are a great alternative in the treatment of diabetes, the disadvantage that can occur is that the mechanism of action is no entirely clear, there are only studies with which a consensus was created in which the inhibitory action present. Among the IAA compounds that have been able to characterize can classify into two big groups: (1) acarbose and its derivatives and (2) polyphenolic compounds.

3.7.1.1 ACARBOSE AND ITS DERIVATIVES

Acarbose (Figure 3.1) is an oligosaccharide obtained from the fermentation of *Actinoplanes uthanesis*, which consists of a polyhydroxylated (valienamine) derived from aminocyclohenene linked through nitrogen atoms to a 6-deoxyglucose, which is bound to maltose (Gyemant, 2003). It is currently used as a complement in the treatment of diabetes when accompanying metformin. It presents a pK_a 11.83, which favors the site of action in the intestinal tract, acts by delaying the digestion and absorption of carbohydrates by blocking the enzymes α-glucosidase and α-amylase (Sales, 2012).

The adverse effects that oral acarbose treatment can present are abdominal discomfort, flatulence, and diarrhea. These effects can cause by the increase of degradation products in the intestine as a result of bacterial fermentation of undigested carbohydrates; they occur very frequently in diabetic patients treated with IAA (Gyemant, 2003; Kwon, 2008). There are also reports of an increase in kidney tumors and liver damage (Murai, 2002). Acarbose may have a synergistic effect in the presence of flavonoids such as baicalein (Zhang, 2017). The enzyme α-amylase binds to the polysaccharides, and acarbose is a pseudotetrasaccharide that contains a nonhydrolyzable nitrogen-<inked bond that suppresses the activity of α-amylase through competition, this inhibition is reversible (Li, 2005; Sales, 2012; Yilmazer-Musa, 2012).

The valienamine ring (Figure 3.3) of acarbose is considered crucial in the mechanism of inhibition of the enzymes α-glucosidase, α-amylase, and other amylolytic enzymes. Brayer et al. (2000) carried out a structural study of the α-amylase/acarbose complex where they describe the inhibition carried out in the active subsites "−3" toward "+2." Kadziola et al. (1998) mentioned that the mechanism of the complex operates between a pair of carboxylic acids at a distance of 5 Å, which is very characteristic of enzyme retention, first the catalyst donates a proton for the formation of the glycosidic bond, followed by the formation of a covalent bond between the rest of the sugar molecule, the second reaction is a hydrolysis by the presence of water from the covalent bond of the glucosilenzime complex creating a displacement with the deprotonated catalytic acid, proposes that the two displacements operate by transition states with substantial character of oxocarbonium ions.

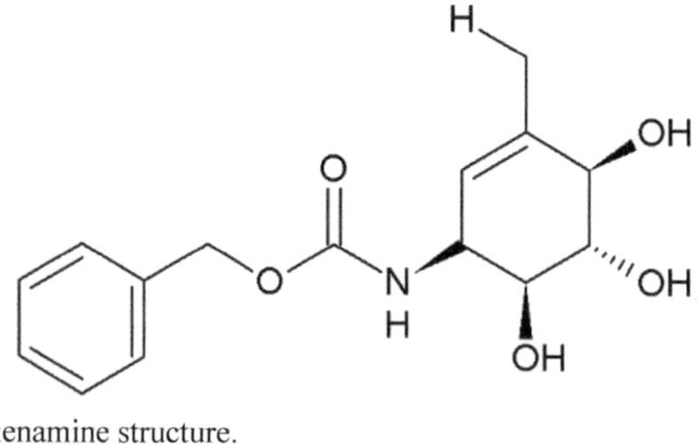

FIGURE 3.3 Valienamine structure.

Rafique et al. (2019) managed to synthesize several compounds (modified *N*-sulfonohydrazide) based on acarbose and compared with it the inhibitory

Natural Extracts and Compounds as Inhibitors

activity of α-amylase in vitro, presenting better performance and thus managed to open a new branch of research where substitutions can be made in the functional groups of acarbose without interfering or in the case of some substitutions which improve inhibition performance. Table 3.1 shows the structures and mean inhibitory concentration in vitro tests.

TABLE 3.1 Compounds Derived from Acarbose and Their Average Inhibitory Concentration

Compound	Structure	IC_{50}
Acarbose	Figure 7.1	1.20 ± 0.09
N-sulfonohydrazide modificado		4.50 ± 0.03
		3.91 ± 0.05
		3.52 ± 0.03
		2.92 ± 0.08

TABLE 3.1 *(Continued)*

Compound	Structure	IC$_{50}$
	(isopropylsulfonyl hydrazone of 6,6-dimethyl-1-phenyl-4,5,6,7-tetrahydro-1H-indazol-4-one)	1.99 ± 0.12
	(trifluoromethylsulfonyl hydrazone of 6,6-dimethyl-1-phenyl-4,5,6,7-tetrahydro-1H-indazol-4-one)	1.23 ± 0.06
	(phenylsulfonyl hydrazone of 6,6-dimethyl-1-phenyl-4,5,6,7-tetrahydro-1H-indazol-4-one)	3.97 ± 0.08
	(p-tolylsulfonyl hydrazone of 6,6-dimethyl-1-phenyl-4,5,6,7-tetrahydro-1H-indazol-4-one)	3.72 ± 0.11

TABLE 3.1 *(Continued)*

Compound	Structure	IC$_{50}$
		3.63 ± 0.12
		3.83 ± 0.11
		3.41 ± 0.16
		1.66 ± 0.08

TABLE 3.1 *(Continued)*

Compound	Structure	IC$_{50}$
	(4-nitrobenzenesulfonohydrazide derivative of 6,6-dimethyl-1-phenyl-4,5,6,7-tetrahydro-1H-indazol-4-one)	1.37 ± 0.03
	(2-methyl-5-nitrobenzenesulfonohydrazide derivative of 6,6-dimethyl-1-phenyl-4,5,6,7-tetrahydro-1H-indazol-4-one)	1.93 ± 0.09
	(2,4-dichlorobenzenesulfonohydrazide derivative of 6,6-dimethyl-1-phenyl-4,5,6,7-tetrahydro-1H-indazol-4-one)	1.38 ± 0.11

TABLE 3.1 *(Continued)*

Compound	Structure	IC$_{50}$
	2,5-dichlorophenylsulfonyl hydrazone of 6,6-dimethyl-1-phenyl-4,5,6,7-tetrahydro-1H-indazol-4-one	1.91 ± 0.11
	3,5-dichloro-2-hydroxyphenylsulfonyl hydrazone of 6,6-dimethyl-1-phenyl-4,5,6,7-tetrahydro-1H-indazol-4-one	1.52 ± 0.09
	5-chloro-2-methoxyphenylsulfonyl hydrazone of 6,6-dimethyl-1-phenyl-4,5,6,7-tetrahydro-1H-indazol-4-one	2.51 ± 0.09
	4-methoxyphenylsulfonyl hydrazone of 6,6-dimethyl-1-phenyl-4,5,6,7-tetrahydro-1H-indazol-4-one	1.97 ± 0.03

3.7.1.2 POLYPHENOLIC COMPOUNDS AS α-AMYLASE INHIBITORS

The α-amylase is one of the main enzymes responsible for the hydrolysis of starch, breaking the α-1–4 glycosidic bonds generating simple sugars such as glucose, in addition to other sugars such as maltose and dextrins. On the other hand, α-glucosidase is a crucial enzyme for metabolism, transforming oligosaccharides into monosaccharides absorbable by the small intestine. The inhibition of these enzymes can delay the digestion of carbohydrates and reduce the rate of glucose absorption; by authorization, these inhibitors can improve glucose tolerance in patients with diabetes (Rubilar et al., 2011).

Polyphenolic compounds are known to be useful in formulating nutritional supplements or medications for different diseases, due to their antioxidant, antifungal, and antimicrobial properties. In addition, these compounds can inhibit digestive enzymes such as α-amylase and α-glucosidase, which can delay the digestion of carbohydrates ingested through the diet, which causes a reduction in glucose absorption (Rubilar et al., 2011).

Phenolic compounds are the largest group of nonenergy substances present in foods of plant origin. In recent years it has been shown that a diet rich in vegetable polyphenols can improve health and reduce the incidence of diseases (Perez et al., 2009).

The ability of polyphenols to modulate the activity of different enzymes and to consequently interfere with signaling mechanisms and different cellular processes may be, at least in part, due to the physicochemical characteristics of these compounds, which allow them to participate in different reactions (cellular metabolic oxide-reduction). Eight thousand different polyphenols have been described, which grouped into four broad classes: phenolic acids, flavonoids, tannins, and lignins, which are highly polymerized compounds. Tannins have astringent properties and affect the digestibility of nutrients, particularly food nitrogen, by their binding to exogenous and endogenous proteins. Also, it can reduce the activity of certain digestive enzymes such as trypsin, amylase, and lipase (Vázquez et al., 2005).

Among the substances of phenolic character, those that have been an object of more in-depth studies are the flavonoids. They comprise a large group of secondary metabolites that derive from subunits that come from the metabolic pathways of acetate and shikimate. They are found almost exclusively in higher plants and are presented in two very distinctive ways: linked to glycidic units (flavonoids glycosides) or free (flavonoids aglycones), as is the case of flavanones (catechins and proanthocyanidins). Flavonoids have been isolated from many plant drugs because they are popular natural products. Its polyphenols have been attributed to a number of actions ranging from inhibitory enzyme activities (hydrolases, cyclooxygenases, alkaline phosphatase, cAMP phosphodiesterases, ATPase, liases, hydroxylases, transferases, oxidoreductases, and kinases), to antiinflammatory, anticancer action, antibacterial, and antiviral (Paredes et al., 2005).

3.7.2 PLANTS AS A SOURCE OF α-AMYLASE INHIBITORS

Plants are one of the most important sources of chemical compounds that can effectively inhibit α-amylase. The advantage of using plant inhibitors as therapeutic agents in the nutraceutical treatments of DM is their accessibility and a secure method of isolation by extraction. It has been reported that there are more than 1200 types of medicinal plants used to treat diabetes, approximately half of which tested for antihyperglycemic properties and an essential part of which is applicable for food purposes (Wijaya et al., 2011, Ilyina et al., 2017).

Among the studied phytocomponents of these plants, phenolic compounds have the highest inhibitory activity (Table 3.2), such as flavones (luteolin, acacetin, bilobetin, lonericin), flavanones (naringenin, hesperetin), flavonols (quercetin, rutin, kaempferol, hyperin), flavanonols, anthocyanidins, flavanediols, chalcones, isoflavones, and catechins, as well as polysaccharides, hypoglycans, peptidoglycans, guanidines, steroids, glycopeptides, terpenoids, and alkaloids (Table 3.2). A high inhibitory potential demonstrated for flavonoids and flavin groups. Besides, tannins are also known for their ability to form active complexes with enzymes (Table 3.2) (Kunakova et al., 2016).

Zaynullin et al. (2018) described the ability of a powerful antioxidant dihydroquercetin (DHQ), also known as taxifolin ((2R, 3R)-2-(3,4-dihydroxyphenyl)-3,5,7-trihydroxy-2,3-dihydro-4H-chromen-4-one), as an inhibitor of the α-amylase. The IC_{50} of this compound was close to 9 mM. The competitive kinetic mechanism of inhibition demonstrated. Inhibition constant estimated as 2.25 ± 0.22 mM (684 ± 66 ppm). It indicated a high affinity of DHQ to the enzyme. The effect of DHQ as a starch hydrolysis inhibitor was confirmed in *Saccharomyces cerevisiae* biological prototypical. Molecular docking analysis demonstrated the binding of DHQ through hydrogen bonding with Tyr^{193}, Trp^{263}, and His^{327} of the α-amylase. These interactions cause competitive enzyme inhibition. DHQ establishes a potential in the development of new drugs for the control of DM. The authors proposed the functional use of DHQ as a hypoglycemic food additive for the prevention and treatment of type 2 diabetes (Zaynullin et al., 2018).

TABLE 3.2 Chemical Structures of Some Phytochemical Compounds are Known for their Ability to Inhibit the Enzyme Alpha-Amylase

Luteolin	Naringenin
Quercetin	Epigallocatechin
Morphine	Tannin
Strictinin	Isostrictinin
Streptozocin	Phaseolamine

Sallam and Galala (2017) analyzed α-amylase inhibition by phenolic compounds presented in extracts from *Gymnocarpos decandrus* Forssk. Gallocatechin exhibited the highest α-amylase inhibitory activity (68.3 %) in comparison with other test substances. Employing docking analysis, the potential binding sites of gallocatechin were found to be Asp^{300}, Glu^{233}, Trp^{59}, and Gln63 through hydrogen bonding. Catechin and epicatechin demonstrated nearly similar α-amylase inhibitory activities at 67.0% and 66.7%, respectively. The potential binding sites of catechin through hydrogen bonding were Asp ^{300}, Glu^{233}, and Gln^{63}, whereas the binding sites of epicatechin were Gln^{63}, Trp^{59}, Asp^{300}, Asp^{197} through hydrogen bonding and Trp^{59} and Tyr^{62} through hydrophobic (π–π) interaction. The binding sites of afzelechin through hydrogen bonding defined as Asp^{300}, Glu^{233}, Trp^{59}, and Tyr^{62} through hydrophobic (π–π) interaction. It is to note that tryptophan residues frequently mentioned as part of the interaction between phenolic compounds and α-amylase.

Amylase inhibitors have been found in the seeds of at least 20 types of legumes, as well as in wheat and rye grains. Species of plants such as jambolana or yambolan (*Syzygium cumini* L.) and guayaeva (*Psidium guajava* L.) are widely used in the traditional system of DM treatment in India (Table 3.3). It was shown that glycosides from guava, such as strictinin and isocyctrinin (Table 3.2), which are widely used in the clinical DM treatment, have antihyperglycemic properties (Kunakova et al., 2016). For an alcoholic extract of nutty lotus (*Nelumbo nucifera* Gaertn.), a decrease in glucose level shown in experimental rats and it discovered that streptozocin (Table 3.2) plays the very important role (Alagesan et al., 2012). Methanol and ethyl acetate extracts from two varieties of amaranth (*Amaranthus caudatus* L.) showed the inhibitory activity on α-amylase (more than 80%) at a dose of 0.25–1.00 mg/mL (Kunakova et al., 2016).

The buffered extracts from several plant species, namely as fenugreek (*Trigonella foenum-graceum* L.), black pepper (*Piper nigrum* L.), Egyptian balanitis (*Balanitesa aegyptiaca* L.), common rosemary (*Rosmarinus officinalis* L.), Chinese camellia (*Camellia sinensis* L.), kratom (*Mitragyna inermis* L.), tamarind (*Tamarindus indica* L.), and lemon balm (*Melissa officinalis* L.), were proved to have α-amylase inhibitory activity: an average of 45% decrease in activity at 0.2 g/mL of extracts (Kunakova et al., 2016).

Extracts from Aloe vera (*Aloe vera* L.), palm-shaped adansonia (*Adansonia digitata* L.), pink catharanthus (*Catharanthus roseus* L.), cinnamon (*Cinnamomum verum* J. Presl), large coccinia (*Coccinia grandis* L.), Indian mango (*Mangifera indica* L.), white mulberry (*Morus alba* L.), common oleander (*Nerium oleander* L.), thin-layered basil (*Ocimum tenuiflorum* L.), haritaki (*Terminalia chebula* Retz.), tinospore cordy (*Tinospora cordifolia* (Thunb.) Miers) is widely used for DM treatment in western India (Ilyna et al., 2017; Bhat et al., 2007).

TABLE 1.3 Some Plants That Contain Alpha-Amylase Inhibitors

| *Galega officinalis* L. | Medicinal dandelion | Common flax |
| Garlic | Jambolan | Guava |

Some degree of inhibition (nearly 30%) of the enzyme observed for extracts of herbs traditionally used in Mongolian medicine, such as rhodiola (*Rhodiola rosea* L.), currant (*Ribespull chelum* Turcz), and lingonberry (*Vaccinium uliginosum* L.) (Kunakova et al., 2016). Methanol, hexane, and chloroform extracts of Lebanese traditional medicinal plants are recommended in this country as the treatment of diabetes (Alagesan et al., 2012). Water extracts of some Indonesian plants, namely, patanga (*Caesalpinia sappan* L.) and kaffir-lime (*Citrus hystrix* L.) also demonstrated nearly 96% and 41% of enzyme inhibition, respectively. IAA also identified in Indian Azadirachta (*Azadirachta indica* A. Juss) (Bhat et al., 2007), Tulasi (*O. tenuiflorum* L.), and for some plant species in the genus *Teucrium* (Bhat et al., 2007). Inhibitory activity was reported for some plant species from Russia: *Galega officinalis* L., the dandelion medicinal (*Taraxacum officinale* L.), common flax (*Linum usitatisumum* L.), and seed garlic (*Allium sativum* L.) (Table 3.2) (Kunakova et al., 2016). Ilyina et al. (2014) defined in vitro the effect of betulin containing extract from *Betula pendula Roth.* bark on α-amylase activity. Betulin-containing extract was isolated from *Betula pendula Roth.* bark with isopropanol. In vitro assay was carried out using amylase from porcine pancreas (Sigma-Aldrich). The more marked activity decrease (at 20%) was detected with extract concentration increasing to 1000 mg/mL, whereas under higher extract concentrations, the minor decline in activity was observed. Using Dixon and Cornish–Bowden coordinates, the competitive mechanism of inhibition was demonstrated. Calculated kinetic parameters were: K_m equal to 0.6 mg/mL, V_{max} equal to 2.6 and 2.1 mM/min from Lineweaver-Burk and Dixon coordinates, respectively; K_i equal to 3670 ± 230 mg/mL. The partial inhibition of enzyme indicates the existence of a low concentration of active inhibitory form, which reaches a saturation level with increasing extract concentration in applied suspension. Therefore K_i has an apparent constant character. This partial inhibition of amylase activity observed in vitro assay was not reflected at any antinutritional effect on weight gain and meat quality of broiler chickens during in vivo assay. Instead, the tendency to improve the meat quality was observed. This extract can be recommended for the application in poultry farming, as well as provides high potential for further pharmaceutical and pharmacological research.

The pharmacotherapeutic role of IAA is based on the blocking effect of α-amylase. First-generation IAAs were recommended in the 1980s to reduce carbohydrate loading and ease hyperglycemia. Initially, an aqueous extract of white beans (*Phaseolu vulgaris* L.) was used, in which a phaseolamin was identified inhibitor (Table 3.2). However, a low selectivity for binding to enzymes found for this compound, and for this reason many undesirable side effects observed. Later similar structures were tested, among which the most promising was acarbose (Figure 3.1) (Kunakova et al., 2016).

As shown by numerous studies, flavonoids and their derivatives (Table 3.2) can influence the course of a wide variety of physiological processes through various mechanisms. Their distinctive pharmacological property is a decrease in the permeability and fragility of the walls of blood vessels due to the antioxidant and membrane-stabilizing effect. The preparations containing them also have antispasmodic, anti-inflammatory, and diuretic effects (Zaynullin et al., 2018). It was found that flavonoids affect regulatory proteins and various protein structures, including enzymes (Sallam et al., 2017). The effectiveness of their interaction with enzymes, combined with low toxicity, allows them to be used to solve a wide variety of technological, pharmaceutical, and biotechnological problems. These examples give a basis to study the inhibitory abilities of phenolic compounds contained in plant tissues and available for isolation concerning the most studied and used enzymes, such as α-amylases (Zaynullin et al., 2018; Sallam et al., 2017; Alagesan et al., 2012).

Zaynullin et al. (2014) studied in vitro effects of extracts from *Betula pendula Roth* bark, china tea, *Opuntia ficus-indica* root on α-amylase activity, which compared with acarbose inhibitory effect (Zaynullin et al., 2014). The similar inhibition was observed for the watery extract of *Opuntia ficus-indica* root and medicine, whereas other extracts characterized by the less inhibitory effect. The competitive kinetic mechanism of inhibition demonstrated for watery suspension of extracts obtained after solvent evaporation from watery and ethanolic infusions of *Opuntia ficus-indica* root.

Arredondo Valdés et al. (2018) performed the search and characterization of extracts of medicinal plants useful for the design of products that lead to postprandial glucose control, to achieve maximum well-being and quality of lifecycle of patients with diabetes or prediabetes. There is an ethnopharmacological study of medicinal plants available in Saltillo, Coahuila, and screening for the evaluation of the inhibitory activity of a-amylase extracts. Exact half of the 22 plants submitted in the study showed a-amylase inhibitory effect above 10% of initial activity at a concentration of 666.7 ppm. The IC_{50} value for the aqueous suspension of the ethanol extract of *Bidens odorata* was 851 ppm similar to that detected with the acarbose drug. The findings of this study revealed the species of medicinal plants, which were not previously considered as carriers of IAA and can be used in the production of functional hypoglycemic food.

The isolation from plants growing in Mexico of new nontoxic extracts and compounds with inhibitory activity on α-amylase will allow the prevention of diabetes complications by developing functional foods and nutraceuticals based on them and will provide an incentive for the development of high-tech, knowledge-intensive, and import-substituting technologies.

3.8 CONCLUSION

Diabetes and obesity are two of the most common problems among human societies. One of the most current methods for the prevention of these diseases is the reduction of caloric intake by decreasing the food digestion and absorption. Carbohydrates digested with many enzymes such as amylase that break glucose polymer such as starch to a low-molecular-weight product that hydrolyzes. Since carbohydrate is one of the significant parts of the human diet, we tried to review on IAA. α-Amylase can hydrolyze carbohydrates such as starch. By α-amylase inhibition in people with diabetes, lowering blood sugar levels can cause weight reduction in obese individuals. The studies on IAA have done on a broad range of research. In research for reducing the number of digestible carbohydrates and low blood sugar, the idea has been to use carbohydrates with low levels. For this purpose, resistance starch should be used which must be prepared either synthetically or extracted and added to food.

On the other hand, foods are manipulated and maybe have some side effects in a long time. Respect to the structure of the enzyme and recognizing its active site, the researchers have tried to reduce the enzyme activity or inactivate it by manipulation of amino acids in the active site. One method that seems to be useful for producing inactive enzymes is the genetic manipulation of the HPA producing genes including AMY2A and AMY2 and producing of the inactive enzyme. This is due to being irreversible of the enzyme and the lack of complete digestion of starch material is not impossible in human samples. So, using amylase inhibitors can be used as an appropriate method in reducing blood sugar levels. Amylase inhibitors either synthetic or derived from plants are generally categorized into three groups including pseudosaccharides, proteinaceous, and polyphenolic inhibitors. The inhibitory effect of polyphenolic compounds has been studied in a wide range. These compounds are very varied and diverse in shape and molecular structure and exist in different amounts in various fruits, vegetables, and plant products. A significant group of polyphenolic compounds is flavonoids, in which many lines of research have done about their inhibition effects. Most of them have a strong inhibitory effect on enzymes and can utilize as a base designation of inhibitory drugs. With this approach, by application of a lower dose of drugs, the blood sugar levels will be reduced. α-Amylase synthesis inhibitors can cause serious side effects, so it is better to emphasize much more significant on the use of plant inhibitors. Most plant-derived compounds have fewer side effects than synthetic compounds. It seems that using fruits and vegetables naturally, which have polyphenolic compounds especially flavonoids, can inhibit α-amylase, which are the best and most useful substances to lower the blood sugar. In conclusion, inhibition of α-amylase is a successful manner in the prevention and therapy of obesity and diabetes. Traditional medicines, including herbal medicine, possess great potential in this regard. A better method than alternative one to inhibit α-amylase is using amylase inhibitors in reducing blood sugar, but it is necessary to continue with the research to find new treatments or alternative to DM treatment.

ACKNOWLEDGMENTS

The author's acknowledgment of the financial support for the development of research to the National Council of Science and Technology of Mexico (CONACYT) for scholarship: 1010048. Another way author's acknowledgment of the financial support for the development of research to the Secretary Public Education for project UACOAH-PTC-485-2019.

KEYWORDS

- α-amylase
- inhibition
- hyperglycemia
- diabetes
- functional food

REFERENCES

Adda, A., Regagba, Z., Latigui, A., & Merah, O. (2014). Effect of salt stress on a-amylase activity, sugars mobilization and osmotic potential of *Phaseolus vulgaris* l. Seeds var. 'Cocorose' and 'djadida' during germination. Journal of Biological Sciences, 14(5), 370–375. doi:10.3923/jbs.2014.370.375

Ahmed, S. A., Mostafa, F. A., Helmy, W. A., & Abdel-Naby, N. A. (2017). Improvement of bacterial a-amylase production and application using two steps statistical factorial design. Biocatalysis and Agricultural Biotechnology, 10, 224–233. Doi: 10.1016/j.bcab.2017.03.004

Alagesan, K., & Krishnan Raghupathi, P. (2012). Amylase inhibitors: potential source of anti-diabetic drug discovery from medicinal plants. International Journal of Pharmacy & Life Sciences, Feb, -P. 1409.

Allala, F., Bouacem, K., Boucherba, N., Azzouz, Z., Mechri, S., and Sahnoun, M. (2019). Purification, biochemical, and molecular characterization of a novel extracellular thermostable and alkaline α-amylase from *Tepidimonas fonticaldi* strain HB23. International Journal of Biological Macromolecules, 132, 558–574.

Alpers, D. H. (2003). CARBOHYDRATES | Digestion, absorption, and metabolism. Encyclopedia of Food Sciences and Nutrition, 881–887. doi:10.1016/b0-12-227055-x/00168-1

Arbige, M. V, Shetty, J. K., & Chotani, G. K. (2019). Industrial enzymology: the next chapter. Trends in Biotechnology, 37(12), 1355–1366.

Arredondo Valdés R. (2013). Detección in vitro de inhibidores de α – amilasa en extractos de plantas mexicanas con aplicaciones bio-médicas. Tesis de Maestría en Biotecnología, Facultad de Ciencias Químicas de la UA de C.

Arredondo Valdés R., Martínez Hernández J.L., Silva Belmares S.Y., Segura Ceniceros E.P., Sierra Rivera C.A, Paredes Ramírez A.R.,& Iliná A. (2018). Evaluación del potencial hipoglucemiante de infusiones vegetales de plantas empleadas en medicina tradicional mexicana. CienciAcierta, 56, 1–10.

Asgher, M., Asad, M. J., Rahman, S. U., & Legge, R. L. (2007). A thermostable α-amylase from a moderately thermophilic *Bacillus subtilis* strain for starch processing. Journal of Food Engineering, 79(3), 950–955. doi:10.1016/j.jfoodeng.2005.12.053

Awasthi, M. K., Wong, J. W. C., Kumar, S., Awasthi, S. K., Wang, Q., Wang, M., . . . Zhang, Z. (2018). Biodegradation of food waste using microbial cultures producing thermostable alpha-amylase and cellulase under different pH and temperature. *Bioresource Technology*, 248(Pt B), 160–170. doi:10.1016/j.biortech.2017.06.160

Azzopardi, E., Lloyd, C., Teixeira, S. R., Conlan, R. S., & Whitaker. (2016) I. S. Clinical applications of amylase: novel perspectives. Surgery, 160 (1), 26–37. doi:10.1016/j.surg.2016.01.005.

Bertoft, E., Andtfolk, C., and Kulp, S. E. (1984). Effect of pH, temperature, and calcium ions on barley malt α-amylase isoenzymes. Journal of the Institute of Brewing, 90(5), 298–302. doi:10.1002/j.2050-0416.1984.tb04278.x

Bhanja, T., & Banerjee, R. (2015). Purification, biochemical characterization and application of α-amylase produced by *Aspergillus oryzae* IFO-30103. Biocatalysis and Agricultural Biotechnology, 4, 83–90.

Bhanja, T., Kumar, A., Banerjee, R., Chandna, P., & Chander, R. (2016). Improvement of microbial α-amylase stability: strategic approaches. Process Biochemistry, 51, 1380–1390.

Bhat, M., Zinjarde, Smita S., et al. (2011). Antidiabetic Indian Plants: A Good Source of Potent Amylase Inhibitors, Institute of Bioinformatics and Biotechnology, University of Pune, Pune 411 007, India Department of Zoology, University of Pune, Pune 411 007, India Received 21 April 2007; Accepted April 11, 2008 Copyright © 2011, -P. 2-3.

Bisswanger, H. (2014). Enzyme assays. Perspectives in Science, 1(1–6), 41–55. doi:10.1016/j.pisc.2014.02.005

Bonnefond, A., Yengo, L., Dechaume, A., Canouil, M., Castelain, M., Roger, E., & Froguel, P. (2017). Relationship between salivary/pancreatic amylase and body mass index: a systems biology approach. BMC Medicine, 15(1). doi:10.1186/s12916-017-0784-x

Brayer G.D, Luo Y., & Withers S.G. (1995). The structure of human pancreatic alpha amylase at 1.8 A resolution and comparisons with related enzymes. Protein Science 4(9), 1730–1734.

Brayer, G. D., Sidhu, G., Maurus, R., Rydberg, E. H., Braun, C., Wang, Y., Nguyen, N. T., Overall, C. M., & Withers, S. G. (2000). Subsite mapping of the human pancreatic α-amylase active site through structural, kinetic, and mutagenesis techniques. Biochemistry, 39(16), 4778–4791.

Butterworth, P. J., Warren, F. J., & Ellis, P. R. (2011). Human α-amylase and starch digestion: an interesting marriage. Starch, 63(7), 395–405. doi:10.1002/star.201000150

Danson, M. J., Hough, D. W., Russell, R. J., Taylor, G. L., & Pearl, L. (1996). Enzyme thermostability and thermoactivity. Protein Engineering, 9(8), 629-630. doi:10.1093/protein/9.8.629

Dar, G. H., Kamili, A. N., Nazir, R., Bandh, S. A., Jan, T. R., & Chishti, M. Z. (2015). Enhanced production of a-amylase by *Penicillium chrysogenum* in liquid culture by modifying the process parameters. Microbial Pathogenesis, 88, 10–15.

Dastjerdi, Z.M., Namjoyan, F. et al. (2015). Alpha amylase inhibition activity of some plants extract of Teucrium species. European Journal of Biological Sciences 7(1), 26–31.

De Souza, P. M., & de Oliveira Magalhães, P. (2010). Application of microbial α-amylase in industry—a review. Brazilian Journal of Microbiology, 41(4), 850–861. doi:10.1590/S1517-83822010000400004

Dhital S., Warren F.J., Butterworth P.J., Ellis P.R., & Gidley M.J. (2017). Mechanisms of starch digestion by α-amylase—structural basis for kinetic properties, Critical Reviews in Food Science and Nutrition, 57(5), 875–892. doi:10.1080/10408398.2014.922043

Eckersall, P. D., & Beeley, J. A. (1981). Genetic analysis of human salivary α-amylase isozymes by isoelectric focusing. Biochemical Genetics, 19(11–12), 1055–1062. doi:10.1007/bf00484564

Erik, J., & Borchert, T. V. (2000). Protein engineering of bacterial alpha-amylases. Biochimica et Biophysica Acta, 1543(2), 253–274.

Ghobadi, N., Ogino, C., Yamabe, K., & Ohmura, N. (2017). Characterizations of the submerged fermentation of Aspergillus oryzae using a Fullzone impeller in a stirred tank bioreactor. Journal of Bioscience and Bioengineering, 123(1), 101–108. doi:10.1016/j.jbiosc.2016.07.001

Gyémánt, G., Kandra, L., Nagy, V., & Somsák, L. (2003). Inhibition of human salivary α-amylase by glucopyranosylidene-spiro-thiohydantoin. Biochemical and Biophysical Research Communications, 312(2), 334–339.

Hansen, G. H., Lübeck, M., Frisvad, J. C., & Lübeck, P. S. (2015). Production of cellulolytic enzymes from Ascomycetes : comparison of solid state and submerged fermentation. Process Biochemistry, 50, 1327–1341.

Hashemi, M., Mohammad, S., Hadi, S., & Abbas, S. (2013). Comparison of submerged and solid state fermentation systems effects on the catalytic activity of Bacillus Sp . KR-8104 amylase at different pH and temperatures. Industrial Crops and Products, 43, 661–667.

Hiteshi, K., and Gupta, R. (2014). Thermal adaptation of a-amylases: a review. Extremophiles, 18(6), 937–944. doi:10.1007/s00792-014-0674-5

Homaei, A., Ghanbarzadeh, M., & Monsef, F. (2016). Biochemical features and kinetic properties of α-amylase from marine organisms. International Journal of Biological Macromolecules, 83, 306–314.

Ilyina A., Arredondo-Valdés R., Farkhutdinov S., Segura-Ceniceros E.P., Martínez-Hernández J.L., Zaynullin R., & Kunakova R. (2014). Effect of betulin-containing extract from birch tree bark on α-amylase activity in vitro and on weight gain of broiler chickens *in vivo*. Plant Foods for Human Nutrition, 69, 65–70. Doi:10.1007/s11130-014-0404-2. ISSN: 0921-9668 (print version). ISSN: 1573-9104 (electronic version)

Ilyina, A., Ramos-González, R., Segura-Ceniceros, E.P., Vargas-Segura, A., Martínez-Hernández, J.L., Zaynullin, R, Kunakova, R., & Korotina, T. (2017). Alimentary and medicinal plants in functional nutrition. Universidad Autónoma de Coahuila. 314 p. ISBN: 978-607-506-313-3

Jabbour, D., Sorger, A., Sahm, K., and Antranikian, G. (2013). A highly thermoactive and salt-tolerant alpha-amylase isolated from a pilot-plant biogas reactor. Applied Microbiology and Biotechnology, 97(7), 2971–2978. doi:10.1007/s00253-012-4194-x

Jacobsen, N., Melvaer, K. L., & Hensten-Pettersen, A. (1972). Some properties of salivary amylase: a survey of the literature and some observations. Journal of Dental Research, 51(2), 381–388. doi:10.1177/00220345720510022501

Jegannathan, K. R., & Nielsen, P. H (2013). Environmental assessment of enzyme use in industrial production—a literature review. Journal of Cleaner Production, 42, 228–240. doi:10.1016/j.jclepro.2012.11.005.

Kadziola, A., Søgaard, M., Svensson, B., & Haser, R. (1998). Molecular structure of a barley α-amylase-inhibitor complex: implications for starch binding and catalysis. Journal of Molecular Biology, 278(1), 205–217.

Karam, E. A., Abdel, W. A., Saleh, S. A. A., Hassan, M. E., Kansoh, A. L., & Esawy, M. A. (2017). Production, immobilization and thermodynamic studies of free and immobilized *Aspergillus Awamori* amylase. International Journal of Biological Macromolecules, 102, 694–703.

Kathiresan, K., & Manivannan, S. (2006). α-Amylase production by *Pencillium felluttanum* isolated from mangrove rhizosphere soil. African Journal of Biotechnology, 5(10), 829–832.

Kılıç Apar, D., and Özbek, B. (2004). α-Amylase inactivation by temperature during starch hydrolysis. *Process Biochemistry, 39*(9), 1137–1144. doi:10.1016/s0032-9592(03)00224-3

Kunakova R.V., R.A. Zaynullin, E.K. Khusnutdinova, B.I. Yalaev, E.P. Segura, A.D. Ilyina (2016). Plants as a valuable source of amylase inhibitors in creating functional foods for diabetes prevention. Herald of the Academy of Sciences of the Republic of Bashkortostan, 21(1): 6–15. (Artículo en ruso) www/vestnikanrb.ru. ISSN 1728-5283

Kwon, Y.-I., Apostolidis, E., & Shetty, K. (2008). In vitro studies of eggplant (Solanum melongena) phenolics as inhibitors of key enzymes relevant for type 2 diabetes and hypertension. Bioresource Technology, 99(8), 2981–2988.

Li, C., Begum, A., Numao, S., Park, K. H., Withers, S. G., & Brayer, G. D. (2005). Acarbose rearrangement mechanism implied by the kinetic and structural analysis of human pancreatic α-amylase in complex with analogues and their elongated counterparts. Biochemistry, 44(9), 3347–3357.

Li, S., Yang, Q., Tang, B., & Chen, A. (2018). Improvement of enzymatic properties of *Rhizopus Oryzae* α-amylase by site-saturation mutagenesis of histidine 286. Enzyme and Microbial Technology, 117, 96–102.

Loo, J. A., Yan, W., Ramachandran, P., & Wong, D. T. (2010). Comparative human salivary and plasma proteomes. Journal of Dental Research, 89(10), 1016–1023. doi:10.1177/0022034510380414

Mahmood N. (2016). A review of α-amylase inhibitors on weight loss and glycemic control in pathological state such as obesity and diabetes. Comparative Clinical Pathology, 25, 1253–1264 doi:10.1007/s00580-014-1967-x

Mandel, A. L., Peyrot des Gachons, C., Plank, K. L., Alarcon, S., & Breslin, P. A. S. (2010). Individual differences in AMY1 gene copy number, salivary α-amylase levels, and the perception of oral starch. PLoS One, 5(10), e13352. doi:10.1371/journal.pone.0013352

Marc J.E.C. van der Maarel, Bart van der Veen, Joost C.M. Uitdehaag, Hans Leemhuis, & L. Dijkhuizen (2002). Properties and applications of starch-converting enzymes of the α-amylase family. Journal of Biotechnology 94, 137–155.

Maximov V., Zaynullin R., Akhmadiev N., Segura-Ceniceros E.P., Martínez Hernández J.L., Bikbulatova E., Akhmetova V., Kunakova R, Ramos R., & Ilyina A. (2016). Inhibitory effect of 4,4′-[ethane-1,2-diylbis(sulfandiylmethanediyl)]bis (3,5-dimethyl-1H-pyrazole) and its derivatives on alpha-amylase activity. Medicinal Chemistry Research, 25 (7), 1384–1389. doi:10.1007/s00044-016-1574-2. ISSN: 1054–2523 (Print), 1554–8120 (Online)

Michelle de Sales P., Monteiro de Souza P., Simeoni L.A, Oliveira Magalhães P., Silveira D. (2012). α-Amylase inhibitors: a review of raw material and isolated compounds from plant source. Journal of Pharmacy and Pharmaceutical Sciences, 15(1), 141–183.

Mobini-Dehkordi, M., & Afzal Javan, F. (2012). Application of alpha-amylase in biotechnology. The Journal of Biology and Today's World, 1(1), 39–50. doi:10.5325/jmedirelicult.37.1.0060.

Monteiro de Souza, P., & de Oliveira e Magalhaes, P. (2010). Application of microbial α-amylase in industry-a review. Brazilian Journal of Microbiology, 41, 850–861. ISSN 1517-8382.

Mortazavi, H., Jazaeri, M., Baharvand, M., & Abdolsamadi, H. (2014). Salivary alpha-amylase alteration as a possible indicator for diabetes. Journal of Basic and Applied Scientific Research, 4(2), 284–288. URL acces: https://www.textroad.com/pdf/JBASR/J.%20Basic.%20Appl.%20Sci.%20Res.,%204(2)284-288,%202014.pdf

Murai, A., Iwamura, K., Takada, M., Ogawa, K., Usui, T., & Okumura, J. (2002). Control of postprandial hyperglycaemia by galactosyl maltobionolactone and its novel anti-amylase effect in mice. Life Sciences, 71(12), 1405–1415.

Nakamura, A., Haga, K., & Yamane, K., (1993). Three histidine residues in the active center of cyclodextringluconotransferase from alkalophilic Bacillus sp. 1011 effects replacement on pH dependence and transition-state stabilization. Biochemistry, 32, 6624–6631.

Nater, U. M., & Rohleder, N. (2009). Salivary alpha-amylase as a non-invasive biomarker for the sympathetic nervous system: Current state of research. Psychoneuroendocrinology, 34(4), 486–496. doi:10.1016/j.psyneuen.2009.01.014

Nielsen, A. D., Pusey, M. L., Fuglsang, C. C., and Westh, P. (2003). A proposed mechanism for the thermal denaturation of a recombinant *Bacillus halmapalus* alpha-amylase—the effect of calcium ions. Biochimica et Biophysica Acta, *1652*(1), 52–63. doi:10.1016/j.bbapap.2003.08.002

Nielsen, J. E., Borchert, T. V., and Vriend, G. (2001). The determinants of alpha-amylase pH-activity profiles. Protein Engineering, *14*(7), 505–512. doi:10.1093/protein/14.7.505

Nikitkova, A. E., Haase, E. M., & Scannapieco, F. A. (2013). Taking the starch out of oral biofilm formation: molecular basis and functional significance of salivary α-amylase binding to oral streptococci. Applied and Environmental Microbiology, 79(2), 416–423. doi:10.1128/AEM.02581-12

Numao S., Damager I., Li C., Wrodnigg T.M., & Begum A. (2004). In situ extension as an approach for identifying novel alpha amylase inhibitors. Journal of Biological Chemistry, 279(46), 48282.

Ochiai A., Sugai H., Harada K., Tanaka S., Ishiyama Y., Ito K., Tanaka T., Uchiumi T., Taniguchi M., & Mitsui T. (2014). Crystal structure of α-amylase from Oryza sativa: molecular insights into enzyme activity and thermostability. Bioscience, Biotechnology, and Biochemistry. 78(6), 989–997. doi:10.1080/09168451.2014.917261

Østergaard, L. H., & Olsen, H. S. (2011). Industrial applications of fungal enzymes. In Industrial Applications; Berlin, Heidelberg: Springer, pp. 269–290. doi:10.1007/978-3-642-11458-8_13.

Panneerselvam, T., Elavarasi, S., & Nadu, T. (2015). Isolation of α-amylase producing *Bacillus subtilis* from soil. International Journal of Current Microbiology and Applied Sciences, 4(2), 543–552.

Paredes-Salido, F., Clemente-Fernández, A. (2005). Polifenoles de aplicación en farmacia. Vol. 24. Núm. 8, 85–94.

Perez-Vizcaino, F., Duarte, J., Jiménez, R., Santos-Buelga, C., Osuna A. (2009). Antihypertensive effects of the flavonoid quercetin. Pharmacological Reports, 61, 67–75.

Perry G.H., Dominy N.J., Claw K.G., Lee A.S., Fiegler H., Redon R., Werner J., Villanea F.A., Mountain J.L., Misra R., Carter N.P., Lee C., & Stone A.C. (2007). Diet and the evolution of human amylase gene copy number variation. Nature Genetics, 39(10), 1256–1260.

Peyrot des Gachons, C., & Breslin, P. A. (2016). Salivary amylase: digestion and metabolic syndrome. Current Diabetes Reports, 16(10), 102. doi:10.1007/s11892-016-0794-7

Prajapati, V. S., Trivedi, U. B., & Patel, K. C. A. (2015). Statistical approach for the production of thermostable and alklophilic alpha-amylase from *Bacillus amyloliquefaciens* KCP2 under solid-state fermentation. Biotechnology, 5(39), 211–220. doi:10.1007/s13205-014-0213-1.

Prakash, O., and Jaiswal, N. (2010). a-Amylase: an ideal representative of thermostable enzymes. Applied Biochemistry and Biotechnology, 160(8), 2401–2414. doi:10.1007/s12010-009-8735-4

Qureshi, A.S., Khushk, I., Ali, H., Chisti, Y., & Ahmad, A. (2016). Coproduction of protease and amylase by thermophilic Bacillus sp. BBXS-2 using open solid-state fermentation of lignocellulosic biomass. Biocatalysis and Agricultural Biotechnology, 8, 146–151.

Rafique, R., Mohammed Khan, K., Arshia, Chigurupati, S., Wadood, A., Ur Rehman, A., Salar, U., Venugopal, V., Shamin, S., Taha, M., Perveen, S. (2019). Synthesis, in vitro α-amylase inhibitory, and radicals (DPPH & ABTS) scavenging potentials of new N-sulfonohydrazide substituted indazoles. Bioorganic Chemistry, 103410.

Rani, R., Sukumaran, R. K., Kumar, A., Larroche, C., & Pandey, A. (2010). Advancement and comparative profiles in the production technologies using solid-state and submerged fermentation for microbial cellulases. Enzyme and Microbial Technology, 46, 541–549. doi:10.1016/j.enzmictec.2010.03.010.

Rubilar, M., Jara, C., Poo, Y., Acevedo, F., Gutierrez, C., Sineiro, J., Shene, C. (2011). Extrants of Maqui (*Aristotelia chilensis*) and Murta (*Ugnimolinae turcz*): sources of antioxidant compounds and alfa-glucosidase/alfa-amylase inhibitors. Journal of Agricultural and Food Chemistry, 59, 1630–1637.

Sales, P. M., Souza, P. M., Simeoni, L. A., Magalhães, P. O., & Silveira, D. (2012). α-Amylase inhibitors: a review of raw material and isolated compounds from plant source. Journal of Pharmacy & Pharmaceutical Sciences, 15(1), 141.

Sallam A, & Galala A.A. (2017) Inhibition of alpha-amylase activity by *Gymnocarpos decandrus* Forssk. constituents. International Journal of Pharmacognosy and Phytochemical Research, 9, 873–879.

Schwarz A., Brecker L., & Nidetzky B. (2007). Acid-base catalysis in *Leuconostoc mesenteroides* sucrose phosphorylase probed by site directed mutagenesis and detailed kinetic comparison of wild-type and Glu237 to Gln mutant enzymes. Biochemical Journal, 403(Pt 3), 441–446.

Seppälä, S., Elmo, S., Knop, D., Solomon, K. V, & Malley, M. A. O. (2017). The importance of sourcing enzymes from non-conventional fungi for metabolic engineering and biomass breakdown. Metabolic Engineering, 44 (June), 45–59.

Sharma, A., & Satyanarayana, T. (2013). Microbial acid stable α-amylases: characteristics, Genetic Engineering and Applications. 48, 201–211.

Simair, A. A., Qureshi, A. S., Khushk, I., Ali, C. H., Lashari, S., Bhutto, M. A., Mangrio, G. S., & Lu, C. (2017). Production and partial characterization of α-amylase enzyme from Bacillus sp. BCC 01-50 and potential applications. BioMed Research International, 9. doi:10.1155/2017/9173040.

Sindhu, R., Binod, P., Madhavan, A., & Sabeela, U. (2017). Molecular improvements in microbial α-amylases for enhanced stability and catalytic efficiency. Bioresource Technology, 245, 1740–1748.

Soccol, C. R., Ferreira da Costa, E., Junior Letti, L. A., Grace Karp, S., Lorenci Woiciechowski, A., & Porto de Souza Vandenberghe, L. (2017). Recent developments and innovations in solid state fermentation. Biotechnology Research and Innovation, 1, 52–71.

Søgaard M., Abe J., Martin-Eauclaire M.F., & Svensson B. (1993). α-amylases: structure and function. Carbohydrate Polymers, 21(2–3), 137–146. doi:10.1016/0144-8617(93)90008-R

Tallapragada, P., Dikshit, R., Jadhav, A., & Sarah, U. (2017). Partial purification and characterization of amylase enzyme under solid state fermentation from *Monascus sanguineus*. Journal of Genetic Engineering and Biotechnology, 15, 95–101.

Tiwari, M. (2011). Science behind human saliva. Journal of Natural Science, Biology, and Medicine, 2(1), 53–58. doi:10.4103/0976-9668.82322

Uitdehaag, J.C.M., Mosi, R., Kalk, K.H., Van der Veen, B.A., Dijkhuizen, L., Withers, S.G., & Dijkstra, B.W., (1999). X-ray structures along the reaction pathway of cyclodextringlycosyltransferase elucidate catalysis in the α-amylase family. Nature Structural & Molecular Biology, 6, 432–436.

Vázquez, C., De Cos, A., López-Nomdedeu, C. (2005). Alimentación y nutrición 2nd ed. Editorial Días de Santos. Madrid, pág. 477.

Vineetha, R., Pai, K. M., Vengal, M., Gopalakrishna, K., & Narayanakurup, D. (2014). Usefulness of salivary alpha amylase as a biomarker of chronic stress and stress related oral mucosal changes—a pilot study. Journal of Clinical and Experimental Dentistry, 6(2), e132–e137. doi:10.4317/jced.51355

Wang, J., Li, Y., & Lu, F. (2018). Molecular cloning and biochemical characterization of an α-amylase family from *Aspergillus niger*. Electronic Journal of Biotechnology, 32, 55–62.

Wang, X., Kan, G., Shi, C., Xie, Q., Ju, Y., Wang, R., Qiao, Y., & Ren, X. (2019). Purification and characterization of a novel wild-type α-amylase from Antarctic Sea ice bacterium *Pseudoalteromonas* sp. M175. Protein Expression and Purification, 164 (March).

Wang, Y., Hu, H., Ma, J., Yan, Q., & Liu, H. A. (2020). Novel high maltose-forming α-amylase from *Rhizomucor miehei* and its application in the food industry. Food Chemistry, 305, (May 2019).

Wijaya H. C., Rahminiwati M., Wu M.C., & Lo D. (2011). Inhibition of α-glucosidase and α-amylase activities of some Indonesian herbs: in vitro study, BITEC Bangna, Bangkok, Thailand, pp. 285–287.

Wo, H. A. B. (2019). Filamentous fungi for the production of enzymes, chemicals and materials. Current Opinion in Biotechnology, 59, 65–70.

World Health Organization. Health Topics /Diabetics. https://www.who.int/health-topics/diabetes (accessed Dec 30, 2019)

Yadav, J. K., and Prakash, V. (2011). Stabilization of α-amylase, the key enzyme in carbohydrates properties alterations, at low pH. International Journal of Food Properties, 14(6), 1182–1196. doi:10.1080/10942911003592795

Yilmazer-Musa, M., Griffith, A. M., Michels, A. J., Schneider, E., & Frei, B. (2012). Grape seed and tea extracts and catechin 3-gallates are potent inhibitors of α-amylase and α-glucosidase activity. Journal of Agricultural and Food Chemistry, 60(36), 8924–8929.

Zakowski, J. J., & Bruns, D. E. (1985). Biochemistry of human alpha amylase isoenzymes. CRC Critical Reviews in Clinical Laboratory Sciences, 21(4), 283–322. doi:10.3109/10408368509165786

Zaynullin R., Arredondo-Valdés R., Ilyina A, Segura-Ceniceros E.P., Martínez-Hernández J.L., & Kunakova, R. (2014). Effect of extracts from birch tree bark, china tea and the root of cactus *Opuntia ficus-indicata* on α-amylase activity. En libro: Biotecnología—de ciencia a práctica. (Editores: Ibraguimov R.I., Garipova M.I., Veselov M. Yu., Farjutdinov R.G., Shpirnaya I.A., Cvetkov V.O., Grigoriadi A.S., Bashirova R.M., Umarov I.A.) Universidad Estatal de Bashkortostan, pp. 102–105. ISBN: 978-5-7477-3641-2

Zaynullin R.A., Kunakova R.V, Khusnutdinova E.K., Yalaev B.I., Segura-Ceniceros E.P., Chavez-Gonzalez M.L., Martínez-Hernández J.L., Gernet M.V., Batashov E.S., & Ilyina A. (2018). Dihydroquercetin: known antioxidant—new inhibitor of alpha amylase activity. Medicinal Chemistry Research, 27, 966–971. doi:10.1007/s00044-017-2119-z. ISSN 1054-2523

Zhang, Q., Han, Y., and Xiao, H. (2017). Microbial α-amylase: a biomolecular overview. *Process Biochemistry, 53*, 88–101. doi:10.1016/j.procbio.2016.11.012

Zhang, B., Li, X., Sun, W., Xing, Y., Xiu, Z., Zhuang, C., & Dong, Y. (2017). Dietary flavonoids and acarbose synergistically inhibit α-glucosidase and lower postprandial blood glucose. Journal of Agricultural and Food Chemistry, 65(38), 8319–8330.

CHAPTER 4

The Implication of Rambutan, Ginseng, and Green Tea in Energy Drinks Through Their Genetic Characterization, Total Phenolics, and Antioxidant Activity

AXEL FLORES-CUÉLLAR, OCTAVIO TORRES-HERNÁNDEZ, MIRIAM VÁZQUEZ-SEGOVIANO, PAOLA ISABEL ANGULO-BEJARANO[*], and VÍCTOR OLVERA-GARCÍA[*]

Tecnologico de Monterrey, Escuela de Ingeniería y Ciencias, Campus Queretaro, Av. Epigmenio González 500, Fracc. San Pablo, CP 76130 Queretaro, Mexico

[*]Corresponding author. E-mail volverag@tec.mx

ABSTRACT

Lack of energy is a problem that affects daily human efficiency. The actual market offers energetic drinks that can lead to critical diseases, hyperglycemia, tachycardia, upon others. Thus energy beverages based on nutraceutical properties are a feasible solution that at the same time are socially accepted. Ginseng (*Panax ginseng*) and green tea (*Camellia sinensis*) have both demonstrated to activate the sympathetic nervous system, regulation of chemical stress, immune modulation, antitumor activities, glucose metabolism, and cognitive performance. On the other hand, rambutan (*Nephelium lappaceum*) provides vitamins and biocompounds that are beneficial for human consumption. Extracts of rambutan fruit, mainly from the peel, have been shown to possess phytochemical compounds that exhibit antioxidant, antimicrobial, antidiabetic, antiviral, anti-inflammatory, antihypoglycemic, and antiproliferative effects in various in vitro and in vivo tests. Then it is possible to combine these three products in an energy drink to stimulate the nervous system and also with other nutraceutical benefits. The objective of this work was to propose a beverage based on rambutan, ginseng, and green tea.

4.1 INTRODUCTION

4.1.1 TRADITIONAL ASIAN MEDICINE

The World Health Organization (WHO) defines traditional medicine as the sum of knowledge, skills, and practices based on the theories, beliefs, and experiences of different cultures that are used to maintain health, as well as to prevent, diagnose, improve, and treat physical and mental illnesses (Nafiu et al., 2017). East Asian medicine was introduced in Korea and Japan around the 6th century. Nevertheless, it was originated in China about 3000 years ago (Park et al., 2012); it embraces a holistic approach where the human body is viewed as a small universe. Also, Asian medicine seeks to restore energy and have a balance among the different components within the body and its functions. Within Asia, traditional Chinese medicine is the system with the longest history and it stands from the ancient Chinese philosophy of Yin-Yang and the five elements. Yin-Yang is the core of traditional Chinese medicine and comes from the philosophy of Confucianism, Buddhism, and Taoism rather than natural science; Chinese believed light and darkness were the ideal opposites (Wang, 2016). Yin-Yang, as a concept, refers to two elements that maintain an equilibrium condition by interacting with each other and being complementary to each other. On the other hand, wood, earth, fire, water, and metal, are the five elements considered by the ancient Chinese philosophy as the basic material units composing the World, each of these elements have an active and dynamic relationship with one another; thus they act collectively (Yu et al., 2006). Traditional Chinese medicine advocates combinatory therapeutic strategies and encompasses a lot of practices, including herbal medicine, acupuncture, moxibustion, and massage (Wang, 2016).

4.1.1.1 ASIAN MEDICINAL HERBS

Plants have been used for medical and nutritional purposes long before recorded history. The use of herbs is considered an essential part of traditional Chinese medicine (Teschke et al., 2015). The plant kingdom is a rich source of natural chemical substances, many of which are important natural pharmaceutical agents, that is why according to WHO, approximately

80% of the world population currently relies on herbal medicines as standard therapeutic modalities.

Medicinal plants are produced and offered in a wide variety of products, from crude to processed materials to packaged products, including pharmaceutical, herbal remedies, teas, among others (Dzoyem et al., 2013). The total value of herbal medicine in China, manufactured in 1995, reached USD 17.6 billion. This trend has continued, sales of herbal products in this country resulted in USD 14 billion between 2003 and 2004, and it is estimated that the annual worldwide market for these products is near the USD 60 billion (Wee et al., 2011).

4.1.2 FUNCTIONAL FOODS

Functional foods are a combination of science, nutrition, and medicine. The term refers to food products that have been enriched with natural components with a specific physiological health-promoting effect (Vukasović 2017; Kumar et al., 2018). Nowadays, there is a trend to rising the consumption of health functional foods on the level of disease prevention by using the bioactive compounds in medicinal plants, along with the increased interest in maintaining health because of the devastating lifestyle that some people may have (So et al., 2018).

4.1.3 LACK OF ENERGY AND COGNITIVE PERFORMANCE

The lack of energy goes hand in hand with the fatigue that a person presents and is a symptom caused by various factors that make the individual unable to perform their activities in the best possible way. The causes can be from diseases such as sclerosis to other deficiencies like a poor diet, but doctors argue that the principal trigger is prolonged stress; this symptom is directly related to the central nervous system (CNS). In the same way, it is increasing proportionally to the age of the person (Pedraz-Petrozzi, 2018, Gibson-Smith et al., 2018).

4.1.4 ENERGY DRINKS IN THE CURRENT MARKET

One of the many solutions that have been developed is the formulation of energy drinks that have focused on an audience of adolescents and adults, being products that their worldwide commercialization is very viable due to the majority consumption that is had daily, and in recent years the sales of these products are increasing thanks to advertising campaigns and the consumer's search for different products. Reports have shown that only in Latin America, 64.9% of the population consumed energy drinks at least once weekly. In most beverages, the main constituent responsible for the stimulation of the CNS is the alkaloid compound, ingredients such as caffeine, theobromine, or taurine that are well known for food formulation. Unfortunately the high amount of sugar, excessive consumption, and mixing with other substances in the drink cause harmful effects on health, being the best-known example taurine drinks which are associated with tachycardia attacks; therefore the offer in the market already has prejudices on the part of society and does not relate it to a healthy habit. The high levels of sugar in sport beverages, which are in constant consumption by athletes, negatively increase the caloric intake, cause weight gain, and other problems like caries (Sánchez et al., 2015).

4.1.5 *MOST POPULAR ENERGY DRINK ACTIVE COMPOUNDS*

Common active compounds used in energy beverages in the markets are vitamins A, C, D, and E, protein, sodium, potassium, calcium, magnesium, iron, phosphorus, inositol, glucuronolactone, taurine, L-carnitine, caffeine, guarana, Ginkgo biloba, ginseng, and green tea extracts (Corbo et al., 2014). Nowadays, the relationship between nourishment and health is beginning to be more understood by producers and consumers, therefore increasing the production and consumption of functional beverages.

4.2 CAMELLIA sinensis

Camellia sinensis (L.) Kuntze is better known as green tea. Other names by which this species is known are Chha (India), Cha (China), Chai (Russia), Itye (Africa), Tea plant (England), Tea (United States), Te (Italy), and Té (Mexico) (Namita, Mukesh, and Vijay, 2012). The tea plant is part of Theaceae within the *Camellia* genus, being the biggest of the Theaceae with 120–300 species (Le et al., 2017). In 1753 Carl Linnaeus described the tea plant taxonomically for the first time in history, nominated it as *Thea sinensis* (Preedy, 2012). Masters, in 1844, redefined its taxonomy as two different species, *Thea sinensis*, and *Thea assamica*. By the beginning of the 19th century, both taxa were uniited in a single species and named it as *Camellia* L. Later Carl Ernst Otto Kuntze recombined the species epithets and redefined it as *C. sinensis* (L.) Kuntze (Zhen, 2002).

The tea plant is an evergreen tree or shrub that grows up to 15 m in high in the wild and up to 1.5 m in cultivation. The leaf lamina usually is erect, elliptic, oblong, obovate-oblong to oblanceolate, thin-coracious to coracious of papery, from light to dark green, and present from 7 to 14 lateral nerves. Flowers are white, solitary, or in small groups of up to four, terminal, or axillary, and 2.5–8 mm long. The flowers contain around 100–300 stamens, 7–15 mm long with yellow anthers, and produce brownish-red capsules. Ovaries are globose to void with 3–5 ovules. Styles are filiform and 8–12 mm long. Capsules are compressed globose with 1–3 seeds each. Seeds are brown and usually 30–45 mm in diameter. Flowering occurs between October and February, whereas the fruiting occurs between August and October (Zhen, 2002; Ross, 2007).

4.2.1 PRODUCTION AND GEOGRAPHICAL DISTRIBUTION

C. sinensis grows naturally in the understory of broad-leaved evergreen forests at altitudes between 100 and 2200 m. The geographic area in which the species grows is known as "tea belt" that covers the southwestern of China, northern Laos, northern Vietnam, Myanmar, Cambodia, and northern India. Nevertheless, there are other regions also where forest-growing tea plants can be found, such as Japan, South Korea, Thailand, and Taiwan. Currently, tea plant cultivation takes place in tropical, subtropical, and temperate regions around the world (Preedy, 2012).

Statistical projections made in 2018 by the Food and Agriculture Organization indicate that the global production and consumption will keep rising over the next decade, mainly because of the increasing demand of developing countries. The global production of green tea is increasing at a rate of 7.5% per year and it is predicted to achieve 3.6 million tons by 2027, being China the biggest producer, with a yield of 1.5 million tons between 2015 and 2017, it is considered for this country to produce 3.3 million tons by 2027 (FAO, 2018).

4.2.2 HISTORY

According to Chinese mythology, the tea plant has been used for three millennia. It is said on Chinese culture that the Emperor Shen Nung saw leaves falling from a tree, by coincidence those leaves drop into his pot of hot water, curious of the scent, he drank it and named it as a "heavent sent," it was the moment when tea began to be known for its medical effects (Zink and Traidl-Hoffmann, 2015).

4.2.3 TRADITIONAL USES

In China, the extract of the dried leaf is normally drunk as a sedative, antihypertensive, and anti-inflammatory. In India, decoctions of leaves and buds are generally drunk to treat ache and fever. Fresh leaf juice is normally drunk for abortion, contraceptive, and hemostatic. In Turkey, the leaves are chewed to treat diarrhea (Ross, 2007). In Ecuador, hot water extract of the dried leaf is used for diuretic purposes, mainly in urinary tract infections. In Peru, hot water extracts of the dried leaves are used to debloat. It is also used for anti-inflammatory properties. In Mexico, the leaf extract in hot water is consumed by nursing mothers to increase the production of breast milk.

4.2.4 MAIN PHARMACOLOGICAL EFFECTS

For many centuries, green tea has been widely used as a health tonic in different cultures around the World. Recently these bioactive compounds have been categorized and studied through scientific and modern technologies. The most important substances in green tea are caffeine, theanine, theobromine, theophylline, theaflavins, and phenolic acids (Schneider and Segre, 2009). Many in vitro and in vivo studies have demonstrated the high antioxidant capacity of the polyphenols present in the plant, being epigallocatechin gallate, the most common and investigated (Natarajan et al., 2019). This activity is accountable for many other beneficial actions attributed to green tea.

4.2.4.1 ANTIMICROBIAL ACTIVITY

Many studies have demonstrated the antimicrobial capacity of green tea extracts. Aqueous, ethanolic, and methanolic extracts were used to study the inhibition capacity of bacterial growth in *Staphylococcus*, *Streptococcus*, *Pseudomonas*, *Escherichia coli*, *Bacillus*, and *Proteus* (Kumar et al., 2012). The extracts demonstrated an effective antimicrobial capacity, mainly because of the bacterial cell wall damage caused by catechins. In vitro antimicrobial activity has also been tested for skin pathogens, caused by the inhibition of the adherence between the pathogen and the host cell membrane (Sharma et al., 2012).

4.2.4.2 ANTICANCER ACTIVITY

Green tea has demonstrated anticancer properties through numerous studies, mainly because of the presence of catechins. It has demonstrated to act in different stages, for example, initiation, promotion, metastasis, and angiogenesis (Niedzwiecki et al., 2016). Inhibition of colon, prostate, skin, lung, and breast cancer has been attributed to mechanisms such as inhibition of free radicals and activation or inhibition of signaling cascades (Butt et al., 2015). Green tea extract is capable of target stem cells of cancer in various tissues, where tumor reduction was achieved by a mixture of anticancer drugs and green tea extracts (Nayyar et al., 2017). Other combinations have been studied, the synergy of tea catechins and quercetin demonstrated inhibition of human pancreatic cancer stem cells (Fujiki et al., 2015).

4.3 NEPHELIUM lappaceum

Nephelium lappaceum L., better known as rambutan in regions of Asia, Central, and South America, is an exotic fruit native to the Asian continent. Other names by which this species is known are nefelio, "mamón chino," lincha, and achotillo (Morton, 1987). The name of the fruit derives from the Malay language with the word "rambut" whose direct translation is hair, this due to the filaments that the peel presents. Carl Linnaeus made the first taxonomic description in the year of 1767 with the kingdom Plantae, phylum Tracheophyta, class Magnoliopsida, order Sapindales, family Sapindaceae (like lychee and longan fruits), genus *Nephelium* L., and the species *Nephelium lappaceum* L.

The plant has oval scattered leaves with an average size of 10–20 cm of length by 5–10 cm width and cylindrical pedicels. Fruits are fleshy drupes, ovoid of 2.5 cm in diameter, with a minimum weight of 30 g, and reddish or yellow in color, with trichomes which are characteristic in the fruit (Lestari et al., 2014). Mesocarp has a sweet taste and surrounds a large seed that cannot be ingested because of its toxicity. Trees reach a height of 10 m or more, not exceeding 25 m. Fruit production reaches 150 kg on average, per tree in their lifetime, meaning a total of 5000–6000 rambutans. For its cultivation, seeds have to be sown as soon as possible after its extraction from the fruit because of its high probability of losing viability (Osorio-Espinoza et al., 2019). Seeds have an elliptical form with a woody texture and wrapped in an edible coat. In order to determine the quality of a fruit that is going to be consumed, the main characteristics to be observed are color, aroma, water content, and filament length (Vargas, 2003; Lestari et al., 2014).

4.3.1 PRODUCTION AND GEOGRAPHICAL DISTRIBUTION

The largest production source of rambutan is found in the Asian continent, with Thailand being the main producer which commercialize approximately 300 thousand tons per hectare of the fruit annually; however, the largest exporter of fruit in the world is its country of origin, Malaysia, producing approximately 70,000 tons per hectare approximately annually, and the main consumers of *N. lappaceum* are China and the European Union that buy derived food products (Arias-Cruz et al., 2016). In Mexico, fruit trade is very active in the southern part of the country, with the State of Chiapas having the highest production with 95%. There are currently approximately 2000 hectares of rambutan production. It is estimated that in the coming years, rambutan production will increase significantly according to world demand (Osorio-Espinoza et al., 2019).

Distribution of the cultivation areas is according to the optimal conditions for the growth of the tree, being tropical climates with temperatures with a range of 26–32 °C, which is the most desirable for its production. It is suitable for various types of soil that have sufficient moisture and nutrients; Asian, Central America, and South America countries are the geographical locations with the best characteristics for plant development. The main production season is from November to February (Osorio-Espinoza et al., 2019).

4.3.2 HISTORY

The discovery of the plant is still not so clear, but the first records are in the country of Malaysia where it has been commercialized for 800 years. Asian merchants have been in charge of distributing the fruit to other continents, like Africa and South America where the crop was better adapted due to the appropriate climatic conditions for its cultivation. From the year 1920, its commercial distribution expanded until it was a very consumed product in certain countries such as Honduras, Costa Rica, Ecuador, Colombia, among others (Morton and Dowling, 1987). In Mexico, the plant was introduced in the year of 1960 to the State of Veracruz, coming directly from Malaysia, but it was not until the first years of the 21st century when it began to be commercially cultivated in the southern states of the country as Chiapas, Guerrero, and Tabasco. Outside these states, its national consumption is very low due to the lack of knowledge of the fruit, although it continues to expand commercially both nationally and internationally (Osorio-Espinoza et al., 2019).

4.3.3 TRADITIONAL USES

The rambutan is attractive to the consumer for its sweet taste to the palate, due to which it has become an ingredient for the preparation of drinks and sweet foods. Various parts of the fruit can be used to relieve various malaises. According to the description of Morton and Dowling (1987), the seed can secrete oil that is pleasant for the smell, so it is used in the manufacture of soaps and candles, in the same way, the fruit has been used to combat stomach problems mainly diarrhea. Leaves are also used for headache remedies. Fats contained in the seed have become a substitute for animal butter and even to generate biodiesel and cosmetics (Uraiwan and Satirapipathkul, 2016). Due to its exotic flavor, the fruit has been subjected to beverage preparation and thanks goes to consumers that seek to try new flavors that come out of the ordinary. Its introduction to the market has been discreet compared to other fruits that are considered exotic but positive this with the help of formulation that is accompanied by different plants with various flavors.

4.3.4 MAIN PHARMACOLOGICAL EFFECTS

The fruit has become an object of study due to its antimicrobial and antioxidant properties, so further research has been conducted to discover the extent of its properties. *N. lappaceum* has the potential to combat hyperglycemia due to its compounds where geraniin stands out (Palanisamy, 2011). High-performance liquid chromatography (HPLC) analyses of fruit extracts reveal that flavonoids, functional groups of CH_3, tannins, aliphatic CH_3, and carbonyl groups are present; these compounds are functional for the elaboration of different phytopharmaceuticals (Lestari et al., 2013). Phenolic compounds like corilagin, gallic acid, and ellagic acid are found in the peel and are responsible for the antioxidant activity of the fruit. Also, the antibacterial properties are strong in the fruit peel, being *Staphylococcus epidermidis* (bacteria responsible for various diseases in humans) highly sensitive to the fruit extracts (Thitilertdecha et al., 2008; Thitilertdecha et al., 2010).

The part with the greatest potential in terms of medicinal use is the fruit peel that is rich in flavonoids, saponins, and tannins. It is considered a waste when it is consumed, so its use becomes beneficial for various industries. Research has shown the ability to combat obesity disease by the polyphenolic compounds present by reducing the size of adipose tissues present in the human body (Lestari et al., 2014). The content of essential oils in seed has a range from 17% to 39%, being an amount that complements the present in the peel and that in the same way is considered, in most cases, a part of the waste of the total fruit (Solís-Fuentes et al., 2010; Rodriguez et al., 2019).

The caloric content of the fruit turns out to be very low, so it is ideal for including in various diets to avoid affect a person's weight. The content per 100 g of aril is 14.5 g of carbohydrates, 83 g of H_2O, 0.8 g of protein, 3 mg of iron, 25 mg of calcium, and 20–45 mg of vitamin C; all with a total of 63 calories (Osorio et al., 2017). Similarly, the seed of the fruit has significant amounts of vitamin E, which is used for skincare methods (Uraiwan and Satirapipathkul, 2016). The content of fatty acids and triglycerides are varied in the seed oil, where oleic acid and stearic acid stand out (Harahap et al., 2012). Pulp produces a juice with approximately 18.3 Brix grades and a pH of 4.4, which causes the characteristic taste pleasant to the palate (Vargas, 2003).

4.4 PANAX ginseng C.A. MEYER

There are several ginseng species, and thus plants with similar properties of *P. ginseng* C.A. Meyer have already been called ginseng; eventually, the term "ginseng" has encompassed more than 10 species belonging to the genus *Panax*. Nevertheless, ginseng is one of the common names for the species *P. ginseng* along with Korean ginseng (Shin, Kwon, and Park, 2015).

Ginseng is a highly valued medicinal plant belonging to the Araliaceae family from the genus *Panax*. This genus encompasses at least nine species, including *P. ginseng*, *P. quinquefolium* (American ginseng), *P. notoginseng* (Chinese notoginseng or Sachi), *P. japonicus* (Japanese ginseng) that are considered the most commonly used ginseng herbs at present (Leung and Wong 2010, Kim, Kim, and Shin, 2013). Although these species belong to the same genus, they can vary widely in their constituents (Kim and Kim 2017).

Panax ginseng is a short perennial herbaceous plant that consists of berries, leaves, stalk, and root (Lee et al., 2018). It is a slow-growing herb native to the regions of China, Japan, Korea, and Russia. *P. ginseng* is characterized by a pale tuberous root that grows upright each year. The annual growth cycle of ginseng plants covers the leaf expansion during spring, followed by flowering, fruit stage, leaf loss, and, finally, the root growing stage during autumn (Lee and Jeong et al., 2019). It can reach large dimensions and has two to five ramified rootlets that increase in size as the plant matures. The thickness of the rhizome is an important indicator of the quality and value of ginseng (Yang and Wu, 2016). Root age strongly determines root growth, shape, and ginseng quality. The rhizome of 6-year-old ginseng root is thick and about 7–10 cm in length and 3 cm in diameter, having several stout rootlets. Ginseng is a self-pollination plant and starts to bloom at its third-year growth stage (CHOI, 2008). The leaves are verticillate and composed of five leaflets, the three-terminal leaflets are larger than the lateral ones, and the fruit is a small berry which is red when ripe (Bone et al., 2012).

4.4.1 PRODUCTION AND GEOGRAPHICAL DISTRIBUTION

Ginseng is distributed in 35 countries around the world, among them, 19 are both importers and exporters (Xu, Choi, and Huang, 2017), the 4 largest producers are China, South Korea, Canada, and the United States. Despite its popularity, the cultivation of this crop is challenging due to a specific climate, soil conditions, and other growth factors (Walsh et al., 2021).

China is the largest producer of ginseng with 44,749 tons per year, ranking in second place South Korea with 27,480 tons, Canada with 686 tons, and finally, the United States with 1054 tons. The total production of these countries is 79,769 tons, which is approximately 99% of the world's production (Baeg and So, 2013). In the United States and Canada, *Panax quinquefolius* has adapted to the environmental conditions, and it is mostly produced in both countries. Thus *P. ginseng* has to be imported from Korea. In 2000, *P. ginseng* was reported as one of the most frequently purchased herbs in the United States (Kim and Kim 2017; Lu et al., 2009). The global ginseng market at the end of 2013 was worth USD 2.1 billion; it is predicted that by 2025 this market will be worth USD 7.51 billion (Shi, Zeng, and Wong 2019).

4.4.2 HISTORY

P. ginseng is one of the most valued medicinal plants in Eastern Asia, its history began 4500 years ago, and its first record was written 2000 years ago (Buettner et al., 2006; Baeg and So, 2013). The origin of *P. ginseng* dates back to prehistory, and it was mainly used as a remedy for spiritlessness and fatigue, since then, there have been extensive studies to demonstrate its efficacy (Yun, 2001; Kim, 2018). The name *Panax* is derived from the Greek word *panacea* which means "all healing" and stemmed from the traditional belief that the plant could cure all illnesses of the human body, on the other hand, the herbal root is named ginseng which derives from the Chinese words *Jen Sheng* which means "essence of man" because of the humanoid shape of the root (Kim, 2018; Khan et al., 2015).

Western medicine texts from the 1800s and early 1900s described *P. ginseng* as both mild sedative and stimulant used to restore the vital energy, increase the production of body fluids, and promote health and longevity (Bone et al., 2012; Buettner et al., 2006). According to traditional Chinese medicine, *P. quinquefolius* was a "cool" or *yin* tonic used to treat those suffering from "hot" symptoms such as insomnia and stress, whereas *P. ginseng* was a "hot" or *yang* to invigorate the weakened body. Recently *P. ginseng* has gained popularity and has been a renewed interest to investigate its pharmacological effects (Khan, Tosun, and Kim, 2015).

4.4.3 TRADITIONAL USES

The great diversity of pharmacological properties attributed to *P. ginseng* has led to use for its therapeutic effects and nutritional benefits, mostly in Asian countries (Kim and Kim, 2017). Depending on the characteristics of each nation, ginseng was used as food and medicine, among others (Baeg and So 2013). In traditional Korean medicine is recorded that *P. ginseng* is a life-preserving drug possessing both sweet and bitter flavor, which was used to promote energy and to fortify *qi* to stabilize the mind and increase wisdom by stimulating the CNS (So et al., 2018). The traditional view is that *P. ginseng* is above all, a tonic herb which can revitalize the function of the organism, as a component of multiple-ingredient drug preparation to increase its efficacy. *P. ginseng* was mostly used for therapeutic uses such as heart failure, asthma, organ prolapse, cold limbs, neuralgia, convulsions, neurosis, long-term debility, and others (Bone et al., 2012).

Among all parts of ginseng plants, roots were mainly used for medical purposes (Lee et al., 2019). The action mechanism of *P. ginseng* had not been known until ginsenosides; main constituents in ginseng were isolated in 1963 (Leung and Wong, 2010). Although *P. ginseng* is reported to have a wide range of therapeutic and pharmacological effects, researchers are focusing on purifying ginsenosides to reveal the specific mechanism, instead of using whole ginseng root extracts (Lu, Yao, and Chen, 2009).

4.4.4 GINSENOSIDES

Ginseng contains valuable substances; besides carbohydrates and proteins, it contains volatile oils called ginsenosides, which have the highest medicinal value among its components. Isolated in the 1960s for the first time (Park et al., 2018), ginsenosides are saponins that are derivatives of triterpene dammarane, and they are classified into three categories based on the position of its sugar moiety on the dammarane and triterpene rings: oleanolic, panaxadiol, and panaxatriol. The most abundant saponins are protopanaxadiols commonly known as the Rb group or PPD (e.g., Rb1, Rb2, Rc, Rd) and protopanaxatriol is commonly known as Rg group or protopanaxatriol (PPT) (e.g., Rg1, Re, Rf, Rg2) (Murthy et al., 2014)

About 200 ginsenosides have been reported in total (Kim, 2018), Rb1, Rb2, Rc, Rd, Re, Rf, and Rg1 are the most abundant in *P. ginseng* and the most studied because of their effects in humans (Table 4.1). In raw roots, ginsenosides comprise more than 90% of the total saponins. Some studies have shown that the quality and composition of ginsenosides are influenced by a range of factors such as the species, age, part of the plant, and cultivation method (Leung and Wong, 2010). In a 10-year-old *P. ginseng* root, ginsenosides account for over 4.99% of the entire plant, unlike a 4-year-old where ginsenosides account only for 2.60% (Shi et al., 2019; Chen et al., 2008). In contrast, total ginsenosides in the leaf decrease with age (Bone and Mills, 2013) (Table 4.2).

Biologic effects of ginsenosides have been attributed to the steroidal structure, and studies have shown that the mechanisms of action take place at the cellular membrane, inside the cell, or in the nucleus (Buettner et al., 2005).

4.4.5 MAIN PHARMACOLOGICAL EFFECTS

Over the years, *P. ginseng* has shown many beneficial effects in humans. Several studies have confirmed the traditional knowledge that this plant has several pharmacological and medicinal benefits (Kim, 2018). *P. ginseng* is considered a tonic and a nutraceutical that can help to invigorate a weakened body by balancing body functions for optimal health by boosting energy levels and the immune system. Besides, ginseng saponins have shown to provide protective effects against mammalian tumor cell lines and exert various actions such as antidiabetes, anticancer, anti-inflammatory effects, and stimulating the formation of blood vessels which enhance memory and cognitive abilities (Shi et al., 2019).

TABLE 4.1 Distribution of Ginsenosides in *Panax ginseng* Plant

	Content (%)							
	Rg_1	Re	Rf	Rb_2	Rb_1	Rc	Rd	Total
Root hairs	0.376	1.512	0.150	0.780	1.351	1.349	0.381	6.148
Leaves	1.078	1.524	–	0.553	0.184	0.736	1.113	5.188
Lateral roots	0.406	0.668	0.203	0.434	0.850	0.738	0.143	3.532
Main root	0.379	0.153	0.092	0.131	0.342	0.190	0.038	1.348
Leaf stalks	0.327	0.141	–	–	–	0.190	0.107	0.765
Stem	0.292	0.070	–	0.397	–	–	–	0.759

TABLE 4.2 Total Phenolic Compounds and Antiradical Activity of Plant Extracts

Plant	Total Phenols[a]	Antiradical Activity[b]
Camellia sinensis	164.18 ± 15.95	90
Nephelium lappaceum	112.16 ± 8.22	38
Panax ginseng	123.42 ± 30.94	34

[a] *Total phenolics concentration expressed as µg/g db.*
[b] *Antiradical activity by DPPH expressed as percent discoloration.*

4.4.5.1 EFFECTS ON THE CENTRAL NERVOUS SYSTEM

P. ginseng has both stimulatory and inhibitory effects on the CNS, where ginsenosides Rb_1 and Rg_1 play a key role. Results in several animal studies have demonstrated that Rb_1, Rg_1 and Re prevent scopolamine-induced memory deficits. Individually, Rb_1 was shown to increase the uptake of choline in central cholinergic nerve endings and to facilitate the release of acetylcholine from hippocampal slices, from this investigation, it can be inferred that ginsenosides may facilitate learning, memory and the enhancement of nerve growth (Attele et al., 1999; Chen et al., 2008).

Another investigation revealed the efficacy of *P. ginseng* on the brain and nervous system by administering 100 mg of a standard ginseng extract, twice daily to a group of men during 12 weeks; the results showed improvements in the psychomotor functions such as attention, information processing, and auditory reaction compared to the baseline. Several studies are now focused on the role of ginseng and ginsenosides on neurological functions or possible target proteins, including different types of ion channels (Kim et al., 2013).

4.4.5.2 ANTICARCINOGENIC ACTIVITY

One of the promising candidates for cancer prevention is ginseng. Studies have demonstrated that people who consume *P. ginseng* preparations are at a lower risk of stomach, lung, liver, pancreas, ovaries, and colon cancer (Wee et al., 2011). Ginsenosides have shown strong chemoprotective, and chemotherapeutic properties, specially Rg_3 and Rh_2 are recognized as major active anticancer saponins in a wide range of in vivo and in vitro experimental studies (Ahuja et al., 2018).

These significant antitumor properties of ginsenosides are known to be the result of their anti-inflammatory, antiproliferation, antimetastasis, and antiangiogenesis effects. For example, ginsenoside Rh_2 has shown to suppress tumor cell growth in breast, prostate cancer, and leukemia (Shi et al., 2019). Similarly, ginsenoside Rg_3 was found to possess the ability to inhibit the lung metastasis of tumor cells, when administered at a dose of 100–1000 μg/mouse (Park et al., 2005). Another example is ginsenoside Rb_1, which inhibits angiogenesis in vitro and in vivo, a crucial step in tumor growth and metastasis (Lu et al., 2009).

4.4.5.3 IMMUNE SYSTEM EFFECTS

Immune system effects attributed to ginsenosides included studies that employed in vitro tests or in vivo models where ginsenosides were injected or orally administered. For instance, ginsenosides demonstrated an enhancement of B and T lymphocyte activities in mice and the protection of one particular mouse strain against *Candida albicans* infection (Bone et al., 2012). Ginseng saponins can also be considered good drug candidates or supplements for transplantation and autoimmune disorders owing to their immunosuppression ability, although there is growing evidence of its immunomodulatory properties, further investigation is needed to uncover the underlying mechanism of ginsenosides (Shi et al., 2019).

4.4.5.4 CARDIOVASCULAR ACTIVITY

Cardiovascular disease is an important problem around the world; it encompasses a spectrum of diseases including coronary artery disease, peripheral vascular disease, congestive heart failure, and hypertension (Kim, 2012). There have been numerous studies that found a close relationship between *P. ginseng* and the cardiovascular system. It is known that ginsenosides regulate blood pressure to normal (Park et al., 2018), they also suppress the apoptosis in neonatal cardiomyocytes during hypoxia and reperfusion (Lee and Kim, 2014) and have the intrinsic property of controlling reactive oxygen species, nitric oxide (NO) production, and the ability to activate a various receptor in endothelial cells (Mohanan et al., 2018). Besides, according to a recent investigation, it has been reported that the chronic feeding of ginsenosides to rabbits may have enhanced vasodilatation by preventing NO degradation (Bone et al., 2012).

4.4.5.5 OTHER EFFECTS

Ginsenosides are also known for their blood sugar regulation, promoting longevity, metabolism growth of normal cells. Besides, evidence suggests that ginseng is widely used in the adjuvant treatment of diabetes and related complications, one example is ginsenoside Rb1 which has a significant antihyperglycemic effect and increases insulin sensitivity and therefore used clinically to treat diabetes mellitus (Boneet al., 2012; Kim and Kim, 2017; Shi et al., 2019).

4.5 GENETIC IDENTIFICATION THROUGH BARCODING OF GINSENG, RAMBUTAN, AND GREEN TEA

DNA barcoding is a system to aid species recognition and identification through internationally agreed protocols and DNA regions to create a global database of living organisms. For instance, barcoding approaches have been used for the verification of medicinal plants (Newmaster, Fazekas, and Ragupathy 2006). Two locus-combination of *mat*K, which has a high evolutionary rate, and *rbc*L, which is conserved among species, is recommended by The Consortium for the Barcode of Life-Plant Working Group as the best plant barcode with a discriminatory efficiency of 72% (Li et al., 2015). Studies have also reported *trn*H-*psb*A intergenic spacer as a standard barcode because of its highly conserved sequences (Kress and Erickson, 2007) and recent studies have demonstrated *ycf*1 gene as a potential plastid DNA barcode for land plants (Dong et al., 2015).

4.5.1 DNA EXTRACTION

Dry samples of *C. sinensis* (green tea) and *P. ginseng* (ginseng) were obtained at a local market in Queretaro. *N. lappaceum* (rambutan) was recovered from cultivation in Tuxtla Chico, Chiapas, Mexico. Samples were cleaned with sodium hypochlorite (5.40% w/v) and sterile water then grounded with a mortar and pestle. A total 20 mg of ginseng root, green tea, and rambutan leaves powder were used for the extractions. Ginseng DNA extraction followed the protocol reported by Khan et al. (2007) with modifications. Rambutan genomic material isolation was performed according to Karaca (2005) protocol with modifications while PureLink® Plant Total DNA Purification Kit by Invitrogen (2012) was used for green tea.

4.5.2 AMPLIFICATION OF BARCODE REGIONS

Four barcoding regions were selected based on previous studies (*rbc*L, *mat*K, *trn*H-*psb*A, and *ycf*1). PCR analyses were performed in the Bio-Rad T100 Thermal Cycler (Hercules, CA) with primers trnHf_05 (5′-CGCG-CATGGTGGATTCACAATCC-3′) and psbA3_f (5′-GTTATGCATGAAC-GTAATGCTC-3′) for the *trn*H-*psb*A spacer region (Sang et al., 1997; Tate and Simpson, 2003), matK-xf (5′-TAATTTACGATCAATTCATTC-3′) and matK-MALP (5′-ACAAGAAAGTCGAAGTAT-3′) for the *mat*K gene (Ford et al., 2009, Dunning and Savolainen, 2010), rbcLa-F (5′-ATGTCACCA-CAAACAGAGACTAAAGC-3′) and rbcLa-R (5′-GTAAAATCAAGTC-CACCRCG-3′) for the *rbc*L gene (Levin et al., 2003; Kress and Erickson, 2007) and ycf1bF (5′-TCTCGACGAAAATCAGATTGTTGTGAAT-3′) and ycf1bR (5′-ATACATGTCAAAGTGATGGAAAA-3′) for the *ycf*1 gene (Dong et al., 2015). The reaction mixture (10 μL) consisted of 5 μL of Bioline 2x BioMix, 0.2 μL primers forward and reverse, 3.6 μL of water and 1 μL of template DNA. To find out the optimal annealing temperature of each barcode, thermal gradients were carried out. Eight different temperatures, for *mat*K from 50 to 57 °C, *ycf*1 from 48 to 55 °C, *trn*H-*psb*A and *rbc*L from 52 to 59 °C. The amplification profiles were done in 2% agarose gel electrophoresis.

It was possible to amplify the *ycf*1 gene in ginseng with a fragment expected size of 900 bp (at 55 °C) and for rambutan with a fragment expected size of 850 bp (at 52 °C). Similarly, *rbc*L gene was successfully amplified in ginseng, rambutan, and green tea with fragments expected size of 650 bp (all at 56 °C). For green tea, it was not possible to amplify for the barcode region *ycf*1 despite carrying it out at the specific temperature (52 °C) indicated by the gradient, some of the reasons may be a large number of polyphenols in the plant and the change of conditions. For the *mat*K and *trn*H-*psb*A barcode regions, fragments in the temperature gradients were not amplified mainly because of the DNA sample concentration and PCR conditions.

4.6 PHYTOCHEMICAL PROFILING

4.6.1 EXTRACTION OF CAFFEINE FROM CAMELLIA SINENSIS AND GINSENOSIDES FROM PANAX GINSENG

Caffeine was extracted by maceration, where 0.5 g of green tea powder was weighed and placed in a beaker together with 10 mL of water. The mixture was placed in a water bath at a temperature of 80 °C for 30 min and filtered with a filter paper. For the ginseng extraction, the root was ground with mortar and pestle, 1 g of the powder was mixed with 120 mL of methanol in a round-bottom flask and extracted for 24 h at 60 °C through Soxhlet. The recovered extract was placed in a Büchi Labortechnik AG CH-9230 Flawil Switzerland R-210 rotary evaporator at 65 °C until dried.

4.6.2 FOLIN CIOCALTEU AND 2,2-DIPHENYL-1-PICRYLHYDRAZYL (DPPH) ASSAYS

To measure the antioxidant capacity and determinate the content of phenols, Folin Ciocalteu and DPPH assays were carried out. A standard of gallic acid was prepared (0–300 µL) for the Folin Ciocalteu assay, In the procedure, a 96-well plate was used in which in each well 20 µL of the standard and samples were added; subsequently 150 µL of Milli-Q water and 50 µL of the Folin reagent were added, and after 8 min, 50 µL of the 20% carbonate was added. The reaction is allowed to occur for 2 h in the dark. The reading was performed at 760 nm on a BioRad xMark Microplate Spectrophotometer.

For the DPPH assay, 20 µL of the sample and 200 µL of DPPH were put in triplicate in each well. The reading was performed again on a BioRad xMark Microplate Spectrophotometer with a wavelength of 520 nm.

According to these results, we can confirm that green tea is the plant with the highest antioxidant capacity, and with a very acceptable total phenolic concentration, the ginseng and rambutan plants were far below compared with that plant.

4.6.3 HPLC

HPLC was performed to analyze the amount of caffeine in green tea. To perform HPLC, a ZORBAX Extend C18 column, Analytical 4.6 × 150 mm 5 µm with a flow of 0.800 mL/min at 40 °C with a lecture of 280 nm (UV) with the mobile phases being methanol and water. Caffeine concentrations of 0.0001, 0.00004, 0.000008, and 0.0000016 g/mL were used to calculate caffeine content in green tea samples. giving a total of four and a half hours of running. The concentration in tea extract was 1.08 mg caffeine/g of sample.

According to the chromatogram (Figure 4.1), the presence of caffeine by comparing the sample with a standard of caffeine was confirmed by the retention time 3.64 min. The concentration of caffeine in the green tea sample accounts for an acceptable percentage, being optimal for formulations to provide energy.

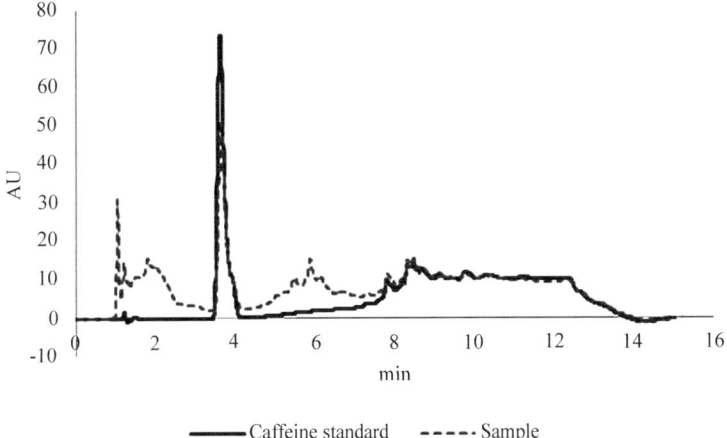

FIGURE 4.1 HPLC chromatogram of green tea and caffeine standard.

4.6.4 GAS CHROMATOGRAPHY-MASS SPECTROMETRY

To identify the compounds of interest of green tea and ginseng, gas chromatography-mass spectrometry (GC-MS) was carried out. GC-MS was performed with an Agilent Technologies 7890 A GC system coupled to an Agilent Technologies 5975C Triple Quadrupole mass spectrometer. Samples were analyzed on a DB-17ht (30 m × 250 μm × 0.15 μm). Analytical conditions were as follows: the oven temperature was held at 50 °C for 1 min, raised to 80 °C for 0 min at 30 °C/min, then increased to 110 °C for 0 min at 10 °C/min, and finally temperature reached 270 °C for 40 min at 6 °C/min. The splitless mode was conducted and helium was used as the carrier gas at a rate of 1.29 mL/min.

The mass spectrometer was operated in the electron impact mode. Tentative identification was done by comparing the mass spectra obtained of the extracts of ginseng and green tea with the NIST 05 library.

The presence of caffeine was confirmed by GC-MS in tea extract (Figure 4.2).

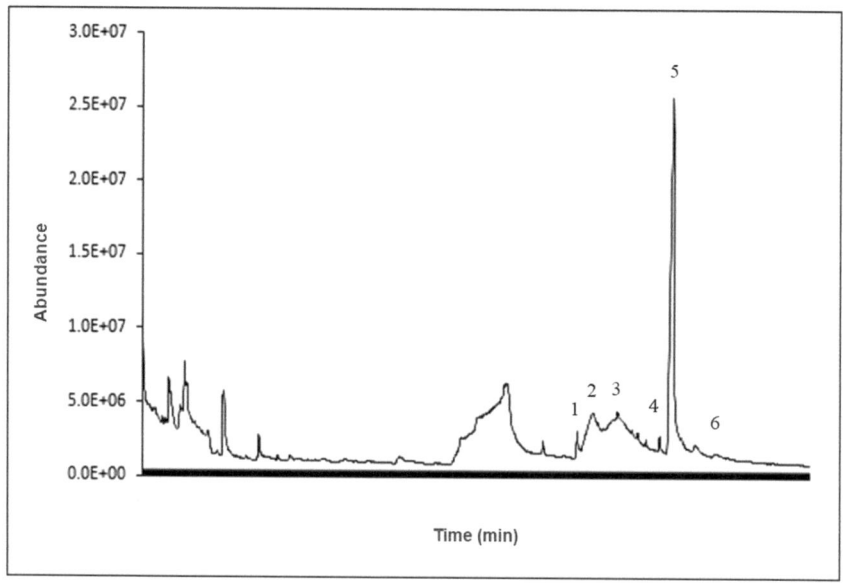

FIGURE 4.2 Gas chromatography–mass spectrometry chromatogram of green tea extract.

In the mass spectra of green tea, compounds with >90% coincidence were marked and numbered. Besides caffeine, theobromine was identified

(Table 4.3). With the GC-MS, it was possible to confirm the presence of caffeine, 19,11 percent of the sample was composed of caffeine; in the same way, it was possible to observe the presence of another alkaloid compound that helps in providing energy such as theobromine.

TABLE 4.3 Compounds in Gas Chromatography-Mass Spectrometry of Green Tea

Sr. No.	Compound	Common Name
1	*n*-Hexadecanoic acid	Palmitic acid
2	Octadecanoic acid, methyl ester	Methyl stearate
3	9,12-Octadecadienoic acid, methyl ester	Linolelaidic acid, methyl ester
4	9,12,15-Octadecatrienoic acid, methyl ester	Alfa-methyl linolenate
5	1H-Purine-2,6-dione, 3,7-dihydro-1,3,7-trimethyl	Caffeine
6	1H-Purine-2,6-dione, 3,7-dihydro-3,7-dimethyl	Theobromine

Ginseng extract showed 13 main peaks in Figure 4.3 and Table 4.4. It was possible to identify various compounds of interest in the food industry, such as maltol, which is a compound that is used to potentiate food flavors. Mannitol is a naturally occurring alcohol found in fruits and vegetables and used as an osmotic diuretic. Mannitol is freely filtered by the glomerulus and poorly reabsorbed from the renal tubule, thereby causing an increase in

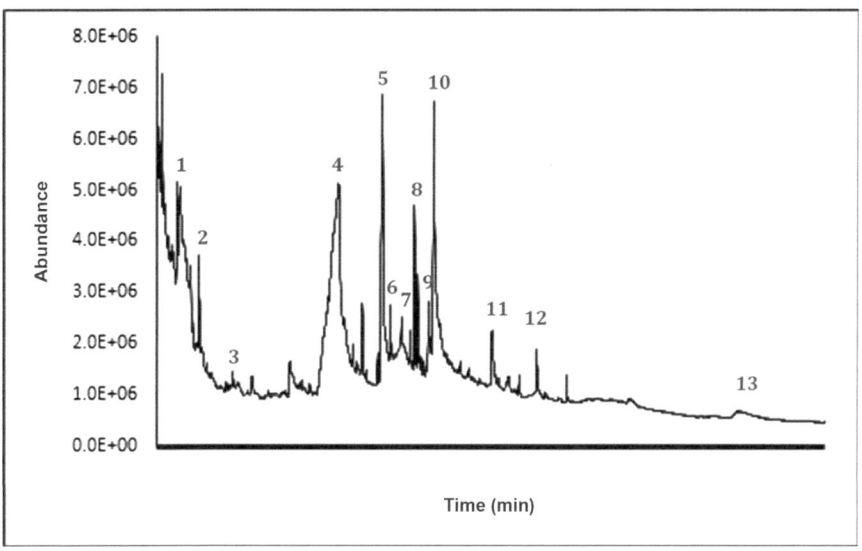

FIGURE 4.3 Gas chromatography-mass spectrometry chromatogram of ginseng extract.

osmolarity of the glomerular filtrate. An increase in osmolarity limits tubular reabsorption of water and inhibits the renal tubular reabsorption of sodium, chloride, and other solutes, thereby promoting diuresis. Also, mannitol elevates blood plasma osmolarity, resulting in enhanced flow of water from tissues into interstitial fluid and plasma (PubChem, 2019).

TABLE 4.4 Compounds in Gas Chromatography-Mass Spectrometry of Ginseng Extract

Sr. No.	Compound	Common Name
1	γ-Hydroxybutyric acid lactone	Butyrolacetone
2	4-Hydroxy-2,5-dimethylfuran-3(2H)-one	Furaneol
3	3-Hydroxy-2-methyl-4-pyrone	Maltol
4	Hexadecanoic acid	Palmitic acid
5	9-Octadecenoic acid (Z)-	Red oil
6	1-Hexadecene	Cetene
7	10,13-Octadecanoid acid	–
8	9,12-Octadecadienoic acid	Leinoleic acid
9	1,2-Benzenedicarboxylic acid, butyl 2-methylpropyl ester	Butyl isobutyl phthalate
10	cis-7-Dodecen-1-yl acetate	Looplure
11	beta-Sitosterol	Sitosterol
12	5-Eicosene	–
13	Phenol, 2,2′-methylenebis(6-(1-1-dimethylethyl)-4-ethyl	Antioxidant 425

^aCompounds with >90% coincidence were marked and numbered.
Relation of compounds and coincidence on the NIST 05 library.

Palmitic acid (PA) whose beneficial uses for health are already reported. PA has been for a long time negatively depicted for its putative detrimental health effects, shadowing its multiple crucial physiological activities. PA tissue content seems to be controlled around a well-defined concentration, and changes in its intake do not influence significantly its tissue concentration because the exogenous source is counterbalanced by PA endogenous biosynthesis (Carta et al., 2017).

The caffeine content was also confirmed. Caffeine is used to restore mental alertness or wakefulness during fatigue or drowsiness (Cappelletti et al., 2015).

4.7 CONCLUSIONS

According to the genetic and phytochemical analysis of each plant, certain important compounds were identified in terms of health benefits. For instance, in the genetic identification of green tea, rambutan, and ginseng, *rbcL* gene was amplified, and *ycf*1 for rambutan and ginseng. There are no previous reported records in gene banks of the amplification of the rambutan plant with the *ycf*1 region. With the amplification of these barcode regions, it will be possible to confirm that the plants used for subsequent analysis are formally the ones that are said to be. For the phytochemical analysis, more enriching results were achieved. The confirmation of the presence of alkaloid compounds such as caffeine helps to be the basis for generating a beverage that stimulated the CNS. Similarly, the use of ginseng and rambutan plants contribute to the product having a higher nutritional value, due to its antioxidant properties that despite not having the same capacity as green tea.

KEYWORDS

- energetic beverages
- ginseng
- nervous system
- rambutan

REFERENCES

Ahuja, Akash, Ji Hye Kim, Jong-Hoon Kim, Young-Su Yi, and Jae Youl Cho. 2018. Functional role of ginseng-derived compounds in cancer. *Journal of Ginseng Research* 42 (3):248–254.

Arias-Cruz, Marco Emilio, Hebert Augusto Vásquez-Ramírez, Diana Mateus-Cagua, Hans Nicolas Chaparro-Zambrano, and Javier Orlando Orduz-Rodríguez. 2016. El rambután (*Nephelium lappaceum*), frutal asiático con potencial para Colombia: avances de la investigación en el Piedemonte del Meta. [The rambutan (*Nephelium lappaceum*), Asian fruit with potential for Colombia: research advances in the Piedemonte del Meta]. *Revista Colombiana de Ciencias Hortícolas* 10 (2):262–272.

Attele, Anoja S, Ji An Wu, and Chun-Su Yuan. 1999. Ginseng pharmacology: multiple constituents and multiple actions. *Biochemical Pharmacology* 58 (11):1685–1693.

Baeg, In-Ho, and Seung-Ho So. 2013. The world ginseng market and the ginseng (Korea). *Journal of Ginseng Research* 37 (1):1.

Bone, Kerry, MCPP Simon Mills, and MA FNIMH. 2012. *Principles and Practice of Phytotherapy: Modern Herbal Medicine*: Elsevier Health Sciences.

Buettner, Catherine, Gloria Y Yeh, Russell S Phillips, Murray A Mittleman, and Ted J Kaptchuk. 2006. Systematic review of the effects of ginseng on cardiovascular risk factors. *Annals of Pharmacotherapy* 40 (1):83–95.

Butt, Masood Sadiq, Rabia Shabir Ahmad, M Tauseef Sultan, Mir M Nasir Qayyum, and Ambreen Naz. 2015. Green tea and anticancer perspectives: updates from last decade. *Critical Reviews in Food Science and Nutrition* 55 (6):792–805.

Cappelletti, Simone, Daria Piacentino, Gabriele Sani, and Mariarosaria Aromatario. 2015. Caffeine: cognitive and physical performance enhancer or psychoactive drug? *Current Neuropharmacology* 13 (1):71–88. doi: 10.2174/1570159X13666141210215655.

Carta, Gianfranca, Elisabetta Murru, Sebastiano Banni, and Claudia Manca. 2017. Palmitic acid: physiological role, metabolism and nutritional implications. *Frontiers in Physiology* 8:902–902. doi: 10.3389/fphys.2017.00902.

Chen, Chieh-fu, Wen-fei Chiou, and Jun-tian Zhang. 2008. Comparison of the pharmacological effects of *Panax ginseng* and *Panax quinquefolium*. *Acta Pharmacologica Sinica* 29 (9):1103.

Corbo, Maria Rosaria, Antonio Bevilacqua, Leonardo Petruzzi, Francesco Pio Casanova, and Milena Sinigaglia. 2014. Functional beverages: the emerging side of functional foods: commercial trends, research, and health implications. *Comprehensive Reviews in Food Science and Food Safety* 13 (6):1192–1206.

Dong, Wenpan, Chao Xu, Changhao Li, Jiahui Sun, Yunjuan Zuo, Shuo Shi, Tao Cheng, Junjie Guo, and Shiliang Zhou. 2015. ycf1, the most promising plastid DNA barcode of land plants. *Scientific Reports* 5:8348.

Dunning, Luke T, and Vincent Savolainen. 2010. Broad-scale amplification of matK for DNA barcoding plants, a technical note. *Botanical Journal of the Linnean Society* 164 (1):1–9.

Dzoyem, Jean P, Emmanuel Tshikalange, and Victor Kuete. 2013. Medicinal plants market and industry in Africa. In *Medicinal Plant Research in Africa*, 859–890. Elsevier.

FAO. 2018. Global tea consumption and production driven by robust demand in China and India. http://www.fao.org/climate-change/news/detail/en/c/1136370/.

Ford, Caroline S, Karen L Ayres, Nicola Toomey, Nadia Haider, Jonathan Van Alphen Stahl, Laura J Kelly, Niklas Wikström, Peter M Hollingsworth, R Joel Duff, and Sarah B Hoot. 2009. Selection of candidate coding DNA barcoding regions for use on land plants. *Botanical Journal of the Linnean Society* 159 (1):1–11.

Fujiki, Hirota, Eisaburo Sueoka, Tatsuro Watanabe, and Masami Suganuma. 2015. Primary cancer prevention by green tea, and tertiary cancer prevention by the combination of green tea catechins and anticancer compounds. *Journal of Cancer Prevention* 20 (1):1–4. doi: 10.15430/JCP.2015.20.1.1.

Gibson-Smith, D, M Bot, I Brouwer, M Visser, and BWJH Penninx. 2018. Diet quality in subjects with and without depressive and anxiety disorders. *Proceedings of the Nutrition Society* 77 (OCE2).

Harahap, Serida Nauli, Nazaruddin Ramli, Nazanin Vafaei, and Mamot Said. 2012. Physicochemical and nutritional composition of rambutan anak sekolah (*Nephelium lappaceum* L.) seed and seed oil. *Pakistan Journal of Nutrition* 11 (11):1073–1077.

Khan, Salman, Alev Tosun, and Yeong Shik Kim. 2015. Ginsenosides as food supplements and their potential role in immunological and neurodegenerative disorders. In *Bioactive Nutraceuticals and Dietary Supplements in Neurological and Brain Disease*, 303–309. Elsevier.

Kim, Hee Jin, Pitna Kim, and Chan Young Shin. 2013. A comprehensive review of the therapeutic and pharmacological effects of ginseng and ginsenosides in central nervous system. *Journal of Ginseng Research* 37 (1):8.

Kim, Jong-Hoon. 2012. Cardiovascular diseases and *Panax ginseng*: a review on molecular mechanisms and medical applications. *Journal of Ginseng Research* 36 (1):16.

Kim, Jong-Hoon. 2018. Pharmacological and medical applications of *Panax ginseng* and ginsenosides: a review for use in cardiovascular diseases. *Journal of Ginseng Research* 42 (3):264–269.

Kim, Min-Hyun, and Hyeyoung Kim. 2017. Ginseng and gastrointestinal protection. In *Gastrointestinal Tissue*, 299–304. Elsevier.

Kress, W John, and David L Erickson. 2007. A two-locus global DNA barcode for land plants: the coding rbcL gene complements the non-coding trnH-psbA spacer region. *PLoS One* 2 (6):e508.

Kumar, Amit, Ajay kumar, Payal Thakur, Sandip Patil, Chandani Payal, Anil Kumar, and Pooja Sharma. 2012. Antibacterial activity of green tea (*Camellia sinensis*) extracts against various bacteria isolated from environmental sources. *Recent Research in Science and Technology* 4 (1):19–23.

Kumar, Chityal Ganesh, Sarada Sripada, and Yedla Poornachandra. 2018. "Status and Future Prospects of Fructooligosaccharides as Nutraceuticals. In *Role of Materials Science in Food Bioengineering*, 45–503. Elsevier.

Le, Ninh Nguyet Hai, Chiyomi Uematsu, Hironori Katayama, Lieu Thi Nguyen, Ninh Tran, Dung Van Luong, and Son Thanh Hoang. 2017. *Camellia tuyenquangensis* (Theaceae), a new species from Vietnam. *Korean Journal of Plant Taxonomy* 47 (2):95–99.

Lee, Chang Ho, and Jong-Hoon Kim. 2014. A review on the medicinal potentials of ginseng and ginsenosides on cardiovascular diseases. *Journal of Ginseng Research* 38 (3):161–166.

Lestari, Sri Rahayu, Muhammad Sasmito Djati, Ahmad Rudijanto, and Fatchiyah Fatchiyah. 2014. Production and potency of local rambutan at East Java as a candidate phytopharmaca. *AGRIVITA, Journal of Agricultural Science* 35 (3):270–276.

Leung, Kar Wah, and Alice Sze-Tsai Wong. 2010. Pharmacology of ginsenosides: a literature review. *Chinese Medicine* 5 (1):20.

Levin, Rachel A, Warren L Wagner, Peter C Hoch, Molly Nepokroeff, J Chris Pires, Elizabeth A Zimmer, and Kenneth J Sytsma. 2003. Family-level relationships of Onagraceae based on chloroplast rbcL and ndhF data. *American Journal of Botany* 90 (1):107-115.

Li, Xiwen, Yang Yang, Robert J Henry, Maurizio Rossetto, Yitao Wang, and Shilin Chen. 2015. Plant DNA barcoding: from gene to genome. *Biological Reviews* 90 (1):157–166.

Lu, Jian-Ming, Qizhi Yao, and Changyi Chen. 2009. Ginseng compounds: an update on their molecular mechanisms and medical applications. *Current Vascular Pharmacology* 7 (3): 293–302.

Mohanan, Padmanaban, Sathiyamoorthy Subramaniyam, Ramya Mathiyalagan, and Deok-Chun Yang. 2018. Molecular signaling of ginsenosides Rb1, Rg1, and Rg3 and their mode of actions. *Journal of Ginseng Research* 42 (2):123–132.

Morton, Julia Frances, and Curtis F Dowling. 1987. *Fruits of Warm Climates*. Vol. 20534: JF Morton: Miami, FL.

Murthy, Hosakatte Niranjana, Milen I Georgiev, Yun-Soo Kim, Cheol-Seung Jeong, Sun-Ja Kim, So-Young Park, and Kee-Yoeup Paek. 2014. Ginsenosides: prospective for sustainable biotechnological production. *Applied Microbiology and Biotechnology* 98 (14):6243–6254.

Nafiu, M. O., A. A. Hamid, H. F. Muritala, and S. B. Adeyemi. 2017. Chapter 7—preparation, standardization, and quality control of medicinal plants in Africa. In *Medicinal Spices and Vegetables from Africa*, edited by Victor Kuete, 171–204. Academic Press.

Namita, Parmar, Rawat Mukesh, and Kumar J Vijay. 2012. Camellia sinensis (green tea): a review. *Global Journal of Pharmacology* 6 (2):52–59.

Natarajan, Satheesh B, Suriyakala Perumal Chandran, Sahar Husain Khan, Packiyaraj Natarajan, and Karthiyaraj Rengarajan. 2019. Versatile health benefits of catechin from green tea (*Camellia sinensis*). *Current Nutrition & Food Science* 15 (1):3–10.

Nayyar, Abhishek Singh, Maneesha Das, Bharat Deosarkar, Soniya Bharat Deosarkar, Abhishek Karan, Pallavi Sinha, and Swati Paraye. 2017. Green tea in medicine: a brief overview. *International Journal of Dental Research* 5 (2):191–194.

Newmaster, SG, AJ Fazekas, and S Ragupathy. 2006. DNA barcoding in land plants: evaluation of rbcL in a multigene tiered approach. *Botany* 84 (3):335–341.

Niedzwiecki, Aleksandra, Mohd Waheed Roomi, Tatiana Kalinovsky, and Matthias Rath. 2016. Anticancer efficacy of polyphenols and their combinations. *Nutrients* 8 (9):552.

Osorio-Espinoza, Humberto, Ángel Leyva-Galan, Ernesto Toledo-Toledo, Francisco Javier Marroquín-Agreda, and Magdiel Gabriel-Hernandez. 2019. La producción de rambután (*Nephelium lappaceum* L.) en Chiapas, México. Oportunidades para una producción agroecológica. [The production of rambutan (*Nephelium lappaceum* L.) in Chiapas, Mexico. Opportunities for agroecological production]. *Cultivos Tropicales* 40 (1).

Park, Hye-Lim, Hun-Soo Lee, Byung-Cheul Shin, Jian-Ping Liu, Qinghua Shang, Hitoshi Yamashita, and Byungmook Lim. 2012. Traditional medicine in China, Korea, and Japan: a brief introduction and comparison. *Evidence-Based Complementary and Alternative Medicine* 2012:9. doi: 10.1155/2012/429103.

Park, Jong Dae, Dong Kwon Rhee, and You Hui Lee. 2005. Biological activities and chemistry of saponins from *Panax ginseng* CA Meyer. *Phytochemistry Reviews* 4 (2–3):159–175.

Park, Sa-Yoon, Ji-Hun Park, Hyo-Su Kim, Choong-Yeol Lee, Hae-Jeung Lee, Ki Sung Kang, and Chang-Eop Kim. 2018. Systems-level mechanisms of action of *Panax ginseng*: a network pharmacological approach. *Journal of Ginseng Research* 42 (1):98–106.

Pedraz-Petrozzi, Bruno. 2018. Fatiga: historia, neuroanatomía y caracteristicas psicopatológicas. Una revisión de la Literatura. [Fatigue: history, neuroanatomy and psychopathological characteristics. A review of the literature]. *Revista de Neuro-Psiquiatría* 81 (3):174–182.

Preedy, Victor R. 2012. *Tea in Health and Disease Prevention*. Academic Press.

PubChem. 2019. National Center for Biotechnology Information. PubChem Database. Mannitol, CID=6251. https://pubchem.ncbi.nlm.nih.gov/compound/Mannitol (accessed on Dec. 31, 2019).

Rodriguez, Evelyn, Oliver Salangad, Ronaniel Almeda, Charisse Reyes, and Kevin Salamanez. 2019. Fatty acid and unsaponifiable composition of ten philippine food plant oils for possible nutraceutical and cosmeceutical applications. *Agriculture & Forestry/Poljoprivreda i Sumarstvo* 65 (3). DOI:10.17707/AgricultForest.65.3.11

Ross, Ivan A. 2007. *Medicinal Plants of the World, Volume 3: Chemical Constituents, Traditional and Modern Medicinal Uses*. Vol. 3: Springer Science & Business Media.

Sánchez, Julio César, César Ramón Romero, Cristhian David Arroyave, Andrés Mauricio García, Fabián David Giraldo, and Leidy Viviana Sánchez. 2015. Bebidas energizantes: efectos benéficos y perjudiciales para la salud. [Energy drinks: beneficial and harmful effects on health]. *Perspectivas en nutrición humana* 17 (1):79–91.

Schneider, Craig, and Tiffany Segre. 2009. Green tea: potential health benefits. *American Family Physician* 79 (7):591–594.

Sharma, Anjali, Sonal Gupta, Indira P Sarethy, Shweta Dang, and Reema Gabrani. 2012. Green tea extract: possible mechanism and antibacterial activity on skin pathogens. *Food Chemistry* 135 (2):672–675.

Shi, Ze-Yu, Jin-Zhang Zeng, and Alice Sze Tsai Wong. 2019. Chemical structures and pharmacological profiles of ginseng saponins. *Molecules* 24 (13):2443.

Shin, Byong-Kyu, Sung Won Kwon, and Jeong Hill Park. 2015. Chemical diversity of ginseng saponins from *Panax ginseng*. *Journal of Ginseng Research* 39 (4):287–298.

So, Seung-Ho, Jong Won Lee, Young-Sook Kim, Sun Hee Hyun, and Chang-Kyun Han. 2018. Red ginseng monograph. *Journal of Ginseng Research* 42 (4):549–561.

Solís-Fuentes, Julio A, Guadalupe Camey-Ortíz, María del Rosario Hernández-Medel, Francisco Pérez-Mendoza, and Carmen Durán-de-Bazúa. 2010. Composition, phase behavior and thermal stability of natural edible fat from rambutan (*Nephelium lappaceum* L.) seed. *Bioresource Technology* 101 (2):799–803.

Teschke, Rolf, Albrecht Wolff, Christian Frenzel, Axel Eickhoff, and Johannes Schulze. 2015. Herbal traditional Chinese medicine and its evidence base in gastrointestinal disorders. *World Journal of Gastroenterology* 21 (15):4466–4490. doi: 10.3748/wjg.v21.i15.4466.

Thitilertdecha, Nont, Aphiwat Teerawutgulrag, Jeremy D Kilburn, and Nuansri Rakariyatham. 2010. Identification of major phenolic compounds from *Nephelium lappaceum* L. and their antioxidant activities. *Molecules* 15 (3):1453–1465.

Thitilertdecha, Nont, Aphiwat Teerawutgulrag, and Nuansri Rakariyatham. 2008.Antioxidant and antibacterial activities of *Nephelium lappaceum* L. extracts. *LWT—Food Science and Technology* 41 (10):2029–2035.

Uraiwan, Kwansiri, and Chutimon Satirapipathkul. 2016. The entrapment of vitamin E in nanostructured lipid carriers of rambutan seed fat for cosmeceutical uses. *Key Engineering Materials* 675–676, 77–80.

Vargas, Alfonso. 2003. Descripción morfológica y nutricional del fruto de rambután *(Nephelium lappaceum)*. [Morphological and nutritional description of rambutan fruit (*Nephelium lappaceum*)]. *Agronomía Mesoamericana* 14 (2):201–206.

Vukasović, Tina. 2017. Functional foods in line with young consumers: Challenges in the marketplace in Slovenia. In *Developing New Functional Food and Nutraceutical Products*, 391–405. Elsevier.

Walsh, Jacob P, Natasha DesRochers, Justin B Renaud, Keith A Seifert, Ken K-C Yeung, and Mark W Sumarah. 2021. Identification of *N,N′,N″*-triacetylfusarinine C as a key metabolite for root rot disease virulence in American ginseng. *Journal of Ginseng Research* 45 (1):156–162.

Wang, Wei. 2016. Genomics and traditional Chinese medicine. In *Genomics and Society*, 293–308. Elsevier.

Wee, Jae Joon, Kyeong Mee Park, and An-Sik Chung. 2011. Biological activities of ginseng and its application to human health. *Herbal Medicine: Biomolecular and Clinical Aspects* 2:157–174.

Xu, Wanqi, Hyung-Kyoon Choi, and Linfang Huang. 2017. State of *Panax ginseng* research: a global analysis. *Molecules* 22 (9):1518.

Yu, F, T Takahashi, J Moriya, K Kawaura, J Yamakawa, K Kusaka, T Itoh, S Morimoto, N Yamaguchi, and T Kanda. 2006. Traditional Chinese medicine and Kampo: a review from the distant past for the future. *Journal of International Medical Research* 34 (3):231–239.

Yun, Taik Koo. 2001. Brief introduction of *Panax ginseng* CA Meyer. *Journal of Korean medical Science* 16 (Suppl):S3.

Zhen, Yong-su. 2002. *Tea: Bioactivity and Therapeutic Potential*. CRC Press.

Zink, Alexander, and Claudia Traidl-Hoffmann. 2015. Green tea in dermatology—myths and facts. *JDDG: Journal der Deutschen Dermatologischen Gesellschaft* 13 (8):768–775.

CHAPTER 5

Innovation and Challenges in the Development of Functional and Medicinal Beverages

DAYANG NORULFAIRUZ ABANG ZAIDEL[*], IDA IDAYU MUHAMAD,
ZANARIAH HASHIM, YANTI MASLINA MOHD JUSOH, and
ERARICAR SALLEH

*Food and Biomaterial Engineering Research Group,
School of Chemical and Energy Engineering, Faculty of Engineering,
Universiti Teknologi Malaysia, 81310 Johor Bahru, Malaysia*

[*]Corresponding author. E-mail: dnorulfairuz@utm.my

ABSTRACT

Recent reports have recognized that functional foods are the most emerging interest in food sectors. In particular, the functional beverage market is one of the fastest-growing segments within this sector. Beverages are categorized as the most dynamic functional food due to their convenience and the high possibility to meet market needs. Functional beverages have a great potential to be incorporated with desirable nutritional ingredients and bioactive compounds such as vitamins, minerals, antioxidants, fatty acids, fiber, prebiotics, and probiotics. This chapter highlighted the recent innovation and development of functional and medicinal beverages including the current processes and methods of developing successful functional and medicinal beverages. The topics include the definition of functional and medicinal beverages and the type of functional beverages available in the market, the market trends and overview, emerging trends in processing technology, quality and safety issues in functional beverages prior to commercialization of products. It has become a challenge to maintain a good quality of functional and medicinal beverages especially in terms of its stability, shelf life, functionality, and safety. These issues must be taken seriously for improving the quality and health benefits of the beverages, as well as for the future of the food industry.

5.1 INTRODUCTION

Recent reports have recognized that functional foods are the most emerging interest in food sectors with functional beverage market, particularly being the fastest-growing segment within this sector. Nowadays, as people are more health conscious, the invention of foods and beverages is focused more on their function toward the prevention and treatment of disease has been growing rapidly. The consumption of functional foods and beverages has been one of the top choices for many people as they target a number of health objectives beyond the basic nutritional functions.

Functional beverages are promoted with many health benefits such as improved immunity and digestion systems, energy-boosting, weight management, improving mental focus and preventing pain that is associated with bone and joint conditions. In addition, some beverages are formulated specifically for age demographic and gender, such as products targeted on kids, women, and seniors. This innovation and diversification in beverages have improved the beverages not only in terms of taste but also nutritional value and functionality.

Beverages have been categorized as the most dynamic functional food category because of the convenience and opportunity to meet consumer demands. The composition, physical attributes, the dimensions of size and shape, and the easiness of delivery, and storage at either ambient or refrigerated temperature are among the acceptable preferences of beverage products. Besides that, it has a great potential to be incorporated with desirable nutritional ingredients and bioactive compounds such as vitamins, minerals, antioxidants, fatty acids, plant extracts and fiber, prebiotics, and probiotics. Functional beverages are a brilliant medium for conveying nutritional ingredients and bioactive compounds to consumers in a natural way.

5.2 DEFINITION OF FUNCTIONAL AND MEDICINAL BEVERAGES

The term "functional food" has no legislative definition in most countries. Mark-Herbert (2004) and Niva (2000) observed that because of the lack of a legitimate definition, food and nutrition experts find it challenging to distinguish between foods that are functional and those that are conventional.

Up until now, academic bodies, national authorities, and even the food industry have defined functional foods differently—some with very simple definitions while others preferring more complex ones. For example, functional food has been simply defined as "foods that provide not only basic nutrition but also other health benefits," whereas a complex definition of functional food is "food that resembles conventional food meant for normal diet consumption, albeit modified to provide more than simple nutrient requirements to serve other physiological roles" (Bech-Larsen and Grunert, 2003).

In the mid-1980s, Japan introduced the first concept of functional foods, namely foods that had ingredients specifically for health functions or health use, also known as FOSHU (Lau et al., 2013). According to Bigliardi and Galati (2013), the specific definition of FOSHU, as given by the Japanese Ministry of Health, Labor and Welfare, is "any food with a particular label and license that promises specific benefits to the health of the persons that consume it for health uses." Diplock et al. (1999) further defined functional foods and drinks as foods that "provide beneficial effects for bodily functions in addition to basic nutrition such that the health of the person consuming it will be improved and the risk of disease is reduced."

In 1998, a report by the Functional Food Science Commission in Europe defined "foods that sufficiently prove to benefit any functions of the body in addition to providing nutritional benefits such that health is improved and the risk of disease is reduced" as functional foods (including drinks). Meanwhile, Pravst (2012) noted the combined definitions of functional foods provided by other studies such as "food that is natural and unmodified; enhanced food with components that have been improved via biotechnological means, special breeding, or unique growing conditions; food that has added beneficial components; food that provides unique benefits as a result of the removal of components via technological or biotechnological means; food with favorable properties as a result of component replacement with another better component; food with a beneficial component modified via technological, chemical, or enzymatic means; food with a component that was modified to become biologically available; or any food with the above combinations."

In addition to the above definition, the International Life Science Institute Europe through the European Commission's Concerted Action on Functional Food Science in Europe defined functional food as "food products that, in addition to providing basic nutrition, also benefit any human bodily function to bring about improved condition, whether general of physical, and causing the risk of disease to reduce."

Diplock et al. (1999) further emphasized that functional food should be consumed as per other food that is naturally consumed as part of one's diet. In other words, functional food is not taken as a capsule or pill but rather in its natural (food) form. Nevertheless, Ohama et al. (2006) disagreed with this definition, stating that Japanese FOSHU products had been consumed as tablets or pills since 2001, whereas other functional foods still retain their normal form. Elsewhere, Yang (2008) contended that China defines functional food as "foods that provide special health benefits or foods that provide minerals or vitamins to the body, regulates the functions of the human body, and are consumed by specific groups of people, but are ultimately not meant for therapeutic purposes, and as a result of consumption, will not cause any acute, subacute, or chronic harm." Canada, on the other hand, defines "food that resembles (or may even be) conventional food that is part of a regular diet consumption that proves to benefit humans physiologically and decreases chronic disease risks in addition to providing basic nutrition" as functional food (Lau et al., 2013).

In the United States, however, functional food is regulated by the Food and Drug Administration but is not defined legally. According to the Academy of Nutrition and Dietetics in the United States, functional foods are "foods that are enhanced, fortified or enriched, or whole foods, that have shown significant evidence of providing effective benefits to health when taken as part of a regular varied diet". Based on the above definitions, functional foods or beverages can be simply defined as "food products that have additional physiological benefits over and above their benefits that occur naturally."

5.3 NUTRITIONAL AND HEALTH CLAIMS

In a regulatory point of view, functional foods do not have an appropriate definition and category. Brown and Chan (2009) reported that any regulatory authority, with the exception of Japan, has classified "functional food" as a distinguishable product. Therefore Sun-Waterhouse (2011) revealed that there is an uncertainty in labeling the new product, either as a food, supplement, or drug. It is important that before a new functional food product enters the market, clear descriptions of functionality or health and nutritional effect of this food should meet the fundamental specifications.

In Europe, it has been stated that "nutrition claim" is made to recommend the specific valuable nutritional composition of food including "source of," "free of," "reduced", whereas "health claim" is documented to propose the synergistic effect within health and food group, a food, or one of its compounds. Dolan (2011) concise that the claims are defining "generic health claim" is associated to the important role of nutrient and other ingredient for growth, development or function of the body; psychological and behavioral performances; weight management, fullness or lessening of existing energy and "health claim" which described the ability of the food in lowering the chance of having disease and children's growth failure.

In the United States, functional food products must follow the guidelines for conventional foods. According to Jackson and Paliyath (2011), it is a requirement to ensure that the product is safe to be advertised as foods, and the incorporated ingredients are "generally recognized as safe" or permitted as food additives. In Hasler (2008), functional food products claims can be classified into three categories: "health claims," "nutrient content claims," and "structure or function claims."

The registered products in the United States which are a claim for dietary supplement ingredients are applicable to be used worldwide. However, the claims of certain US supplement ingredients have not been approved in Europe and vice versa because of the distinct regulations applied. Dolan (2011) reported that the approval of the products needs immense process involving the national authorities, EFSA and the European Commission which are the main restrictions to advertise the product with claims in Europe.

5.4 DEVELOPMENT OF FUNCTIONAL AND MEDICINAL BEVERAGES

Development of a functional beverage includes the extraction of functional ingredient or active compound, formulation of the beverage, processing method or technology used, and quality assessment of the beverage prior to commercialization. Each step in producing functional or medicinal beverages is important to make sure that the quality and safety of the beverages are the top priority. The beverages quality includes shelf life, stability, storage, and packaging that may then contribute to the health of consumers. Different types of functional and medicinal beverages vary from dairy based, vegetable and fruits drinks, energy drinks, sport and performance drinks, soy-based drinks, and herbal drinks. These different types of beverages are formulated based on different functional ingredients and can be produced using different processing methods. As the modern technology has evolved more innovation on functional beverages has been reported to improve the processing and preserve the functional ingredient or active compound in the beverages. Many studies have been reported on the extraction of functional ingredients or active compound from natural sources either from animal or plant for application in beverage production (Ottaway, 2009; Ramachandran and Rao, 2013; Das et al., 2012).

5.5 TYPES OF FUNCTIONAL AND MEDICINAL BEVERAGES

Functional and medicinal beverages appeared to be the most active functional foods category due to convenience, health benefits, ease of distribution and storage as, well as ease of fortification of products with desirable nutrients and bioactive compounds (Ozen et al., 2012). This chapter highlights several types of popular beverages available in this fast-growing sector, namely dairy-based beverages, vegetables and fruit beverages, energy drinks, sport and performance drinks, soy-based beverages, herbal drinks, and ready-to-drink (RTD) tea.

5.5.1 DAIRY-BASED BEVERAGES

Yogurt drinks, fresh milk, and fermented milk are the example of dairy-based beverage products that are often regarded as an excellent vehicle for probiotics. According to FAO/WHO (2006), probiotics are living microorganisms which when consumed in a specific dose will provide beneficial health effects for the host. In fact, the individual probiotics give various health effects including immune response enhancement, cholesterol level reduction, colon cancer prevention, irritable bowel syndrome alleviation, inflammable bowel disease alleviation, and antibiotic-associated diarrhea treatment, and others (Ozer and Kirmaci, 2010).

Today, with regard to the aforementioned health benefits, many dairy manufacturers began to incorporate probiotic bacteria (i.e., *Lactobacillus* spp., *Bifidobacterium* spp., etc.) into their products (Corbo et al., 2014). The main reason behind such action is to satisfy consumers' demand at the same time to gain a competitive advantage in the world market. Besides probiotic bacteria, some producers may fortified/enriched dairy beverages with bioactive components (i.e., Omega-3, phytosterols, fiber, minerals, vitamins, etc.) in order to enhance the products' functional properties (Boroski et al., 2012). Table 5.1 lists some examples of functional dairy-based beverages available in the market.

Functional dairy-based beverages have continued to be as essential as the primary food worldwide, particularly in the Western parts. However, several health risks are associated with these products. According to Kumar et al.

(2015), the major drawback of dairy beverage intake appears to be lactose intolerance. The condition when lactose, a milk sugar contains abundantly in dairy products, is unable to digest because of the lack of the lactase enzyme in the small intestine is called lactose intolerance (Deng et al., 2015). People with lactose intolerance are unable to fully digest the sugar present in dairy products. As a result, they have symptoms like flatulence, bloating, loose stool, and cramping (Szilagyi, 2015). In this case, the market size for dairy companies will be reduced. Thus the companies are launching more alternatives that could provide the same nutrients and health benefits as dairy products so that the range of their targeted customers will become larger.

TABLE 5.1 Examples of Commercially Available Functional Dairy-Based Beverages (Corbo et al., 2014; Ozer and Kirmaci, 2010; Prado et al., 2008)

Brand	Producer	Active Compounds
Probiotics		
Yakult®	Yakult Honsha Co (Japan)	• *Lactobacillus casei* Shirota
Vitagen®	Malaysia Milk Sdn. Bhd. (Malaysia)	• *Lactobacillus acidophilus* • *L. casei*
Chamyto®	Nestle (France)	• *Lactobacillus johnsonii* • *Lactobacillus helveticus*
Enriched Beverages		
Evolus®	Valio Ltd. (Finland)	• Bioactive peptides
Heart Plus®	PB Food (Australia)	• Omega-3
Benecol®	Mc Neil Nutritionals (United Kingdom)	• Phytosterol
Dairyland Milk-2-Go®	Saputo (Canada)	• Omega-3 • Calcium • Vitamins
Night-Time Milk®	Cricketer Farm (United Kingdom)	• Melatonin
Zen®	Danone (Belgium)	• Magnesium
Natural Linea®	Corporacion Alimentaria Penanata SA (Spain)	• Conjugated linoleic acid

5.6 VEGETABLES AND FRUIT BEVERAGES

Dairy-based beverages have long been considered as an ideal delivery vehicle of nutrition for human. However, lactose intolerance appears to be the major drawback of functional dairy products. According to Lomer et al. (2008), lactase nonpersistence is suffered by up to 70% of the total population of the world that may negatively affect the overall market of those products.

Unlike dairy-based products, vegetable- and fruit-based functional beverages are regarded as safe and healthy drinks by consumers particularly to whom are sensitive to dairy products. Additionally, the vital nutrients of vegetables and fruit juices, such as vitamins, minerals, phytochemicals, antioxidants, could provide an excellent medium for probiotic (Granato et al., 2010). The global market for fortified vegetables and fruit juices is expected to reach 65.9 billion liters by 2022, driven by the growing awareness over the health benefits of juicing fruits and vegetables (Global Industry Analysts Inc., 2016). Blueberry, strawberries, cranberry, grapes, blackcurrant, cherries, pomegranate, mango, apple, acai, guarana, kiwifruits, peach, and plums are examples of fruits that are often utilized in industrial productions (Sun-Waterhouse, 2011). Based on the reports from Gaanappriya et al. (2013), orange and watermelon juices were proven to be a suitable carrier of lactobacilli (for the purpose of functional beverage preparation) for individuals who are sensitive of dairy products. Table 5.2 shows several instances of vegetable and fruit-based functional beverages that are accessible in the food industry, which are fortified with probiotics and enriched with vitamins and minerals.

Apart from nutrients that contribute to the health of mankind, most commercialized vegetables and fruit juices contain a high amount of sugar for preservation purpose as well as to enhance the taste of products (Kumar et al., 2015). Nevertheless, this raised the health concern of consumers as long-term consumption of high-calorie product may increase the risk of diabetes and heart-related complications. Thus manufacturers nowadays are attempting to adapt modified or other new approaches in order to develop a novel, low-calorie fortified vegetable or fruit juice containing high-nutritional value at the same time maintaining the taste that able to satisfy consumers' preferences.

5.7 ENERGY DRINKS

Beverages loaded with stimulant drugs such as caffeine are well known as energy drink and often targeted to young people aged within 21 and 35 years as to provide mental and physical stimulation (Corbo et al., 2014). According to Duncan and Hankey (2013), more than 40% of athletes consume energy drinks to boost their athletic performance. Besides caffeine, vitamin B complex and taurine are the most common ingredients for such product. Taurine that can be found naturally in breast milk, meat, and fish is an amino acid that can be obtained naturally in the body. Several investigations found that supplementation with taurine could enhance athletic performance whereas when mixed with caffeine could enhance mental performance (Ripps and Shen, 2012). Examples of commercialized energy drinks are Red Bull® (Red Bull GmbH, Austria), Monster Energy® (Hansen Natural Corp., United States), and Full Throttle® (Coca-Cola Co., United States) (Corbo et al., 2014).

TABLE 5.2 Examples of Commercially Available Vegetable and Fruit-Based Functional Beverages (Corbo et al., 2014; Ozer and Kirmaci, 2010)

Brand	Producer	Active Compounds
	Probiotics	
Whole Grain Probiotic Liquid®	Grainfields (Australia)	• Vitamins
		• Amino acids
		• Enzymes
		• Probiotics:
		• *Lactobacillus acidophilus*
		• *Lactobacillus delbrueckii*
		• *Saccharomyces cerevisiae* var. *boulardii*
		• *S. cerevisiae*
Biola®	Tine BA (Norway)	• Probiotics:
		• *Lactobacillus rhamnosus* GG
Gefilus®	Valio Ltd. (Finland)	• Vitamins C and D
		• Probiotics:
		• *L. rhamnosus* GG
	Enriched Beverages	
Tomato Juice Plus®	Langer Juice Co., Inc. (United States)	• Vitamins
		• Minerals
Tropicana Essentials Orange Juice & Calcium®	Tropicana (United States)	• Calcium
Welch's 100% Grape Juice with Calcium®	Welch Foods Inc. (United States)	• Calcium

Some energy drinks may include herbal extracts as an ingredient to enhance the effect of their products. Examples of herbal extract usually used are ginseng, ginkgo biloba, guarana, and others. Ginseng is a type of herb that has been used for over 2000 years in China, Korea, and Japan, as a remedy for promoting longevity as well as enhancing physical and mental conditions (Kim et al., 2013). In addition, ginseng can be used as an antistress and antioxidant agent. However, the effect might be depending on the dose and frequency of consumption since this herb has multiple drug interactions (Gunja and Brown, 2012).

It has been reported that energy drinks might also be associated with health risks, for an instance the raised risk of alcohol-related harm, and immoderate drinking may lead to psychiatric and cardiac conditions (Sanchis-Gomar et al., 2015). Adults' physical performance has been revealed to be boost when taking caffeine by improving magnitude of response to aerobic resistance and strength, slow down exhaustion, and improving reaction time (Cappelletti et al., 2015). Nevertheless, these effects are extremely inconsistence and depend on the intensity of caffeine, whereas these effects are not being investigated on children and adolescents (Committee on Nutrition and the Council on Sports Medicine and Fitness, 2011).

5.8 SPORTS AND PERFORMANCE DRINKS

Sports drinks are flavored beverages that are particularly formulated to help consumers/athletes to replenish water, electrolytes (minerals such as potassium, sodium, magnesium, chloride, and calcium), and energy, particularly after exercise. Unlike energy drinks, sport and performance drinks do not contain caffeine (Committee on Nutrition and the Council on Sports Medicine and Fitness, 2011). Generally, sports and performance drinks can be categorized into three major types, namely, isotonic, hypertonic, and hypotonic, all of which comprise of different amount of fluid, electrolytes and carbohydrate, and different functions as shown in Table 5.3.

Glucose, sucrose, fructose, and maltodextrin/glucose polymers, or a combination of these appear to be the common ingredients for sports and performance drinks. According to Campbell (2013), an advantage of adding maltodextrin/glucose polymers is that the lower sweetness of these compounds increases the concentration of carbohydrate without increasing the sweetness of the product. Nevertheless, when high carbohydrate-containing drinks are taking in immoderate, it may increase an individual's intake of total daily caloric that will eventually be gaining extra weight, dental caries, and low quality of diet. Table 5.4 illustrated several examples of commercially available sport and performance drinks.

Development of Functional and Medicinal Beverages

TABLE 5.3 Difference Between Three Major Types of Sports and Performance Drinks Available in the Market (Diabetes.co.uk, 2017)

Types of Sports and Performance Drinks	Isotonic	Hypertonic	Hypotonic
Level of fluid, electrolytes, and carbohydrate	Same concentrations of salt and sugar as in the human body	The higher concentration of salt and sugar than the human body	Lower concentration of salt and sugar than the human body
Function	To replace fluids lost through sweating quickly To supply a boost of carbohydrate	To supplement daily carbohydrate intake. Top-up muscle glycogen stores	To replace fluids lost through sweating
Recommended to	Most athletes, including middle and long-distance running. Individuals involve in team sports	Individuals involve in ultra-distance events, but must be used in conjunction with isotonic drinks to replace lost fluids	Athletes who require fluid without a carbohydrate boost, such as gymnasts

TABLE 5.4 Examples of Commercially Available Sport and Performance Drinks (Corbo et al., 2014; Heckman et al., 2010)

Brand	Producer	Active Compounds
Powerade®	Coca-Cola Co. (United States)	Sodium, Iron
Powerade Zero®	Coca-Cola Co. (United States)	Sodium, Potassium, B-vitamins
Accelerade®	Pacific Health Laboratories Inc. (United States)	Sodium, Potassium, Calcium, Vitamin E, Protein
All Sport Body Quencher®	All Sport., Inc. (United States)	Sodium, Potassium, Vitamin C
Gatorade Endurance®	PepsiCo Inc. (United States)	Sodium, Potassium Calcium, Magnesium

5.9 SOY-BASED BEVERAGES

Soy-based beverages are aqueous extracts of whole soybeans (Glicine max) and a significant source of high-quality protein. They are frequently used as a dairy substitute, especially by the individuals who are allergic to milk or having severe intolerance to lactose because soy-based beverages contain almost similar nutrients and health benefits as in milk and dairy products. Although soy extracts lack calcium, the present soy-based beverages have this problem solved by adding artificial calcium to them where this will further increase their nutritional values (Granato et al., 2010). Besides that, these plant-based products are also a suitable alternative to be offered to vegetarians. Soy-based beverages have a wide applicability in the food industry, such as they can be used directly as a beverage or as an ingredient in desserts, infant formulas, and yogurt (Chen et al., 2012).

Products of this category have become increasingly popular among consumers ever since the development of new technologies that can reduce the usual astringent taste in soy products (Childs et al., 2007). According to the statistics provided by Statisca (2017), the global sales volume of soy-based beverages amounted to 13.48 billion liters in 2015 and this was expected increase to 16.29 billion liters in 2018. Thus many food manufacturers are constantly launching new products in the soy beverage category nowadays, thereby widening the product varieties available for consumers in the market. Some of the key players of the soy food products market are Whole Soy & Co., Northern Soy Inc., DuPont Solae, ADM Inc. Solbar Ltd., the Scoular Company, Cargill Inc., and Linyi Shansong Biological Products Co. Ltd. (Granato et al., 2010).

Soy-based beverages have relatively less calories and various health benefits. Consumption of these products can reduce the risk of heart failure as soy extracts reduce the cholesterol level in the blood (Sala-Vila et al., 2015). Other potential health benefits recently discovered in various studies are related to osteoporosis, menopausal symptoms, breast cancer, and prostate cancer (Omoni and Aluko, 2005; Tripathi and Misra, 2005). In addition, soy-based beverages also provide minerals (i.e., zinc and iron) and vitamins (i.e., B6 and B12). Vitamins B6 and B12 are essential for increasing hair growth as they nourish the hair follicles (Rodriguez-Roque et al., 2013). Nevertheless, individuals with kidney stones containing oxalates should avoid consuming these beverages as they are a rich source of oxalates (Food Insight, 2009).

5.10 HERBAL DRINKS

Herbal drinks are beverages made from the infusion or decoction of medicinal plants/herbs in hot water. Various parts of the plant can be used to prepare such beverages; for instance, roots, stems, bark, leaves, fruits, or flowers.

One popular product currently available in the market that falls under this category is herbal tea.

Herbal teas, also called tisanes, are the second largest consumed beverage in the world after water. Some of the major players in herbal tea market are Nestea, Global Herbitech, Tata Global Beverages, Green Earth Products Pvt Ltd., AB Food and Beverages, and Buddha's Herbs (Euromonitor, 2015). Herbal teas are free of caffeine. They are often made from one main or a combination of several herbal ingredients, designed to bring about a specific therapeutic and energizing effect, such as promoting a good night's sleep, relieving stress, supplying antioxidants to the body, supporting heart health, helping to prevent colds, strengthening the immune system, providing cleansing properties for the body (Ravikumar, 2014). Table 5.5 shows various types of herbal teas and their specific health benefits.

TABLE 5.5 Different Types of Herbal Teas and Their Specific Health Benefits (Chen et al., 2008; Ravikumar, 2014; Srivastava et al., 2010)

Types of Herbal Tea	Health Benefits
Ginger tea	• Stimulates and soothes digestive system.
	• Contains anti-inflammatory properties which are helpful for arthritic people.
	• Aids people experiencing nausea of any kind, such as motion sickness, morning sickness.
Peppermint tea	• Treats upset stomach effectively when combined with chamomile.
Chamomile	• Promotes a good night's sleep.
	• Soothes stomach pains and acts as a gentle laxative.
	• Alleviates menstrual cramps.
Cardamom tea	• Relieves coughs.
	• Provides calming effect to women who experience mood swings during their menstrual period.
	• Treats indigestion, relieves flatulence, and prevents stomach pain.
	• Reduces nauseous feeling.
Green tea	• Provides antiaging benefits.
	• Prevents certain types of cancers, such as breast, prostate, stomach, skin, throat, and lung.
	• Aids in weight loss.
	• Fights tooth decay and gum diseases.
	• Reduces the LDL (bad) cholesterol level in the body.

While these products offer many beneficial health effects, there are several risks associated with the consumption of herbal teas. Studies revealed that sassafras tea contains 80% of the toxic compound known as safrole where large amount of safrole intake may lead to liver damage (Elvin-Lewis, 2001). Additionally, Rodriguez-Fragoso et al. (2008) has found that long-term consumption of peppermint tea might lower the testosterone level in adult males, thereby affecting their fertility. Ernst (1998) has proven that pyrrolizidine alkaloids detected in coltsfoot tea (Tussilago farfara) are toxic, and consumption of this tea may increase the risk of liver toxicity and a person's chance of developing cancers.

Nevertheless, the safety of many herbal drinks remains unknown as toxicological studies on herbal teas are very limited. More studies on the toxicity and herb–drug interactions of medicinal should be performed. Thus it is recommended that widespread consumption of herbal infusions should be minimized until data on the levels and varieties of toxicants carcinogens and mutagens are made available.

5.11 READY-TO-DRINK TEA

RTD tea refers to tea-based beverages that are sold in a prepared form and are ready for consumption. It is one of the world's fastest-growing soft drinks categories, driven by the rising consumers' demand for healthy, low-sugar beverages. According to Fortitech (2011), the fortified RTD tea market has grown from $1.4 billion to $2.6 billion from 2004 to 2009. The market continues to increase steadily ever since then. In 2016, RTD tea appeared to be one of the most popular beverages in Malaysia where an estimated of 15% off-trade volume growth was achieved (Euromonitor, 2017a). Additionally, in the same year, RTD tea performed strongly in Australia as the retail volume sales increased up to 7% (Euromonitor, 2017b). The key vendors in the global RTD tea market include Sunny Delight Beverages Co., Unilever, Talking Rain and TeaZazz, whereas Fraser & Neave Holdings Bhd. dominates the RTD tea market in Malaysia (Euromonitor, 2017a, 2017b).

RTD teas, regardless of the types, are convenient and most offer a healthy alternative to other RTD beverages on the market. Consumption of these products may aid in a person's quest to manage a myriad of health conditions ranging from weight loss to cognitive function and heart health. Besides that, RTD teas are desirable vehicles for fortification where even more health benefits can be supplemented to this timeless drink.

Matcha, one of the latest consumer trends in RTD tea products, is a powder ground from raw green leaves called tencha and is made into a nutrient-packed form of green tea. Matcha has been promoted to contain polyphenol, particularly flavonols and flavanols, which represents 30% of fresh leaf dry weight (McKay and Blumberg, 2002; Naghma and Hasan, 2007). Some of the possible health benefits gained from matcha intake are the prevention of cancer and cardiovascular diseases, antioxidative, anti-inflammatory, antibacterial, antiarthritic, and others (Suzuki et al., 2012). Other examples of RTD teas are jasmine hibiscus RTD tea, passion fruit iced green tea, and green tea with cactus water, among others (Bae et al., 2016).

5.12 FUNCTIONAL INGREDIENTS IN BEVERAGES

Functional ingredients are defined as ingredients that can provide health benefits for increased well-being. The micronutrients including antioxidant, probiotics, prebiotics, vitamins, and minerals that are included in a healthy beverage are reported to have physiological benefits and are able to reduce the risk of chronic diseases more advanced than a macronutrients function (Rifnaz et al., 2016). Regularly, functional ingredients are loaded in a beverage with a small amount of less healthy ingredients such as sugar and fats to produce functional beverages. This is done in order to obtain health benefits and at the same time to develop a stable beverage that has appealing flavors throughout the shelf life (Fallourd and Viscione, 2009). The functional properties and benefits of these ingredients can influence the consumers' awareness when they are purchasing the functional beverages. Each consumer had their own different reasons on why they are consuming the beverages and these because of either to treat and prevent certain diseases or just to absorb nutrients for keeping and maintaining a healthy eating pattern (Bomkessel et al., 2014).

5.12.1 ANTIOXIDANT (A, C, E, Z)

Antioxidants are molecules that are able to prevent the oxidation of other molecules. There are increasing consumer demands for food and beverage containing high antioxidants in the food and beverage sector. Research shows that food containing antioxidants are associated with better long-term health advantages including the improvement of immunity performance and the reduced risk of cancer and heart disease. Drinking high antioxidant beverage may help to protect against ageing, Alzheimer disease, and other chronic diseases (Lal, 2015; Nanasombat et al., 2015). Vitamins A, C, E, and zinc are a good source of antioxidants (Lal, 2015).

Vitamin A is a type of fat-soluble vitamin and plays an important role in maintaining healthy vision and skin, neurological functionality, and reducing inflammation through fighting free radical damage. Nutritionally, vitamin A required by the human body can be obtained from two primary sources which are from animal sources and plant sources (Chauhan et al, 2013). The animal-derived sources can be obtained in a form of retinol (active vitamin A), which can be used directly by the body. For plant-derived vitamin A, which can be obtained from fruits and vegetable, it is in the form of provitamin A such as beta-carotene and related carotenoids, which have to be converted to retinol after being consumed in order to be fully utilized by the body (Ottaway, 2009).

Vitamin C or L-ascorbate is a water-soluble antioxidant that is widely distributed in nature. It is highly abundant in some fruits and vegetables and is also found in animal organs such as liver and kidney (Ottaway, 2009; Farbstein et al., 2010). Usually, fruits and vegetables serve as the main sources of vitamin C for the human body. Farbstein et al. (2010) had explained that vitamin C is needed for many physiological processes and has shown a great potential in improving immunity performances, as well as lowering the risk of cardiovascular disease and cancer. Vitamin C plays an important role in human health as an antioxidant that can protect the body against oxidative stress (Ogundele et al., 2016).

Vitamin E is a fat-soluble vitamin and consists of eight groups of lipophilic antioxidant compounds, which are comprised of four tocopherols compound and another four tocotrienols compound. The vitamin can be easily obtained from green vegetables, grains, nuts, and various vegetable oils, such as wheat germ, sunflower seed, and maize oil. Vitamin E can also be found in eggs and milk. Tocopherols and tocotrienols expressed variety biological properties, such as antioxidant activity, the potential to modulate protein function, and gene expression, which helps in forming blood cells and boost the immune system (Ottaway, 2009; Farbstein et al., 2010; Chauhan et al, 2013).

5.12.2 NUTRACEUTICAL INGREDIENTS

Nutraceuticals terms are used to define health-promoting foods, or components that are extracted from them. The term "nutraceutical" was first derived from "nutrition" and "pharmaceutical," and defined as "a food or part of a food that can provide medical or health benefits, including the prevention and/or treatment of a disease" (Lal, 2015; Chauhan et al., 2013). Nutraceutical products may be available in a form of individual, isolated nutrients, herbal products, and dietary supplements. Examples of nutraceutical ingredients are dietary fibers, polyunsaturated fatty acids, and polyphenols such as carotenoids (Chauhan et al., 2013).

Dietary fiber is the part of plant material that is unsusceptible to the enzymatic digestion in the digestive tract but is able to be digested by microflora in the gut. Nonstarch polysaccharides is a common source of dietary fiber, for example, celluloses, lignin, hemicelluloses, gums and pectins, resistant dextrins, and resistant starches. Foods that are rich in soluble fibers include fruits, oats, barley, and beans (Das et al., 2012; Dhingra et al., 2012).

Polyunsaturated fatty acids are crucial to the body to function effectively and these are introduced externally through the diet. PUFAs can be divided into two classes of essential fatty acids which are omega-3 fatty acids and omega-6 fatty acids. PUFAs can be found in fishes such as mackerel and salmon, flaxseed, soybeans, canola, vegetable oils such as sunflower and corn oil, meat, poultry, and eggs (Das et al., 2012). PUFAs especially omega-3 fatty acids provide benefits in many areas of health including premature infant health, bipolar and depressive disorder, and inflammatory diseases such as asthma (Gulati et al., 2016). PUFAs are also being associated with cardiovascular protection and able to reduce morbidity and mortality from cardiovascular diseases due to its anti-inflammatory and antiatherogenic properties (Massaro et al., 2010).

Phytochemicals are available in plant and contain many types of polyphenols whereas will be detected as secondary metabolites and also must be protected from photosynthetic stress. Polyphenols can be distinguished according to approximately 8000 groups and the well-known polyphenols such as flavonols, flavones, flavan-3-ols, flavanones, and anthocyanins had been widely investigated. Polyphenols could promote many functional properties including antioxidant, anti-inflammatory, antimicrobial, cardioprotective activities, as well as a protective effect on neurodegenerative diseases (Das et al., 2012). Some of the medicinal plants such as feverfew (*Tanacetum parthenium*) has been used as a source of nutraceuticals in the production of a functional beverage with anti-inflammatory properties (Carbo et al., 2014). Nutraceuticals have been claimed to have a physiological benefit and provide protection against chronic diseases (Lal, 2015).

5.12.3 VITAMINS (B, D, K)

Vitamin can be described as a group of organic, heterogenous nutrients that are required by the human body in a very small amount (Ottaway, 2009). Vitamin can be categorized into two groups which are either fat-soluble vitamins or water-soluble vitamins. Other than vitamin A, C, and E which can provide antioxidant benefits, another group of vitamins such as vitamin B, D, and K can also provide health benefits to human body. Different types of vitamins and the sources are presented in Table 5.6.

TABLE 5.6 Commonly Used Synonyms and Example of Sources of Vitamin B, D, and K (Berry Ottaway, 2009; Chauhan et al., 2013)

Vitamin	Synonyms	Example of Sources
Fat-soluble vitamins		
Vitamin D_2	Ergocalciferol	Fish liver oil, wheat germ oil, egg yolk, milk, butter
Vitamin D_3	Cholecalciferol	
Vitamin K_1	Phylloquinone, phytomenadione	Cabbage, cauliflower, tomatoes
Vitamin K_2	Menaquinone, farnoquinone	
Vitamin K_3	Menadione	
Water-soluble vitamins		
Vitamin B_1	Thiamin	Cereals, pulses
Vitamin B_2	Riboflavin	Nuts, yeast
Vitamin B_3	Pantothenic	Liver, meat, yeast
Vitamin B_5	Nicotinic acid	Yeast, egg, milk
Vitamin B_6	Pyridoxal, pyridoxine, pyridoxamine	Red meat, liver, cod roe and liver, milk, and green vegetables

5.12.3.1 VITAMIN B

An emerging interest toward vitamin B is observed as it is advertised as a source of nutrient for energy which usually found in energy drinks (Lal, 2015). One of the examples of vitamin B is vitamin B1 which is also known as thiamine and widely distributed in living tissue. Mostly, it could be found in a form of the phosphorylated group in animal products and it is identified as nonphosphorylated form in plant products (Ottaway, 2009). Vitamin B2 or riboflavin is widely distributed in all plant and animal cells. This type of vitamin occurs naturally in two types of form which are riboflavin mononucleotide and flavin adenine dinucleotide. Plants and many bacteria can synthesize riboflavin and it is available as dietary amounts in dairy products (Ottaway, 2009).

Another type of vitamin B is vitamin B6. This vitamin's activity is shown by pyridoxine that consists of three different compounds: pyridoxol, pyridoxal, and pyridoxamine. Generally, B vitamins are essential for energy production because they convert food into energy and help in maintaining the brain function. B vitamins are important in neurologic and for various nervous system function and also help to produce essential protein (Chauhan et al, 2013).

5.12.3.2 VITAMIN D

Vitamin D is abundant in dairy foods and beverages. It presents in nature in several forms, which is vitamin D2 and vitamin D3. The dietary vitamin D occurs predominantly in animal products with a very small amount gained from plant sources. Vitamin D2 or ergocalciferol is produced by the ultraviolet irradiation of ergosterol. This vitamin is widely distributed in plants and fungi. Vitamin D3, also known as cholecalciferol, is derived from ultraviolet irradiation of 7-dehydrocholesterol and can be found in the skin of animals, including humans. Human requirements for this vitamin are obtained both from the endogenous production in the skin and from dietary sources (Ottaway, 2009). Vitamin D is essential for the formation of teeth and bones. It helps the body to absorb and use calcium in an effective manner (Chauhan et al., 2013).

12.5.3.3 VITAMIN K

Vitamin K occurs in a number of forms. Vitamin K1 (phytomenadione or phylloquinone) is found in green plants and vegetables, potatoes, and fruits, whereas vitamin K2 (menaquinone) can be found in animal and microbial materials. Vitamin K is essential in blood clotting and bone metabolism (Chauhan et al., 2013).

5.13 BOTANICALS INGREDIENTS

Botanicals can be defined as fresh or dried plants, plant parts, or collective chemicals isolated from plants, which often obtained in the form of concentrated extracts. They are commonly added and used in the food and beverage manufacturing for flavor, fragrance, or other technical properties, such as coloring, thickening, or preservative (How the Food Industry Defines Botanicals, 2012). Examples of botanical extracts are green tea and aloe vera extracts.

5.13.1 GREEN TEA EXTRACT

Tea is one of the most popular beverages that are consumed by most of the people around the world. Tea that is produced from plant *Camellia sinensis* is consumed daily as green, black, or Oolong tea. Among these three types of tea, green tea has been observed to provide the most significant effects on the human health (Chacko et al., 2010). Green tea comprises of polyphenols and four catechins which are epigallocatechin (EGC), gallocatechin (GC), epicatechin (EC), and epigallocatechin gallate (EGCG). Catechins contain a huge amount of antioxidant properties and are also recognized as tea flavonoids (Lal, 2015). These catechins are reported to provide various health impacts that involve to decrease the risk of Alzheimer disease, to prevent many types of cancers including stomach, lung, pancreas, and kidney cancer and also can act as antitumorigenic agents. Other than that, the consumption of green tea catechins can also prevent cardiovascular and oral diseases and can enhance the metabolism rate. The popularity of green tea extract in nonalcoholic beverages other than tea had been increasing due to it may supply a natural source of caffeine (Lal, 2015; Chacko et al., 2010).

5.13.2 ALOE vera

Aloe vera (*Aloe barbadensis*) has long been used in health foods and beverage as well as for medical and cosmetic purposes. It has been widely used due to its aromatic properties, bitter taste, and other pharmacological activities such as reduction of inflammation and wound healing (Ramachandran and Rao, 2013; Kumar, 2015). Besides providing medicinal values, aloe vera also contains various nutrients such as vitamins, amino acids, mineral, enzymes, sugars, saponin, and fatty acids. These nutrients are able to provide positive effects on human health and can be used in food formulation as a functional ingredient for health benefits. Examples of aloe vera beverage products are aloe soft drink, diet drink with soluble fiber, hangover drink, tropical fruit juice with aloe vera, and yogurt drinks. Other than that, Aloe vera can be turned into aloe vera desserts with chunks of aloe, instant aloe vera granules, aloe vera gums for sore or bleeding gums, aloe vera candy, and aloe vera smoothies (Ramachandran and Rao, 2013).

5.14 PREBIOTICS AND PROBIOTICS

Prebiotic and probiotic can be used to further improve the functionality and quality of the functional beverage. A nondigestible food ingredient is called prebiotics which beneficially affects the host by selectively stimulating the growth or activities of useful intestinal bacteria in the colon, by providing growth enhancers and nutrients to probiotic bacteria. Basically, prebiotics are mainly obtained by extraction from plants, followed by enzymatic hydrolysis. Compounds such as lactulose and lactitol, fructooligosaccharides, inulin, galactooligosaccharides, and oligosaccharides from soy and levans are grouped as prebiotics (Corbo et al., 2014; Vieira da Silva et al., 2016).

Probiotics are described as "a microbial preparation which contains live and/or dead cells including their metabolites which is intended to improve the microbial or enzymatic balance at mucosal surfaces or to simulate immune mechanisms" (Lal, 2015). Probiotic cultures are beneficial for human health because it helps to improve intestinal microbial balance, which results in the inhibition of bacterial pathogens, improving the immune system, reducing the risk of colon cancer, lowering serum cholesterol levels, alleviation of lactose intolerance, and nutritional enhancement (Kumar, 2015). Probiotics are commonly incorporated in dairy-based beverages that includes many drinks—milk-based, yogurt, and cultured buttermilk. In 2008 to 2013, nondairy products especially probiotic drinks are emerged in the functional beverages market. Table 5.7 shows some of the examples of commercially available probiotic beverages (Lal, 2015; Saarela, 2009).

5.14 MINERALS

Human's body needs minerals in order to build strong bone and muscles (Salvia-Trujillo et al., 2017). Besides, this inorganic compound is also important to heal wound, repair cellular damage, and enhance immune system (Kawashima and Valente Soares, 2003). There are two types of minerals which are major minerals (macrominerals) and trace minerals (microminerals). For major minerals, it comprises of sodium, potassium, chloride, calcium, phosphorus, magnesium, and sulfur. As for trace minerals, it consists of iron, zinc, iodine, copper, manganese, and fluoride (Pinto et al., 2014). Fruits and vegetables can be a good source of minerals thus can be the best functional ingredient in beverage (Pereira et al., 2017). Beverages comprising of milk and fruit juices contain great sources of minerals. Study done by Salvia-Trujillo et al. (2017) has highlighted that by using high-intensity pulsed electric field (HIPEF) the mineral concentration in the fruit juices stayed highly stable during storage, regardless of the processing applied. The macromolecule calcium was the major mineral recognized in the fruit juice-whole and fruit juice-skimmed milk beverages. Hence, HIPEF process is a possible treatment for advance functional beverages with high elements of health correlated substances.

TABLE 5.7 Example of Commercially Available Probiotic Beverages

Probiotic Content	Brand	Producer
Lactobacillus acidophilus, Lactobacillus casei, Bifidobacterium bifidum	Vita Bosa®	Biola Inc., Canada
Lactobacillus plantarum 299v	Proviva®	Skane Dairy, Sweden
	Goodbelly®	NextFood, United States
L. acidophilus, Lactobacillus delbruekii, Saccharomyces cerevisiae var. *boulardii*	Whole Grain Probiotic Liquid®	Grainfields, Australia
Lactobacillus rhamnosus GG	Gefilus®	Valio Ltd., Finland
	Biola®	Tine BA, Norway

Adapted and modified from Lal (2015) and Saarela (2009).

In general, fermented coconut water consists significantly higher quantity of sodium as well as other minerals, such as magnesium, calcium, and potassium (Lee et al., 2013). The high mineral content in coconut water has made it a popular substitute to sports drinks that contain high electrolytes (Kalman et al., 2012). The probiotic fermentation does affect toward the mineral content formerly exist in the coconut water (Lee et al., 2013). Their findings have shown that the nutritive advantages and hydrating properties of fermented coconut water can be estimated to be as good as to those of the fresh coconut water. Normally, drinks that are specifically made for rehydration must be able to restore the volume of water lost and recover the electrolytes elements, particularly sodium. Sodium level must be in between 20 and 30 mmol/L for a rehydration beverage to provide fluid replacement and to stimulate absorption (Mitchell et al., 1994). Hence, this shows that sodium gives better hydration for utmost people.

Maple syrup is one of the popular food and drink ingredients which is made of maple sap. Without pasteurization or sterilization process the maple sap is not suitable for human consumption. There are about seven minerals detected in the sterilized maple sap that can be applied as a functional beverage. This includes sodium (7.70 ± 2.07), magnesium (1.14 ± 0.64), aluminum (1.19 ± 0.35), ferum (0.68 ± 0.85), zinc (0.25 ± 0.52), calcium (4.38 ± 1.57), and potassium (5.74 ± 2.90) calculated by mg/100 g (Yuan et al., 2013).

Functional coffee containing ginger has been reported to have an excellent source of selenium, zinc, manganese, copper, and iron but rather modest in sodium, calcium, potassium, and magnesium. This property has promoted a great competitiveness of the functional coffee compared to common coffee types that comprised of traditional, instant, and espresso (Švarc-Gajić et al., 2017). In pure origin coffee country that produce espresso beverages like Brazil, Ethiopia, Honduras, Guatemala, Papua New Guinea Colombia, India, Mexico, Kenya, Cuba, Timor, Mussulo, and China, a comparison of mineral characterization has been done. The results show that calcium and manganese were found to be the best chemical description of origin. Nonetheless, South American pure origin coffees are richer in the analyzed elements except for calcium, whereas samples from Central America have mostly lower mineral quantities, excluding for manganese (Oliveira et al., 2015). According to Costa et al. (2018), coffee silverskin which is the major coffee roasting by-product comprises large amounts of potassium (5 g/100 g), magnesium (2 g/100 g), and calcium (0.6 g/100 g). It has the ability to inhibit induced hemolysis and act in cell protection against oxidative injuries. This suggests that beverages made from coffee silverskin have a higher potential to be one of the functional beverages (Costa et al., 2018).

5.15 SUPERFRUIT EXTRACT

The term "superfruit" is a comparatively recent term used in the food and beverages industries. It is considered a new marketing strategy in order to promote common or rare fruits that can be consumed as food or drinks, or used as ingredients in the production of functional foods, beverages, and nutraceuticals. Superfruits possessed high nutritional values due to their richness in nutrients, antioxidants, and taste appeal. Superfruits have also been proved to have potential health benefits to the consumers. Although, there are no scientific criteria for defining which fruits can be considered as superfruits and the scientific society did not come out with any specific description for "superfruits," however, antioxidants rich fruits that may possesses various health benefits are likely to be included in this category (Felzenszwalb et al., 2013). Fruits such as pomegranate, blueberries, acaí, noni, acerola, and mangosteen are appropriate to be called as "superfruits" and can be obtained at fruit juices section in beverage market (Lal, 2015; Felzenszwalb et al., 2013).

Acai (*Euterpe oleracea*) is a type of palms that have an edible palm heart and the fruit is known as acai berry. Initially, acai berry has been consumed in the form of juice and pulp. It has a high antioxidant capacity and its extract has been incorporated as an ingredient in food and beverages. Several studies have demonstrated that acai berry possesses anticancer, hypocholesterolemic effects, and anti-inflammatory activity (Felzenszwalb et al., 2013; Kubitz, 2016). Besides antioxidants, it also contains other nutrients such as vitamins, mineral, dietary fiber, and complex carbohydrates (Enos, 2011).

Another example of superfruit is pomegranate. Pomegranate (*Punicagranatum* L.) is one of the oldest edible fruits and is usually consumed fresh as a whole fruit or transformed into juice, jellies, and coloring agents (Hmid et al., 2017). It contains important sources of antioxidants such as anthocyanin and the 3-glucosides and 3,5-diglucosides of delphinidin, cyanidin, and pelargonidin. Studies have shown that nutrients from the pomegranate possess antioxidant and antitumor activity and can help to protect the body from ailments like heart problems, diabetes, and osteoarthritis (Gil et al., 2000; Hmid et. al., 2017; Enos, 2011).

5.16 AMINO ACIDS

Amino acid is vital in practically all biological progressions in the human body (Morell and Fiszman, 2017). This is because amino acids give a structure to the cells and act as building blocks of it (Foegeding et al., 2017). A large number of cells, muscles, and tissue are made up of amino acids, meaning they perform many important bodily tasks such as carrying substances, combating pathogens, regulating processes, sending signals, and act as a catalyst for biological reactions (Wang et al., 2012). There are many types of amino acids, for instance, histidine, methionine, phenylalanine, threonine, isoleucine, leucine, lysine, tryptophan, and valine (Bihuniak and Insogna, 2015). Human body cannot produce its own amino acid and must be obtained through food or beverage that contains high with protein (Stroup et al., 2017).

Amino acid level is very important constituents to determine the quality of tea as a functional beverage. Mulberry leaves especially *Morus alba* are rich with amino acid specifically leucine, phenylalanine, lysine, histidine, arginine, threonine, valine, and methionine (Chan et al., 2016). In the Xinjiang province of China, Russian mulberry contains the highest essential amino acids followed by white mulberry and black mulberry (Jiang and Nie, 2015). That is why there are lots of mulberry tea products in the functional beverage market (Gryn-Rynko et al., 2016). Another type of tea leaves that contain high amino acids are AnJiBaiCha albino mutant. The verdicts in this research have provided new insights that lead to the albino phenotype and amino acid enhancement of AnJiBaiCha white tea (Yuan et al., 2015). Kocadağli et al., (2013) has found that brewing conditions intensely influence the presence of free amino acid in tea. There are 20 amino acids which comprise of 18 proteinogenic and 2 nonproteinogenic that were identified in green and black tea samples. After 2 min of brewing procedure, the quantity of free amino acids in tea infusion was 220.53 mg/L and 211.21 mg/L for black and green tea, respectively.

A normal milk beverage can be upgraded into a functional milk beverage by adding phenolic compounds (100 and 200 mg/L). The phenolic was extricated from olive vegetable water, and later on being fermented with γ-amino butyric acid-producing (*Lactobacillus plantarum* C48) and human gastrointestinal (*Lactobacillus paracasei* 15N). During fermentation occur, the total concentration of free amino acids elevated from ca. 190±8.3 mg/L to 303.7±11.8–319.5±9.1 mg/L (Servili et al., 2011).

5.17 EMERGING TECHNOLOGIES IN FUNCTIONAL AND MEDICINAL BEVERAGES PROCESSING

5.17.1 HIGH-PRESSURE PROCESSING TECHNOLOGY

The preservation of fruit juices by thermal blanching above 80 °C is a common process. Nevertheless, products subjected to these process conditions will typically lose their natural flavor as well as some of their nutrient content (Rossi et al., 2003). The development of nonthermal approach such as high-pressure processing (HPP) is due to the market need which is a request for minimally processed foods (Hendrickx and Knorr, 2002).

HPP or known as high hydrostatic pressure (HHP) is a nonthermal method that exposed foods to high pressure between 100 and 1000 MPa for 1–20 min with pressure-transmitting condition using water. Hydrostatic pressure is operated by heighten the free energy which can be accomplished through physical compression during pressure treatment in a closed system by decreasing the mechanical volume. HPP is typically operated by a moderate increase in temperature (adiabatic heating), which this temperature increase due to the composition of the food product (Knorr, 2002; Knorr et al., 2006; Naik et al., 2013; Liepa et al., 2016).

In general, two types of HPP devices are available, horizontal and vertical. The horizontal type of HHP tool is commonly used for commercial purpose since this type of design facilitates the loading and unloading of containers during the production process (Marketsandmarkets, 2013). Figure 5.1 shows the horizontal type of HPP tool from Multivac Inc., Germany (Huang et al., 2017).

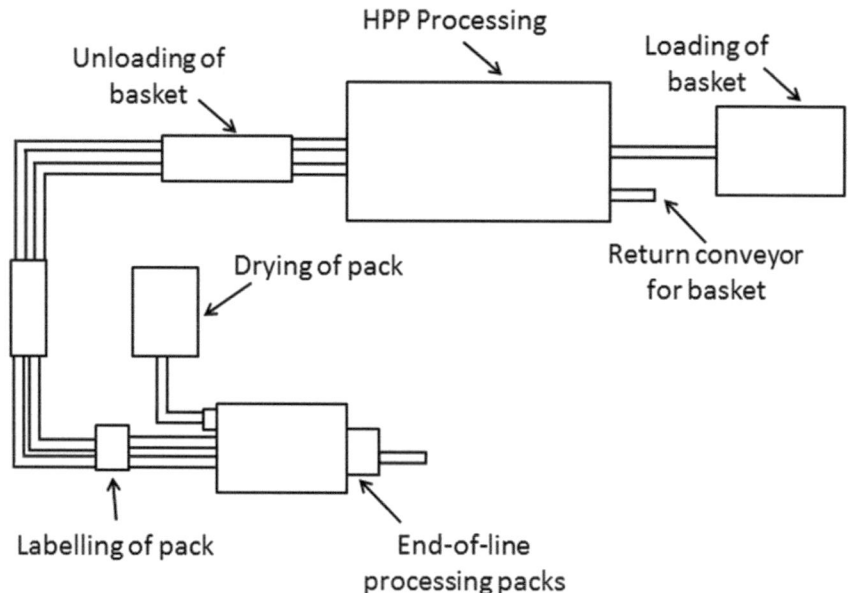

FIGURE 5.1 A horizontal HPP production systems (MULTIVAC, Germany) [Adapted and modified from Huang et al. (2017)].

HPP is able to preserve the nutritional quality of the product with minimal effect to quality which is associated to its little effect on the covalent bonds of low-molecular-mass compounds such as color and flavor compounds (Butz et al., 2003; Oey et al., 2008). HPP inactivates microorganisms and enzymes; therefore it assists in enhancing the safety and shelf-life of perishable foods, whereas the thermal influence on nutritional and quality parameters are considered low in HHP since this procedure is operated at room temperature or moderately above (Knorr, 1993; Butz et al., 2003; Boye, 2015).

In the treatment of HPP, individual phenolic compounds, total phenols, and the bioactivity of a food product are regularly deteriorate. This might be happen because of the higher residual activity of polyphenol oxidase (PPO) and peroxidase (POD), two enzymes that are responsible in catalyzing the phenols oxidation. For example, these enzymes are proficient at oxidizing (+)-catechin, which causes deterioration and then brown polymers will be generated (Lopez-Serrano and Barceloä, 2002).

Over the years, the observation of the application of HPP in the area of food and beverage processing has been done and many fruit juices products, such as HHP treated mandarin, grapefruit, apple, orange, carrot juices, and broccoli–apple juice mixture, were found commercially in the market (Buzrul et al., 2008). The advantages of HHP on the orange juice–milk beverage were observed in terms of nutrition, quality, and energy conservation (Barba et al., 2012). The result showed that 5-log inactivation of *L. plantarum* in the orange juice–milk beverage had been achieved by the minimum HPP treatment. This minimum HHP treatment had shown a greater effect on sustaining the ascorbic acid level than the heat treatment. In another study by Barba et al. (2013), blueberry juice subjected to HPP had shown more than 92% vitamin C retention and increased in total phenolic content. HHP also exert the least influence on the quality of the treated juice. Nutrients and flavors of the product might be eliminated by the thermal processing, HPP benefited the quality of coconut water which the quality was observed to be similar with fresh coconut water (Huang et al., 2017). In yogurt production, just before the fermentation, the HPP treatment had been applied on milk. As a result, HPP could produce fat set-type yogurt that contained 12% total solids with a creamy uniformity, without a need to add polysaccharides (Udabage et al., 2010). The additives used in the yogurt processing can influence the taste, flavor, aroma, and mouthfeel; however, HPP is advisable to be applied on the milk used for yogurt making as it could substitute the food additives (Sfakianakis and Tzia, 2014).

5.17.2 MEMBRANE TECHNOLOGY

Membrane filtration technologies are increasingly employed in the functional beverage industry. The minimal beverage processing techniques could be generated by improving the product quality and stimulating energy saving. Membrane processes technique benefited a low-cost and nonthermal separation effect as well as include no phase change or chemical agents, however, traditional juice processing technique demonstrated a distinctive performance. These benefits prompted an initiative of manufacturing a new fruit juice containing natural fresh tastes and additives free initiates (Cassano et al., 2011).

Membrane technology is a separation technique that involves a hydrostatic pressure gradient using semipermeable membranes which depends on the specific permeability of the membrane for individual or mixed compound of the fluid. In general, molecules in the ionic size range will be collected by reverse osmosis (RO) membranes, the dispersion of macromolecules are obtained by ultrafiltration (UF) membranes while micron size of particle is attained by microfiltration (MF). The definite osmotic pressure is applied to each of membranes, RO, UF, and MF due to the membranes separated the molecules at the different sizes (Cassano et al., 2014).

All three types of membrane filtration involve a cross-flow filtration process (Figure 5.2). In this process, the liquid stream is treated at the feed section and then it flows loosely throughout the membrane surface, therefore two streams are occurred. Wolf et al. (2005) described some particles in the solution will pass through the membrane and ready to be collected or the concentrated solids will be trapped at the membrane. Cross-flow filtration operates a self-cleaning that means the process needs extensive process times without cleaning compared to the traditional filtration and this is related to the feed stream that flows parallelly to the membrane not vertically (Cassano et al., 2014).

Ultrafiltration is commonly used in the juice industry to clarify and stabilize fruit juices and industrial-scale commercial processes have already been applied in apple and orange juice industry (Girard and Fukumoto, 2000). Microorganisms, lipids, proteins, and colloids are a large species that could be collected through UF technique, however, vitamins, salts, and sugars are a small compound will flow together with water. Fukumoto et al. (1998) had listed some of the advantages of fruit juice processed by UF rather than conventional technique such as increase juice yield, increase chances for single-step operating, reduce the chances of using gelatins, adsorbents and other filtration aids, as well as enzyme, easy cleaning and maintenance of the equipment, promote the exclusion of pasteurization, better juice clarity, reduce the filtration times and waste products. Besides, UF is also a commonly applied unit operation in dairy processing. It is used for the standardization of the protein content in milk.

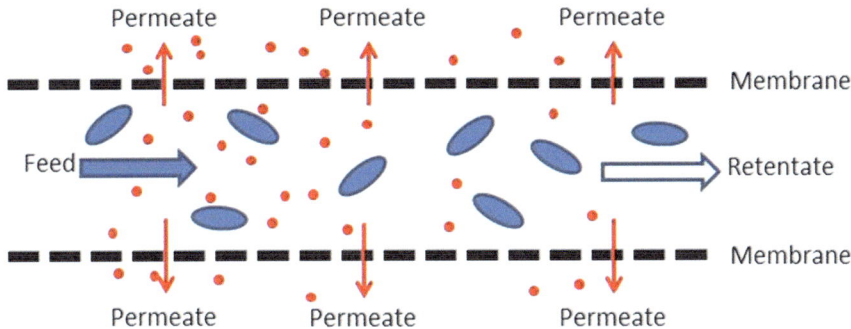

FIGURE 5.2 Cross-flow filtration process [adapted and modified from Cassano et al. (2014)].

Another membrane technology for juice clarification is microfiltration. MF is a membrane separation technique that involves a pressure system and the fruit juice can be clarified without the need for enzymatic treatment. A melon juice had been clarified using cross-flow MF and the juice was found to have a comparable quality with the fresh juice, but with exception for insoluble solids and carotenoids (Vaillant et al., 2005). Acerola juice was produced and clarified using MF, thus the investigation toward consumer preferences had been done and 84% of consumers preferred the MF juice (Matta et al., 2004). Another investigation by Mirsaeedghazi et al. (2010) had reported that the microfiltration is able to clarify pomegranate juice without causing significant chemical changes in its chemical properties when compared to enzymatic methods. MF is also principally used in dairy processing to reduce the bacterial loading in a dairy fluid by retaining bacteria and spores. MF is a preferred method since it minimizes the use of heat treatment, which is beneficial for products which are sensitive to high temperature (Lawrence et al., 2008).

Reverse osmosis is a technique for concentrating liquid which is normally used in juice production. Fruit juices are concentrated to reduce water activity and also provide economic impact for storage, transportation, distribution, and commercial operations. RO operation allows the separation of water from the juice solids; however, it is restricted by high osmotic pressures. Therefore the technique is applied for preconcentration treatment that permits concentration values of about 30 g total suspended solids/100 g, corresponding to osmotic pressures of about 50 bar (Boye, 2015). No phase change involves in RO process and this process is normally performed at room temperature which results in considerable energy saving to the industry. In addition, the RO process compact installation is also easily conducted.

The RO quality of treated juice is better than the quality of the juice produced via conventional thermal treatment because heat treatment affects sensory and nutritional properties of the juice (Álvarez et al., 2000). Furthermore, this method permits the retention of compounds by permeation of water and concentration of solutes such as salts and sugars (Cheryan, 1998) while the retention of water could be appropriate for recycling within the same plant (Balannec et al., 2005; Galambos et al., 2004).

5.17.3 MICROWAVE HEATING TECHNOLOGY

Microwave heating technology is applied in many parts of the functional or medicinal beverage production such as in pretreatment, extraction, and pasteurization or sterilization. In microwave heating, heat is being generated within the material, which is known as volumetric heating. This volumetric heating occurs due to dipolar rotation and ionic conduction of the constituents in the materials when exposing to an oscillating electric field in the microwave (Venkatesh and Raghavan, 2004).

In conventional juice processing, juice is commonly obtained via mechanical pressing. In order to maximize juice yield, in some fruit juice processes, enzymes are usually added into the fruit mash to assist cell walls degradation. This enzymatic cell walls degradation assists in increasing juice production. Alternatively, applying microwave heating on fruit mash before mechanical pressing also produces a higher yield of juice compared to mechanical pressing alone. The study conducted by Gerard and Roberts (2004) on microwave heating of Fuji and McIntosh apple mash prior to pressing process significantly increases apple juice yield. In this study, the apple mash was heated around 40–70 °C using a 2450 MHz microwave at 1500 W. The increase in juice production is due to the liberation of juice or bioactive compounds via cell wall breakage from rapid heating by microwave. Microwave heating destructive effect on orange peel tissue was demonstrated in the study by Kratchanova et al. (2004).

Besides increasing juice yield, microwave technology has also been applied in the area of juice extraction particularly for "hard to press" fruits. "Hard to press" fruits are fruits that not suitable for mechanical pressing such as plum and apricot. This technology is known as microwave hydrodiffusion and gravity (MHG) wherein fruit juice is forced out via hydrodiffusion and the juice produced flow into the collector compartment via gravitational force (Cendres et al., 2011). No requirement of mechanical pressing and solvent and reduction in the number of juice processing steps are the advantages offered by MHG. MHG technique has successfully shown able to extract plum juice (Cendres et al., 2012) and strawberry juice (Turk et al., 2017). Even though the yield from microwave hydrodiffusion is low, additional pretreatment method such as enzymatic degradation helps in increasing the juice yield.

Microwave-assisted extraction has long been used for extraction of various plant bioactive compounds such as from fruits, vegetables, spices, or agriculture by-products. Internal superheating, caused by microwave energy absorption by water within the plant matrix, promotes cell disruption and subsequently improve the nutraceuticals recovery (Wang and Weller, 2006). Microwave extraction of fruit juice by-product showed that more bioactive compounds can be extracted from fruit by-products, and then these compounds can be added into the juice to make the juice more valuable. Apple juice produced from microwave heating of apple mash has higher total phenolic and flavonoid content (Gerard and Roberts, 2004). An innovative grape juice high in polyphenols was produced by Al Bittar et al. (2013) using microwave-assisted extraction of grape by-products.

Another application of microwave technology in juice processing is in the area of pasteurization. Microwave has been studied as an alternative to conventional pasteurization method. High temperature, pressure and long processing duration in conventional pasteurization processes have been associated to the quality loss in fruit juice. Continuous flow microwave pasteurization has been proven to be effective in inactivating of *Escherichia coli* and *Listeria* in apple juice (Siguemoto et al., 2017). Microwave pasteurization shows better retention of Kava juice quality compared to the conventional method (Abdullah et al, 2013). Microwave pasteurization has shown to retain bromelain and vitamin c in pineapple juice (Abd Aziz et al., 2017). In a comparative study between microwave and conventional pasteurization of tomato juice by Stratakos et al. (2016), microwave pasteurized tomato juice showed to have similar physicochemical and microbiological properties as the conventional pasteurized juice, which means that microwave continuous system can be used to replace conventional pasteurization system, since microwave treatment offers shorter processing time. Another plus point for microwave pasteurization is that the tomato juice produced via this technique has higher antioxidant capacity, which is an added value to the product.

Studies in microwave heating also showed that microwave heating is capable of inactivating endogenous enzymes in fresh fruits and vegetables that cause degradation, PPO, and pectin methyesterase (PME). PPO is the enzyme responsible for the loss of quality in fresh fruits and vegetables, whereas PME is the enzyme responsible for the loss of cloudiness in fresh juice (Sampedro et al., 2009). PPO in PME activities in Kava juice was significantly reduced by microwave heating in (Abdullah et al., 2013).

5.17.4 SONIFICATION TECHNOLOGY

Sonication or ultrasound is a nonthermal process that can be applied in many parts of juice processing such as pretreatment, extraction, and pasteurization. The sonication frequencies generally used in power ultrasound region 20–25 kHz. The production and collapse of cavitation bubbles in sonication process promote mechanical, chemical, and biochemical effects in the liquids (Paniwynk, 2017). The cavitation phenomenon occurs in sonication due to the interaction between ultrasonic waves, liquid medium, and dissolved gas in the liquid. Ultrasonic waves form bubbles and when the bubbles grow into unstable size, they will burst and release high heat and pressure that break the compounds in the liquid which induce particle dispersion and cell disruption effect into the system (Figure 5.3) (Abdullah and Chin, 2014). The efficiency of the ultrasonic treatment can be increased by combining this treatment with temperature treatment, pressure treatment, and both, which are known as thermosonication, manosonication, or manothermosonication, respectively (Jambrak et al., 2017).

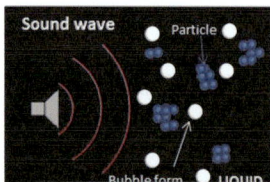
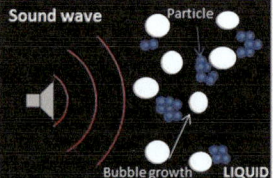
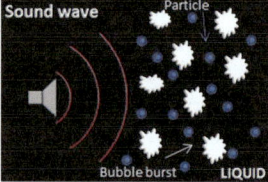

FIGURE 5.3 Cavitation phenomenon in liquid [adapted and modified from Abdullah and Chin (2014)].

Beverage rich in valuable bioactive compounds can be obtained through ultrasonic extraction because the cavitation process in ultrasound treatment breaks down plant matrices to facilitate extraction of bioactive compounds. The cavitation phenomenon during ultrasonic treatment assists in producing juice with higher yield, better nutritional, and sensory quality with shorter processing time. Bora et al. (2017) studied the effect of sonication on the banana juice yield. Higher juice yield was obtained from ultrasonic treatment combined with enzymatic treatment, compared to juice obtained from conventional extraction. In addition to that, the sonication technique in juice processing increases the functional properties of the juice. This can be seen from the work by Abid et al. (2014), where apple juice treated with ultrasound contained higher phenolic compounds and sugar contents and sonicated for 30 min and higher of content of carotenoids and minerals after 60 min of sonication. The total phenolics, carotenoids, and retinol activity equivalent in ultrasonically treated Cape gooseberry juice were found to be higher than conventional extracted juice in the study conducted by Ordonez-Santos et al. (2017).

Besides increasing functional properties of the juice, with minimal loss of on sensory, color, and bioactive compounds of the juice. Significant loss of functional properties in a functional and medicinal drink is normally associated to the application of high temperature during the conventional fruit drink production process. Sonication treatment offers treatment with a lower temperature than conventional thermal process, therefore it is able to retain the quality of the juice. A study performed by Khandpur and Gogate (2015) on orange juice showed ultrasonically treated orange juice has the similar taste to the fresh untreated orange juice and it was accepted by consumers.

In addition, ultrasound treatment is also found to be effective in inactivating microorganisms at a lower temperature, therefore ultrasonic is an alternative treatment for conventional pasteurization. Complete inactivation of microorganisms (total plate counts, yeast and mold counts) in apple juice treated with thermosonication bath (25 kHz) and probe (20 kHz) at 60 °C was achieved in the study done by Abid et al. (2014). Another study by Jambrak et al. (2017) on thermosonication treatment of apple, cranberry, and blueberry juices and nectars at temperature 60 °C, amplitude between 60 and 120 um, and exposure time between 3 and 9 min demonstrated complete inactivation of *Aspergillus ochraceus* 318, *Penicillium expansum* 565, *Rhodotorula* sp. 74, and *Saccharomyces cerevisiae* 5 in the juice samples.

Sonication is also used in juice processing to inactivate degrading enzymes PPO, POD, and PME. PPO and POD reactions cause discoloration and off-flavor in fruits and vegetable produce. Thermosonication using an ultrasonic bath (25 kHz) and probe (20 kHz) have been studied to inactivate PPO, POD, and PME in apple juice and the study result showed that these enzymes were completely inactivated using ultrasonic probe treatment at 60 °C (Abid et al., 2014).

5.18 PHYSICAL AND CHEMICAL PROPERTIES OF FUNCTIONAL BEVERAGES

5.18.1 TEXTURE AND STABILITY

It is challenging to modify or create a beverage so that it can fulfill its claim of improving health or a functional property because the "functional" ingredients added to achieve this aim may change the stability (e.g., proteins or minerals) or texture (e.g., fibers) of the beverage. For example, a beverage with reduced sugar levels to deliver on its promise of having "no/lower/reduced" health claims is often affected texturally and stability-wise. The very low viscosity matrix of the finished product further compound the challenges in product development. Generally, compared to more viscous products, beverages are not as robust when its formulation or processing is changed. It is crucial to consider the composition of the beverage when selecting the type of ingredients to add, as this directly impacts the stability and texture of the beverage. Besides that, other key factors affecting texture and stability include process parameters such as filling temperature, heat treatment (time, temperature), and homogenization. To offset the issues in stability and texture of the final product, ingredients such as emulsifiers and hydrocolloids are often used.

5.18.1.1 FUNCTIONAL BEVERAGE TEXTURE AND STABILIZATION DEFINITION

Viscosity measurements are often used to measure or quantify the texture of a beverage. However, since viscosity is measured at a specific shear rate, this parameter only describes one attribute, but in reality, the texture is a very broad term—encompassing sensory descriptors and rheological properties as well. In other words, the texture is essentially the sensation and appearance of the beverage as a person drinks it, specifically:

1. What does it look like when poured?
2. What does the glaze look like on the bottle?
3. What sensation does it give in the mouth at first sip?
4. How does it coat the mouth?
5. What does the residue look like?

According to Glicksman (1982), the feel of the beverage in the mouth (i.e., mouthfeel) is largely dependent on the hydrocolloids (the flow properties of the gum), as it displays rheological behavior that affects the textural properties of gum. The term "stabilization" is a broad, general term that covers various exceedingly complex biophysicochemical mechanisms. Therefore to address the issue of stability, each of its parameters must be categorized and described appropriately. In general, beverages that are smooth and homogenous are known as stabilized beverages. Stabilization can further be defined as follows, based on the targeted formulation and functionalities of the final beverage product:

1. Stabilization of particles: The beverage must have an even suspension of particles (e.g., cacao particles, minerals, or pulps) throughout.
2. Stabilization of emulsion: The top of the beverage bottle or container must be devoid of fat or oil rings.
3. Stabilization of proteins: The beverage must exhibit a nonsandy and smooth feel in the mouth with no flocculation or sedimentation of proteins.
4. Textural stabilization: The above functionalities all indicate texture stabilization. Besides, the beverage must have a homogenous-looking appearance and viscosity, specifically:

   - there are no lumps or gel points;
   - there is no gradient in viscosity;
   - there are no layers that have formed;
   - there are no separations in the phase;
   - there is no clarification;
   - there is no flocculation.

Depending on the process and formulation of the beverage, each parameter will require a different ingredient solution.

5.18.1.2 FUNCTIONAL BEVERAGE TEXTURE AND STABILIZATION MECHANISMS

5.18.1.2.1 *Hydrocolloids as Gelling and Thickening Agents*

Hydrocolloids mainly help thicken or gel beverages, so they are generally divided into gelling or thickening agents. A beverage with a thickening agent will have the desired texture but its particulates will not be suspended. According to Hoefler (2004), thickening agents help to prevent particles from settling down too quickly and prevents oil droplets from rising rapidly, but separation will still occur. To address this issue, gelling agents are used, as they can prevent oil droplets or particles from separating by permanently suspending them in a matrix, by building a 3D network through the formation of links between molecules. However, the yield value that the network creates must be higher than the density of the particulates for the agent to work effectively.

Acidic beverages contain food proteins such as casein that usually agglomerate and form sediments, so these proteins must be stabilized. However, applying heat to the proteins can dehydrate them easily and make them sandy. Therefore specific hydrocolloids must be formulated to directly interact with the proteins via electrostatic interactions to prevent a chalky texture and sedimentation due to protein agglomeration. Complex coacervates can be formed when food proteins (isoelectric point around 5) react with anionic polysaccharides such as propylene glycol alginate (isoelectric point around 3.5) and high ester pectins, such as carboxymethylcellulose, as long as the pH is below the isoelectric point of the protein and above the isoelectric point of the polysaccharide such that the reaction occurs in the intermediate region where the two macromolecules carry opposite net charges. Several factors determine the strength of the interaction between the polysaccharide and protein, namely the surface distribution of the ionizable groups, the three-dimensional structure of the protein, and the backbone distribution of the carboxyl groups. Futo (1993) added that the presence of fat or sugars, pH, and ionic strength also affect the interaction as a whole. The flow curve of a beverage is an ideal approach for characterizing beverages.

Extensive research has shown that a shear rate ranging from 1 to $100s^{-1}$ correlated to good stimuli when evaluating viscosity orally. Sworn (2007) explained that the reason is that at this range, the flow behavior of the beverage and choice of hydrocolloid could be tweaked to meet specific sensory and processing requirements. However, Glicksmann (1982) reported a much narrower shear range of 20–50 s^{-1}. Therefore beverage development should always be evaluated based on its sensory attributes and flow curves. In this case, Glicksmann (1982) highlighted two main benefits of conducting rheological measurements based on the beverage flow curves (i.e., viscosity vs shear rate): (1) the best beverage out of several beverages containing different gums could be chosen based on the evaluation and characterization of the best mouthfeel using the flow curves, and (2) existing beverages can be qualitatively measured and characterized to be used as targets for duplication.

Nevertheless, because rheological methods require specific methods and rheometers, the process could be very time-consuming. Besides, because of their low viscosity and small particle content, beverages are extra difficult

Development of Functional and Medicinal Beverages   173

to rheologically evaluate. Therefore besides rheological evaluation, sensory evaluation is also commonly carried out to characterize beverages. Because beverages tend to have low viscosities than other food matrices, their matrices have a polymer concentration that is around or a little above the critical concentration (C*). Therefore the main challenge in formulating beverages is to ensure that there is minimal interaction between their polymers so that the particles in the 3-D network of the beverage are suspended, whereas still maintaining the low viscosity of the beverage.

5.18.1.2.2 Emulsifying Agents

The interfacial tension of two normally immiscible phases can be reduced with the use of emulsifiers. These substances prevent the creaming, coalescence, and flocculation of the different immiscible phases (that could lead to emulsion breakdown) during storage by allowing them to mix and form a stable emulsion. Becher and Walstra (1983) observed that turbulent conditions almost always cause oil-in-water (o/w) emulsions to emulsify. Besides, compared to the contribution of the applied homogenization energy to oil droplet disruption, the effect of the interfacial tension is relatively negligible. Beverages that contain o/w emulsions such as juice beverages with essential oil-based or omega-3 oil-based flavors and high-fat milk beverages usually benefit from the use of emulsifiers, as these agents can stabilize the emulsion and ensure that it remains stable throughout its shelf life.

5.19 SENSORY QUALITY

The sensory quality of functional beverages is assessed based on the properties and attributes listed below:

1. Properties involving the use of organs (organoleptic properties): The consumer does not decide his beverage of choice based on tactile, auditory, visual, or olfactory elements, so these properties are not touched on in this section.
2. Experiences after ingesting the drink (digestive properties): pleasure, fullness, and/or heaviness.

According to Hwang and Hong (2015), the positive attributes that define a good quality beverage are tactile properties such as the absence of contaminants (strange flavors or odors), body, and mouthfeel; olfactory properties such as odor (orthonasal and retronasal) and aroma; taste properties such as aftertaste, mouth persistence, and flavor; and color and overall appearance. Meanwhile, negative attributes include gas, sedimentation, astringency and bitterness, discoloration, foaming, and unpleasant smell production (vinegary or ketone notes).

5.19.1 THE ACCEPTABILITY OF FUNCTIONAL BEVERAGES

The acceptability and availability of a beverage/drink among consumers are usually evaluated before it is released to the market. A hedonic scale is used to evaluate the attributes of a beverage, whether positive or negative. The scale usually varies from 5 to 11 points indicating maximum displeasure to maximum pleasure. For adults, verbal evaluation is used while facial hedonic scale evaluation is used for infants. The tests to measure whether or not a drink is accepted usually employ scales with reliability, good predictability, and discriminative power correlating to eating habits. Most tests employ the nine-point hedonic scale, but others also use seven-, five-, or three-point hedonic scales. Normally, acceptance tests involve consumers, where the level of confidence desired for decision-making will determine the number of participants involved. The commonly evaluated attributes of natural drinks and the associated sensory tests are listed in Table 5.8. Usually, the attributes of taste and smell and aroma and smell are evaluated based on the qualities mentioned below:

1. The feelings that occur on the tongue and in the mouth, such as sourness, bitterness, saltiness, or sweetness (and most recently umami) all fall under taste, indicating whether or not a drink is delicious.
2. Drinks that evoke mixed and pleasant olfactory sensations or are perfumed or fragrant usually indicate a nice aroma.
3. Usually, unpleasant olfactory sensations that refer to a specific type of aroma are indicated by the odor of the product.
4. The viscosity or body of a drink indicates the oral sensation when ingesting the drink.
5. Taste, viscosity, and aromatic sensations combine to denote flavor, such that the brain understands flavor by evaluating these components.

The time it takes for the flavor of a drink to disappear from the mouth is referred to as the flavor persistence in the mouth. This attribute is synonymous with wine tasting, but with the recent developments of juices and teas, this attribute has become an important one to evaluate. Usually, one's palate

Development of Functional and Medicinal Beverages 175

TABLE 5.8 Attributes Commonly Evaluated in Natural Drinks and the Sensory Tests Used

Product Evaluated	Sensory Evaluation	Sensory Panel	Scale	Palate Cleansing	Reference
Amarone red wine	Development of sensory descriptors for aroma, taste, flavor, and mouthfeel	12	Nine points	Undefined	Pagliarini et al. (2004)
Apple and apple-whey beverages	(a) Appearance (color, sediments and suspension); odor desirability, odor intensity and flavor (b) Descriptive flavor analysis	(a) 15 (b) 18	Five points	Undefined	Jaworska et al. (2014)
Arjuna-ginger medicinal mix blended	Appearance, color, taste, flavor, brightness, and strength	20	Nine points	Undefined	Verma and Singh (2013)
Assai and passion fruit with banana pulp	Affective tests of ordering and preference, purchase intention tests.	Undefined	Five point	Undefined	Camargo et al. (2012)
Black cherry, concord grape, and pomegranate juices blend	(a) Consumers: overall liking, appearance, Just about right (JAR) attributes (b) Descriptive analysis: development of sensory descriptors for flavor, mouthfeel, and strange flavors	(a) 100 consumers (b) 10	JAR scale nine points	Unsalted crackers and water	Lawless et al. (2013)
Cloudy plum juices	Color, odor, taste, consistency, and overall sensory impression	10–15	Nine points	Undefined	Levaj et al. (2012)
Custard apple leaf extract fortified sweetened aonla juice	Color and appearance, flavor, taste and mouthfeel, and overall acceptability	Undefined	Nine points	Undefined	Sushilkumar et al. (2016)
Drink lemon juice and honey ready to serve	Color/appearance, flavor/aroma, body, taste, and overall acceptability	Undefined	Nine points	Undefined	Sharma et al. (2016)
Gowe beverages	Perception of the basic tastes (sweet and sour) and familiarity with the product	22 panelist	Descriptive analysis	Undefined	(Adinsi et al., 2015)
Grape juice mixed with apple, pear, and peach juices	Color, odor, aroma, sweet taste, persistence in the mouth, overall pleasantness	50 consumers	Five points	Natural water	Chiusano et al. (2015)

TABLE 5.8 (Continued)

Product Evaluated	Sensory Evaluation	Sensory Panel	Scale	Palate Cleansing	Reference
Antioxidant Rich Drink	Appearance, color, texture, aroma, taste, and overall acceptability	10	Nine points	Undefined	Puri et al. (2017)
Mugwort tea (*Artemisia argyi* H. Lev & Vaniot)	Color acceptability, flavor acceptability, saltiness, bitterness, sourness, astringency, sweetness and overall preference	15	Labeled affective magnitude (LAM) scale 15 points	Undefined	Jae et al. (2015)
Plum nectar (*Prunus domestica*)	Quantitative descriptive analysis (QDA): color, odor, taste, consistency, and overall sensory impression	15	Intensity scale 0–10	Salt-free bread and water	Hruškar et al. (2012)
Wine	Aroma, tactile attributes	8	10 points	Water	Xing Chen et al. (2017)

Adapted and modified from Saucedo-Pompa et al. (2018).

must be cleansed from bitter, fatty, spicy, or sweet tastes to allow for the experience of new sensory attributes. In this case, the ideal complement for cleaning the mouth is water. Since it is a given that water is used in the evaluation of sweet beverages and natural beverages, this combination of products is often not reported in publications.

5.19.1.1 FUNCTIONAL BEVERAGE STATISTICAL TESTS AND ANALYSES FOR SENSORY EVALUATION

Qualitative tests, discriminative tests, and acceptance tests are the tests used to evaluate the sensory attributes of beverages. The hedonic scale, namely Just About Right (JAR) and Labeled Affective Magnitude (LAM), are used in discriminative tests and acceptance tests. Meanwhile, the intensity grade scale is used in qualitative tests, to evaluate the taste profile and to conduct qualitative descriptive analysis (QDA) of the beverage.

The sequence of sensory tests used for beverage development is shown in the yellow box of Figure 5.4. Based on the optimization, substitution, or acceptance of ingredients in the test of acceptance, alternative tests are presented. Based on the review of the literature presented in this section, the sensory tests provide the researcher with different strength points to discriminate or accept a product. Hence, the questions that are to be answered during the beverage development product will determine whether or not the product is rejected or accepted.

Consumer acceptance tests normally use a nine-point verbal hedonic scale indicating degrees of taste from "extremely disliked" to "extremely liked." Following that, the verbal responses are transformed into numerical values to further analyze statistically. A study attempted to discover whether the words on the nine-point hedonic scale could be changed to numbers (Nicolas et al., 2009). They found that scales that used "only words" or "only numbers" usually induced the customers to respond differentially. In particular, 79%–100% of the customers would give different results when given either scale. The study concluded that the researcher should apply caution when transforming one scale to the other, as the numerical data derived from these two scales are not interchangeable. Lawless et al. (2009) mentioned the LAM scale, which is another scale used in consumer acceptance tests that has higher discriminative power than the nine-point scale. The LAM scale also has good spacing between anchor words for foods and well-liked beverages. Lawless et al. (2009) further compared the nine-point hedonic scale and the LAM scale in food acceptance tests.

They found that both discriminated products well based on the tastes of the consumer. Also, they found that consumption patterns had a strong effect on consumer acceptance ratings. The study concluded that the nine-point hedonic scale was comparable to the LAM scale. They suggested using more comprehensive scales to ensure that the true differences between products could be detected, and thus preventing Type II error.

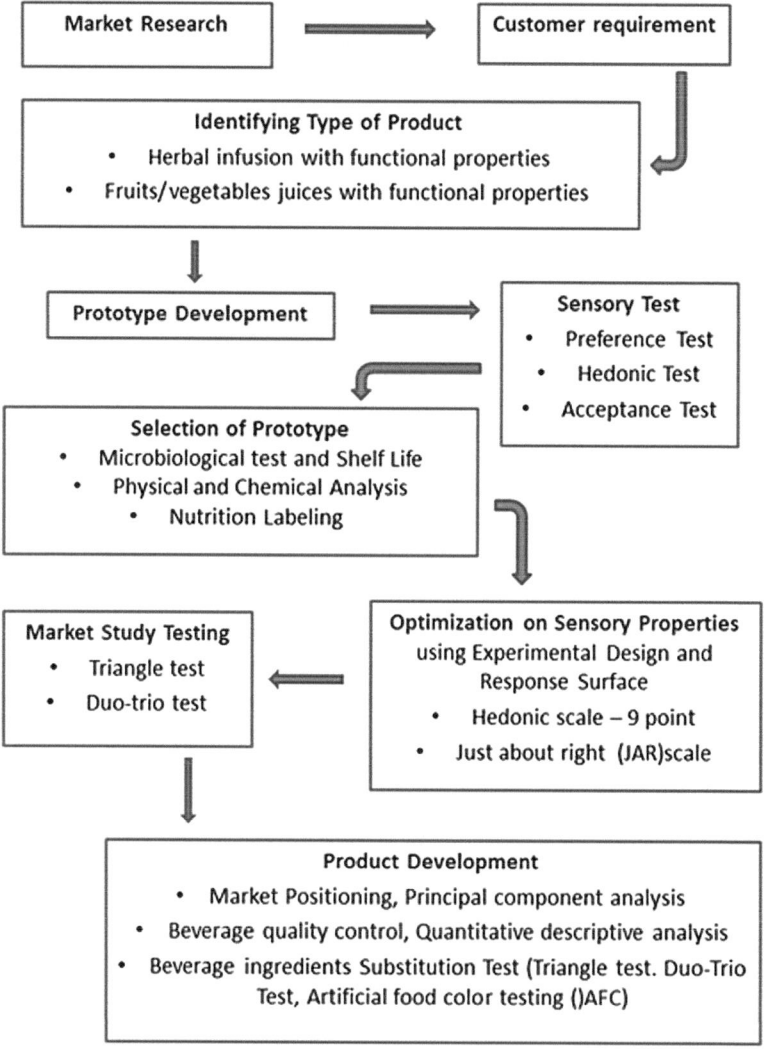

FIGURE 5.4 Sequence of sensory tests that are used in the development of natural beverages [adapted and modified from Saucedo-Pompa et al. (2018)].

The LAM scale was used in a previous study to determine the acceptability of two species of Artemis tea based on general preferences, bitterness, acidity, color, taste, and salinity (Jae et al., 2015). The range for the LAM scale was 0–15, correlating to greatest imaginable dislike and greatest imaginable like. The study found that the differences in the volatile compounds in the tea, namely the terpenic compounds, contributed to significant differences in general preferences and taste acceptability while the rest of the attributes showed no significant difference.

The JAR scale is normally used to identify attributes that need improvement or to optimize a product attribute. The scale shows that the ideal value of an attribute is close or near to the most-liked attribute. In other words, after evaluation using the JAR scale, products that qualify as "just right" products indicate the most-preferred or most-liked products. Nevertheless, studies that have employed the JAR scale often discover very different optimal values of attributes compared to the attributes of current products on the market.

For example, to optimize the sweetness of lemonade, Epler et al. (1998) assessed the box and line JAR scales against the hedonic scale. Several formulations of the lemonade were developed with different sugar content (6%–14%) and the taste and the "optimal" level of sweetness of each were predicted. Both the line JAR scale and the box JAR scale produced similar optimum values (9.2% and 9.4% sucrose), but this was lower than the results of the hedonic scale, which showed 10.3% sucrose as the optimum sweetness. Then a preference test was conducted, showing that consumers preferred the lemonade with 10.3% sugar content compared to the 9.3% lemonade. Therefore the JAR scale is not as good at predicting sweetness as the hedonic scale.

The optimal sweetness concentration of a beverage could be assessed using three- and nine-point JAR scales. The data from both scales were compared via a survival statistical analysis and then a regression analysis. The study used orange juice with three different sucrose concentrations. It was found that the three-point JAR scale showed an optimum sucrose concentration of 13.1%, higher than the 8.2% optimum sucrose concentration produced by the nine-point scale. Later, additional tests showed that the sample with 13.1% sucrose concentration was preferred by 70% of subjects rather than the sample with 8.2% concentration. The study, therefore, showed that combining statistical analysis with the three-point JAR scale yielded more real optimal sucrose concentration compared to the nine-point scale. The Box and Wilson methodology (also known as response surface) has also been used to optimize drinks. This method combines regression analysis with experimental design. This method enables the determination of the optimal value of an independent variable (e.g., sugar level) that could maximize the response variable (e.g., flavor). One study used the response surface methodology to find the optimal level of sweetness for a soybean extract fermented beverage. They obtained predictive response models that indicated the ideal sweetness, flavor, and generally accepted purchase attitude, with the ideal sweetness being 11 g of sucrose per 100 mL.

Elsewhere, Pagliarini et al. (2004) differentiated wines from different harvests based on their taste and aroma, also with the help of response surface methodology. A few studies have conducted similar tests, but the above studies better illustrated the use of the method for developing beverages. Figure 5.5 compares the different tests that have been developed and widely used.

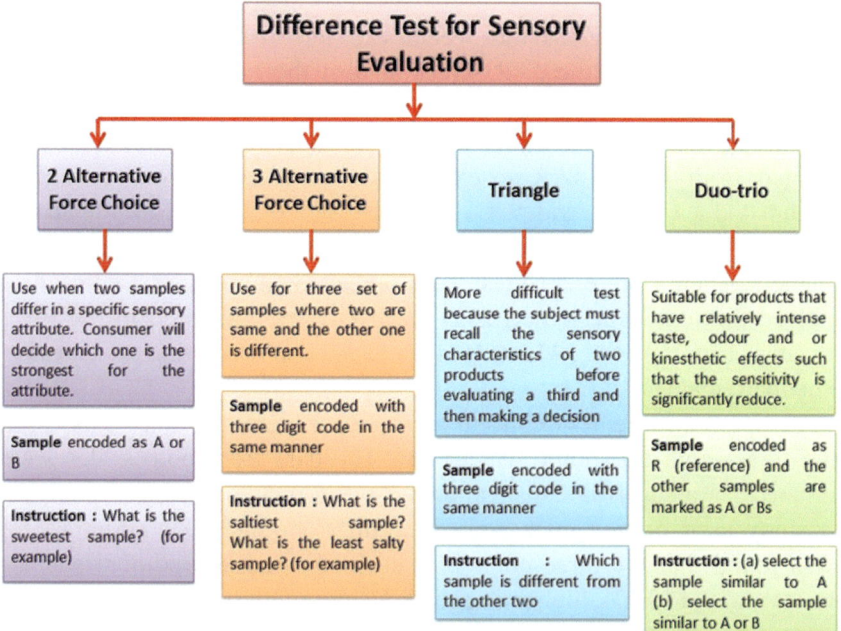

FIGURE 5.5 Type of difference sensorial test [adapted and modified from Fallourd and Viscione (2009)].

Difference tests are usually used to compare the attributes of a new beverage or to replace the ingredient in a beverage. The differences tests are divided into two categories: (1) duo-trio and triangular tests (when the cause of the difference is not asked, and other attributes are focused on), and (2) 2AFC and 3AFC tests (when the cause of the difference is asked, focusing on a specific attribute). Normally, these tests are conducted on consumers and trained judges. Simpler tests are conducted with consumers where the reliability of the results can be increased with an increase in the number of participants.

The Chi-square statistical method is used in the differences test. Because this test requires repeated sample testing such that adaptation to stimulus could occur, the judge can become fatigued. Hence, one session only allows the evaluation of six samples at most. The results of this test are interpreted using statistical tables, depending on the required significance level, sample size, and the minimum number of correct answers. A new drink that is about to launch in the market must first have an established flavor profile, defined based on its sensory properties. Besides, the properties must also be monitored in the process of quality control. In this case, the sensory attributes of the product, such as appearance, aftertaste, and mouthfeel, but no more than seven attributes, all in all, can be determined via a QDA involving 10–12 trained judges. The QDA is conducted to quantitatively specify and establish the nature and intensity of a product's sensory attributes.

In a QDA, the intensity of the attributes of some beverages is evaluated based on the intensity level on a vertical line. Once converted into numerical values, the distances between the vertical lines are analyzed using Analysis of Variance (ANOVA). In one study, a QDA was conducted on nectars and 10 descriptive terms were generated to describe the products, relating to consistency, smell, taste, color, and overall sensory impression (Hruškar et al., 2012). Based on the ANOVA results, the overall sensory impression, sour and sweet intensity, and color intensity of the nectars were identified as significantly different, but the addition of acid or sugar did not yield any significant differences.

To establish the position of a beverage in the market, principal component analysis (PCA) can be used. This analysis leverages on the beverage's sensory attributes, namely body, color, flavor, and aroma. Using PCA, the sensory attributes of a beverage could be quantitatively analyzed based on the correlation between the attributes. It also enables the calculation of new variables by grouping attributes such that the distance between groups and attributes can be observed. Therefore the product with better market positioning among consumers can be identified, and its consequences on the market assessed.

5.20 FLAVOR RETENTION AND RELEASE

Flavor retention and release are mostly studied in functional beverages (including drinkable meal replacers or sports supplements), to design healthier food products (low-fat milk, alcohol-free beer, etc.) without compromising on traditional product acceptability, and beverages with exotic features (exotic fruit tastes, cocktails, fusions, etc.). In skim milk, loss/lack of hydrophobic flavors challenges consumer's acceptability compared to that of high-fat milk, whereas the potential health benefit of soy milk suffers from a beany off-flavor (Kinsella and Damodaran,1980) originating from lipoxygenase activity (Mana Kialengila et al., 2013). Flavor release or retention is generally affected by the intrinsic chemical properties of the flavor (hydrophobicity, hydrophilicity (log P value), and volatility), the composition of the medium (lipid, protein, salt, sugar, etc.), and finally environmental conditions (temperature, pH). In other words, the interaction between flavor compounds and other food ingredients under given environmental conditions determines the intensity of flavor retention or release from a product.

5.21 QUALITY AND SAFETY ISSUES IN FUNCTIONAL AND MEDICINAL BEVERAGES

Maintaining bioactivity of the active ingredient and quality of the functional and medicinal beverages after the processing stage is very important. It has always been a challenge to provide a fresh and perfect functional beverage to the consumers because as the time goes by, the quality of the product may deteriorate and their functionality may reduce or lost due to several reasons, such as degradation of the functional ingredient during production, short shelf life, and unsuitable packaging and storage condition (Andersson et al., 2015).

5.21.1 SHELF-LIFE OF FUNCTIONAL AND MEDICINAL BEVERAGES

During the extended storage of the functional and medicinal beverage in refrigerated conditions, the changes in quality of the product especially in the sensory and nutritional value are unavoidable. This change limits the shelf-life of the product. According to Rysstad and Johnstone (2009), the changes in the quality of certain beverages can be due to (1) the activity of enzymes or microbes that are present in the packaged product, (2) chemical changes in the product such as oxidation, and (3) interactions of the beverage with the packaging material. A huge challenge is faced by the researcher in improving the shelf life of functional beverages. Therefore many investigations are being done to enhance the viability and the efficacy of functional ingredient used in the beverage as it should be sustained in the food product until it reaches the end user. One of the methods includes the usage of electromagnetic heating such as ohmic heating due to electrical resistance (50–60 Hz); radio frequency (RF), heating (10–60 MHz), and microwave heating (1–3 GHz) (Zhao et al., 2000; Rysstad and Johnstone, 2009). Conventionally, heat or thermal treatment, such as low temperature long time and high temperature short time, has been used in pasteurization to eliminate microbial contamination in the beverage (Rupasinghe and Yu, 2012). Although this treatment is able to be used to extend the shelf life by reducing the activity of the enzymes present in the product, it can cause adverse reactions such as nutrient degradation and reduce the freshness of the beverage (Awuah et al., 2007; Rupasinghe and Yu, 2012). Therefore in order to overcome this problem, electromagnetic heating has been employed by transferring the direct heat from the source into the food without going through a heat transfer surface. These methods are implemented homogeneous and quick heating and cooling to promote less damage from thermal treatment thus increase the efficacy of the functional beverages.

According to the research done by Benlloch-Tinoco et al. (2015), microwave pasteurization of formulated fruit juice is expected to bear a longer shelf life than formulated fruit juice treated by conventional thermal pasteurization. This is due to greater penetration depths of 2450 MHz microwave radiation, which results in the better inactivation of spoilage microorganisms present in fruit juices. An excellent quality of fruit-based products might be accomplished through microwave heating due to greater potentials in prolonging the shelf-life which then can substitute many conventional preservation methods. This method provides faster heating rates, higher thermal efficiency, higher penetrative power, and shorter processing times due to the specific way of heating that takes place during microwave processing, thus results in better organoleptic, functional, and nutritional properties preservation, especially on the color.

Besides that, natural preservatives such as bacteriocins, organic acids, lactoperoxidase, essential oils, carotenoids, and phenolic compounds have shown a great potential in extending the shelf life (Rupasinghe and Yu, 2012; Aneja et al., 2014). For instance, natural beverage preservatives could be produced from fruit by-products containing phenols and carotenoids as they are able to prolong the shelf life of the product by slowing down the occurrence of off-flavors and rancidity. Besides, they also exhibit well-known health benefits (Carbo et al., 2014). Other than electromagnetic heating and natural preservatives, there are also another methods that can be used to preserve and prolong the shelf life of the beverage, such as pulsed electric fields (Oziemblowski and Kopec, 2005), reverse osmosis, and ultrasonication (Benlloch-Tinoco et al., 2015).

5.21.2 STABILITY OF FUNCTIONAL AND MEDICINAL BEVERAGES

The suitable concentration of a functional ingredient incorporated into a beverage to achieve the optimum efficacy of the specific functional compound is often very challenging as these functional compounds may directly influence on the physical properties and the stability of the product. Besides that, process parameters such as homogenization, filling temperature, and heat treatment such as time and temperature also play an important role to provide stability to the final product.

Commonly, a stabilized beverage defines a well uniform and smooth beverage. Referring to the beverage formulation and specific final functionalities, stabilization can be particularly defined into three categories: (1) particle stabilization where the particles such as pulps, cacao particles, and minerals are uniformly suspended in the beverage, (2) emulsion stabilization in which fat or oil ring does not appear on the top of the beverage packaging, and (3) protein stabilization where no flocculation or sedimentation of proteins occurs in the beverage or the beverage exhibits a smooth and nonsandy mouthfeel (Fallourd and Viscione, 2009). To overcome these challenges, the use of ingredients such as hydrocolloids, polyphenols, emulsifiers, and enzymes can help in ensuring the stability and texture of the final product. (Fallourd and Viscione 2009; Andersson et al., 2015).

5.22 STORAGE AND PACKAGING

The quality and the longevity of the final product can be directly or indirectly affected by the packaging material and the storage condition (Andersson et al., 2015). The choice of the packaging materials can influence the interaction between the product and its environment. There are four types of interaction that are involved that are (1) egress and scalping that causes aroma loss; (2) sorption; (3) migration of monomers and additives from the packaging materials; and (4) ingress of oxygen and light (Figure 5.6).

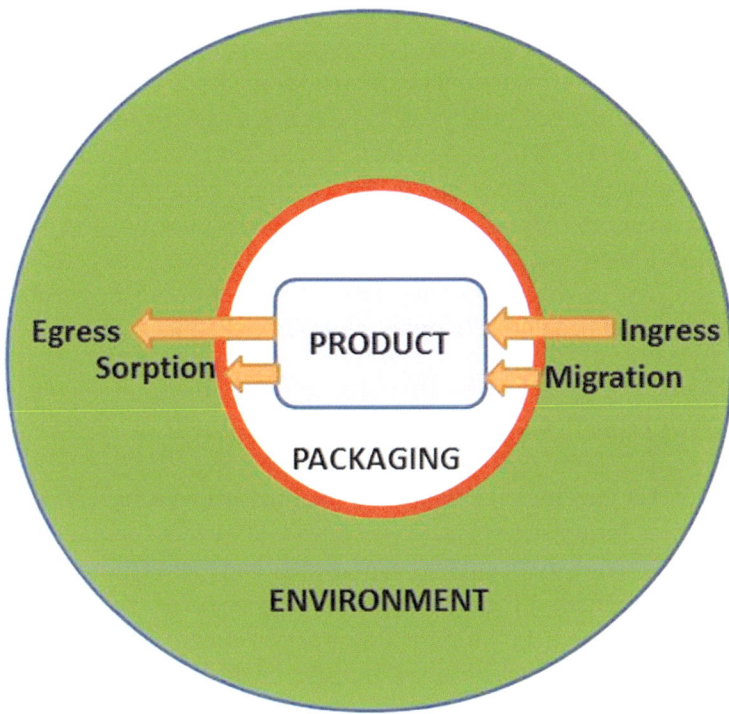

FIGURE 5.6 Influence of food-packaging interaction on the product shelf life [adapted and modified from Rysstad and Johnstone (2009)].

Functional beverages can also be damaged when exposed to light. For example, bioactive compounds such as antioxidants, vitamins, or other light-sensitive compounds that incorporated in functional beverages must be shielded from light exposure. Functional drinks may contain special antioxidants, vitamins, or other light-sensitive ingredients that must be protected from light exposure during storage. When the products are exposed to the light, it may result in accelerated vitamin C degradation as well as photobleaching (Rysstad and Johnstone, 2009). To overcome this, aluminium-foil lined cartons can be used to have a complete light protection.

Another method that can be used to improve the quality of the product is by sterilization of the packaging. The most common packaging utilizes the combination of concentrated hydrogen peroxide, heat, and UV treatment in cartons which encourages bacterial spores to be inactivated synergistically, hence produces a sterile condition (Ansari and Datta, 2003; Rysstad and Johnstone, 2009). Another potential method in the sterilization of product packaging includes pulsed light (Gomez-Lopez et al., 2007), plasma, and UV-excimer (Warriner et al 2004).

Another important factor that needs to be considered in ensuring the best condition of the finished product is the storage temperature. A study done by Andersson et al. (2015) shows that storage temperature influenced the quality of the bioactive compound in the beverage. Higher storage temperature (20 °C) causes rapid changes in the level of bioactive component as compared to lower storage temperature (4 °C). Therefore it is crucial for a rapid cooling and lower storage temperature to be applied to the product after the packaging had been done in order to preserve the functional nutrients.

5.23 CONCLUSIONS

Food and beverages manufacturers are continuously facing challenges in producing new formulations and development of new beverages especially beverages with added functional ingredients that could promote health. The challenges include the quality and safety aspects of the beverages while maintaining or improving the taste for consumers benefit. Functional beverages that could benefit health, improve energy and immunity system especially that are relatively easy to consume would be very high in demand as the consumers are looking more for convenient due to busy lifestyle. Furthermore, before promoting its health-promoting effect to the consumers, a quality functional ingredient should be analyzed and tested for clinical evidence for its safety and efficacy. It is easy to claim that the products are having health-promoting effect but it would be unfair for consumers to get the side effect if the issues of safety are not taken seriously by the manufacturers. The market trend in functional and medicinal beverages is increasing and it would be beneficial for the manufacturers and researchers to collaborate in improving the quality and producing more innovative products in the future.

ACKNOWLEDGMENTS

The authors would like to acknowledge Ministry of Education Malaysia and Universiti Teknologi Malaysia for the financial support (Research Grant No. R.J130000.7851.4F993 and Q.J130000.3551.05G90). Special thanks to Nurul Hazirah Hamidon, Nabilah Abdul Samad, Farhani Ahmad, Chin Chooi Li, and Hazelynna Makerly for the contributions in this chapter.

KEYWORDS

- **functional beverages**
- **healthy drinks**
- **processing**
- **quality**
- **safety issues**

REFERENCES

Abd Aziz, N.A., Jusoh, Mohd Jusoh, Y.M., Nik Mahmood, N. A., Yunus, N.A., and Endut, A. Quality evaluation of microwave and conventional pasteurised pineapple juice. International Food Research Conference Proceeding, Universiti Putra Malaysia, 2017.

Abdullah, S.A., Lee, S.H., Cho, I.K., Li, Q.L., Jun, S., and Choi, W. Pasteurization of Kava juice using novel continuous flow microwave heating technique. Food Science and Biotechnology, 2013, 22(4), 961–966.

Abdullah, N. and Chin, N.L. Application of thermosonication treatment in processing and production of high quality and safe-to-drink fruit juices. Agriculture and Agricultural Science Procedia, 2014, 2, 320–327.

Abid, M., Jabbar, S., Wu, T., Hashim, M.M., Hu, B., Lei, S. and Zeng, X. Sonication enhances polyphenolic compounds, sugars, carotenoids and mineral elements of apple juice. Ultrasonic Sonochemistry, 2014, 21, 93–97.

Adinsi, L., Akissoé, N. H., Dalodé-Vieira, G., Anihouvi, V. B., Fliedel, G., Mestres, C. and Hounhouigan, J. D. Sensory evaluation and consumer acceptability of a beverage made from malted and fermented cereal: case of gowe from Benin. Food Science and Nutrition. 2015, 3(1), 1–9.

Al Bittar, S., Perino-Issartier, S., Dangles, O. and Chemat, F. An innovative grape juice enriched in polyphenols by microwave-assisted extraction. Food Chemistry, 2013, 141, 3268–3272.

Alvarez, S., Riera, F. A., Alvarez, R., Coca, J., Cuperus, F. P., Th Bouwer, S., Boswinkel, G., van Gemert, R.W., Veldsink, J.W., Giorno, L., Donato, L., Todisco, S., Drioli, E., Olsson, J., Tragardh, G., Gaeta, S.N., Panyor, L. A new integrated membrane process for producing clarified apple juice and apple juice aroma concentrate. Journal of Food Engineering, 2000, 46, 109–125.

Andersson, S.C., Ekholm, A., Johansson, E., Olsson, M.E., Sjoholm, I., Nyberg, L, Nilsson, A. and Rumpunen, K. Effect of storage time and temperature on stability of bioactive compounds in aseptically packed beverages prepared from rose hips and sea buckthorn berries. Agriculture and Food Science. 2015, 24, 273–288.

Aneja, K.A, Dhiman, R.D, Aggarwal, N.K. and Aneja, A. Emerging preservation techniques for controlling spoilage and pathogenic microorganisms in fruit juices. International Journal of Microbiology, 2014, 1–14.

Ansari, I.A., and Datta, A.K. An overview of sterilization methods for packaging materials used in aseptic packaging systems. Food and Bioproducts Processing, 2003, 81(1), 57–65. https://doi.org/10.1205/096030803765208670

Awuah, G.B., Ramaswamy, H.S and Economides, A. Thermal processing and quality: Principles and overview. Chemical Engineering and Processing, 2007, 46, 584–602.

Balannec, B., Vourch, M., Rabiller-Baudry, M., and Chaufer, B. Comparative study of different nanofiltration and reverse osmosis membranes for dairy effluent treatment by dead-end filtration. Separation and Purification Technology, 2005, 42(2), 195–200.

Bae, Y. W., Lee, S. M., and Kim, K.O. Age and gender differences in the influence of extrinsic product information on acceptability for RTD green tea beverages. Journal of Science and Food Agriculture, 2016, 96(4), 1362–1372.

Barba, F. J., Jäger, H., Meneses, N., Esteve, M. J., Frígola, A., and Knorr, D. Evaluation of quality changes of blueberry juice during refrigerated storage after high-pressure and pulsed electric fields processing. Innovative Food Science & Emerging Technologies, 2012, 14, 18–24.

Barba, F. J., Esteve, M. J., and Frigola, A. Physicochemical and nutritional characteristics of blueberry juice after high pressure processing. Food Research International, 2013, 50(2), 545–549.

Bech-Larsen, T., and Grunert, K.G. The perceived healthiness of functional foods: A conjoint study of Danish, Finnish and American consumers' perception of functional foods. Appetite, 2003, 40(1), 9–14.

Becher, P. and Walstra, P. Encyclopedia of Emulsion Technology, Marcel Dekker Inc., 1983, 57.

Benlloch-Tinoco, M., Igual, M., Rodrigo, D., and Martínez-Navarrete, N. Superiority of microwaves over conventional heating to preserve shelf-life and quality of kiwifruit puree. Food Control, 2015, 50, 620–629.

Bigliardi, B. and Galati, F. Innovation trends in the food industry: the case of functional foods. Trends in Food Science & Technology, 2013, 31, 118–129.

Bihuniak, J.D. and Insogna, K.L. The effects of dietary protein and amino acids on skeletal metabolism. Molecular and Cellular Endocrinology, 2015, 410, 78–86.

Bomkessel, S., Broring, S., Omta, S.W.F., and van Trijp, H. What determines ingredient awareness of consumers? A study on ten functional food ingredients. Food Quality and Preference, 2014, 32, 330–339.

Bora, S. J., Handique, J., and Sit, N. Effect of ultrasound and enzymatic pre-treatment on yield and properties of banana juice. Ultrasonics Sonochemistry, 2017, 37, 445–451.

Boroski, M., Giroux, H.J., Sabik, H., Petit, H.V., Visentainer, J.V., Matumoto-Pintro, P.T., and Britten, M. Use of oregano extract and oregano essential oil as antioxidants in functional dairy beverage formulations. Journal of Food Science and Technology, 2012, 47, 167–174.

Boye, J. I. Nutraceutical and functional food processing technology. Statewide Agricultural Land Use Baseline, 2015 (Vol. 1).

Brown, P. N. and Chan, M. An overview of functional food regulation in North America, European Union, Japan and Australia. In: Smith, J., Charter, E., (Eds.), Functional Food Product Development. New York: John Wiley & Sons, 2009, 257–292.

Butz, P., Fernández García, A., Lindauer, R., Dieterich, S., Bognár, A., and Tauscher, B. Influence of ultra high pressure processing on fruit and vegetable products. Journal of Food Engineering, 2003, 56, 233–236.

Buzrul, S., Alpas, H., Largeteau, A., and Demazeau, G. Modeling high pressure inactivation of *Escherichia coli* and *Listeria innocua* in whole milk. European Food Research International, 2008, 227, 443–448.

Camargo, G.A., Mieli, J., Prati, P., Ormenese, R.S.C. and Schmidt F. L. Quality and sensory evaluation of beverage of assai and passion fruit enriched with unripe banana pulp. RETEC, Ourinhos, 2012, 5(01), 80–92.

Campbell, B. Dietary carbohydrate strategies for performance enhancement. In: Campbell, B. (Ed.), Sports Nutrition-Enhancing Athletic Performance. Florence, SC: Taylor & Francis, 2013, 75–124.

Cassano, A., Conidi, C., and Drioli, E. Clarification and concentration of pomegranate juice (*Punica granatum* L.) using membrane processes. Journal of Food Engineering, 2011, 107 (3–4), 366–373.

Cassano, A., Conidi, C., and Drioli, E. Membrane processing. In: Conventional and Advanced Food Processing Technologies, 2014, 537–566.

Cappelletti, S., Daria, P., Sani, G., and Aromatario, M. Caffeine: cognitive and physical performance enhancer or psychoactive drug? Current Neuropharmacology, 2015, 13(1), 71–88.

Cendres, A., Chemat, F., Page, D., Le Bourvellec, C., Markowski, J., Zbrzezniak, M., Renard, C.M.G.C., and Plocharski, W. Comparison between microwave hydrodiffusion and pressing for plum juice extraction. LWT—Food Science and Technology, 2012, 49, 229–237.

Chacko, S.M., Thambi, P.T., Kuttan, R., and Nishigaki I. Beneficial effects of green tea: a literature review. Chinese Medicine, 2010, 5(13), 1–9

Chauhan, B., Kumar, G., Kalam, N., and Ansari, S.H. Current concepts and prospects of herbal nutraceutical: a review. Journal of Advanced Pharmaceutical Technology & Research, 2013, 4(1), 4–8.

Chen, K.I., Erh, M.H., Su, N.W., and Chou, C.C. Soyfoods and soybean products: from traditional use to modern applications. Applied Microbiology and Biotechnology, 2012, 96, 9–22.

Chen, Z.Y., Huang, Y., Leung, F.P., Tian, X.Y., Wong, W.T., Yung, L.M., Yung, L. H., and Yao, X.Q. Tea polyphenols benefit vascular function. Inflammopharmacology, 2008, 16, 230–234.

Cheryan, M. Ultrafiltration and Microfiltration. Lancaster: Technomic Publishing Co Inc., 1998.

Childs, J.L., Yates, M.D., and Drake, M.A. Sensory properties of meal replacement bars and beverages made from whey and soy proteins. Journal of Food Science, 2007, 72, S425–S434.

Chiusano, L., Cravero, M.C., Borsa, D., Tsolakis, C, Zeppa, G, Gerbi, V. Effect of the addition of fruit juices on grape must for natural beverage production. Italian Journal of Food Science, 2015, 27, 375–384.

Committee on Nutrition and the Council on Sports Medicine and Fitness. Clinical report—sport drinks and energy drinks for children and adolescents: are they appropriate? Pediatrics, 2011, 127, 1182–1189.

Coppens, P., Fernandes Da Silva, M., and Pettman, S. European regulations on nutraceuticals, dietary supplements and functional foods: a framework based on safety. Toxicology, 2006, 221, 59–74.

Corbo, M.R., Bevilacqua, A., Petruzzi, L., Casanova, F. P., and Sinigaglia, M. Functional beverages: the emerging side of functional foods. Comprehensive Reviews in Food Science and Food Safety, 2014, 13(6), 1192–1206.

Costa, A. S.G., Alves, R.C., Vinha, A.F., Costa, E., Costa, C.S.G., Nunes, M.A., Almeida, A.A., Santos-Silva, A., Oliveira, M.B.P.P. Nutritional, chemical and antioxidant/pro-oxidant profiles of silverskin, a coffee roasting by-product. Food Chemistry, 2018, 267, 28–35.

Das, L., Bhaumik, E., Raychaudhuri, U., and Chakraborty. Role of nutraceuticals in human health. Journal of Food Science and Technology, 2012, 49(2), 173–183

Deng, Y., Misselwitz, B., Dai, N., and Fox, M. Lactose intolerance in adults: biological mechanism and dietary management. Nutrients, 2015, 7(9), 8020–8035.

Dhingra, D., Michael, M., Rajput, H. and Patil, R.T. Dietary fibre in foods: a review. Journal of Food Science and Technology. 2012, 49(3), 255–266.

Diplock, A. T., Aggett, P. J., Ashwell, M., Bornet, F., Fern, E. B., and Roberfroid, M. B. Scientific concepts of functional foods in Europe: consensus document. British Journal of Nutrition, 1999, 81(Suppl. 1), S1–S27.

Dolan, L.C. Claims: the United States and European perspective. Agro FOOD Industry Hi Tech, 2011, 22, 5–7.

Duncan, M.J., and Hankey, J. The effect of a caffeinated energy drink on various psychological measures during submaximal cycling. Physiology & Behavior, 2013, 116–117, 60–65.

EC. Regulation (EC) No. 1924/2006 of the European Parliament and of the Council of 20 December 2006 on nutrition and health claims made on foods. Official Journal of the European Union, L 2006, 12, 3–18.

Elvin-Lewis, M. Should we be concerned about herbal remedies. Journal of Ethnopharmacology, 2001, 75(2), 141–164.

Ernst, E. Harmless herbs? A review of the recent literature. American Journal of Medicine, 1998, 104(2), 170–178.

Epler S., Chambers Iv E., Kemp, K.E. Hedonic scales are a better predictor than scales of optimal sweetness in lemonade. Journal of Sensory Studies, 1998, 13, 191–197.

Europe, I. Scientific concepts of functional foods in Europe consensus document. British Journal of Nutrition, 1999, 81(4), S1–S27.

Fallourd, M.J. and Viscione, L. Ingredient selection for stabilisation and texture optimisation of functional beverages and the inclusion of dietary fibre. In: Paquin, P. (Ed.), Functional and Speciality Beverage Technology, Woodhead Publishing Limited, 2009, 3–38.

Farbstein, D., Kozak-Blickstein, A., and Levy, A.P. Antioxidant vitamins and their use in preventing cardiovascular disease. Molecules, 2010, 15(11), 8098–8110.

Felzenszwalb, I., da Costa Marques, M.R., Mazzei, J.L., and Aiub, C.A. Toxicological evaluation of *Euterpe edulis*: a potential superfruit to be considered. Food and Chemical Toxicology, 2013, 58, 536–544.

Foegeding, E.A., Plundrich, N., Schneider, M., Campbell, C., and Lila, M.A. Protein-polyphenol particles for delivering structural and health functionality. Food Hydrocolloids, 2017, 72, 163–173.

Futo, l. Pectin food use, In Encyclopaedia of Food Science and Nutrition, 1993, 2nd ed., Vol. 7, Benjamin Cabillero, 4445, 4448, 4449.

Fukumoto, L.R., Delaquis, P., and Girard, B., Microfiltration and ultrafiltration ceramic membranes for apple juice clarification. Journal of Food Science, 1998, 63, 845–850.

Gaanappriya, M., Guhankuma, R.P., Kiruththica, V., Santhiya, N., and Anita, S. Probiotication of fruit juices by *Lactobacillus acidophilus*. International Journal of Advanced Biotechnology and Research, 2013, 4, 72–77.

Galambos, I., Mora, J., Járay, P., Vatai, G., and Bekássy-Molnár, E. High organic content industrial wastewater treatment by membrane filtration. Desalination, 2004, 162, 117–120.

Gerard, K.A. and Roberts, J.S. Microwave heating of apple mash to improve juice yield and quality. LWT—Food Science and Technology, 2004, 37 (5), 551–557.

Gil, M.I., Tomas-Barberan, F.A., Hess-Pierce, B., Holcroft, D.M., and Kader, A.A. Antioxidant activity of pomegranate juice and its relationship with phenolic composition and processing. Journal of Agricultural and Food Chemistry, 2000, 48, 4581–4589.

Girard, B. and Fukumoto, L.R. Membrane processing of fruit juices and beverages: a review. Critical Reviews in Food Science and Nutrition, 2000, 40, 91–157.

Glicksman, M. Functional properties of hydrocolloids. In: Food Hydrocolloids, Vol 1, Boca Raton, FL: CRC Press Inc., 1982, 59, 61.

Gomez-Lopez, V.M, Ragaert, P., Debevere, J., and Devlieghere, F. Pulsed light for food decontamination: a review. Trends in Food Science & Technology, 2001, 18, 464–473.

Granato, D., Branco, G.F., Nazzaro, F., Cruz, A.G., and Faria, J.A.F. Functional foods and nondairy probiotic food development: trends, concepts, and products. Comprehensive Reviews in Food Science and Food Safety, 2010, 9, 292–302.

Gryn-Rynko, A., Bazylak, G., and Olszewska-Slonina, D. New potential phytotherapeutics obtained from white mulberry (*Morus alba* L.) leaves. Biomedicine & Pharmacotherapy, 2016, 84, 628–636.

Gulati, K., Rai, N., Chaudhary, S., and Ray, A. Nutraceuticals in respiratory disorders. In: Gupta, R.C. (Ed.), Nutraceutical: Efficacy, Safety and Toxicity, Academic Press, 2016.

Gunja, N., and Brown, J. A. Energy drinks: health risks and toxicity. Medical Journal of Australia, 2012, 196, 46–49.

Hasler, C.M. Health claims in the United States: an aid to the public or a source of confusion? Journal of Nutrition, 2008, 138(6), 1216S–1220S.

Heckman, M.A., Sherry, K., and Gonzalez de M.E. Energy drinks: an assessment of their market size, consumer demographics, ingredient profile, functionality, and regulations in the United States. Comprehensive Reviews in Food Science and Food Safety, 2010, 9, 303–317.

Hendrickx, M.E.G., and Knorr, D.W. Ultra High Pressure Treatments of Foods. Kluwer Academic/Plenum Publishers, 2002.

Hmid, I., Elothmani, D., Hanine, H., Oukabli, A., and Mehinagic, E. Comparative study of phenolic compounds and their antioxidant attributes of eighteen pomegranate (*Punica granatum* L.) cultivars grown in Morocco. Arabian Journal of Chemistry, 2017, 10(2), 2675–2684.

Hoefler A.C. Hydrocolloids: Practical Guides for Food Industry, Eagan Press Handbook Series, Saint Paul, MN: Cereals & Grains Assn, 2004, 33–34.

Huang, H.W., Wu, S.J., Lu, J.K., Shyu, Y.T., and Wang, C.Y. Current status and future trends of high-pressure processing in food industry. Food Control, 2017, 72, 1–8.

Hruškar, M., Levaj, B., Kovačević, D.B., Kićanović, S. Sensory profile of plum nectars. Croatian Journal of Food Technology, Biotechnology and Nutrition; 2012, 7(Special Issue), 28–33.

Hwang, S.H. and Hong, J.H. Determining the most influential sensory attributes of nuttiness in soymilk: a trial with Korean consumers using model soymilk systems. Journal of Sensory Studies, 2015, 30(5), 425–437.

Jackson, C-J. C., Paliyath, G. Functional foods and nutraceuticals. In: Paliyath, G., Bakovic, M., and Shetty, K. (Eds.), Functional Foods, Nutraceuticals and Degenerative Disease Prevention. Oxford Unite Kingdom: Wiley-Blackwell, 2011, 11–43.

Jae, K.K., Shin, E.C., Lim, H.J., Choi, S.J., Kim, C.R., Suh, S.H., Kim, C.J., Park, G.G., Park, C.S., Kim, H.K., Choi, J.H., Song, S.W., Shin, D.H., Characterization of nutritional composition, antioxidative capacity, and sensory attributes of Seomae Mugwort, a native Korean variety of *Artemisia argyi* H. Lév. & Vaniot. Journal of Analytical Methods in Chemistry, 2015, 1–9.

Jambrak, A. R., Šimunek, M., Petrović, M., Bedić, H., Herceg, Z., and Juretić, H. Aromatic profile and sensory characterisation of ultrasound treated cranberry juice and nectar. Ultrasonics Sonochemistry, 2017, 38, 783–793.

Jaworska, G., Grega, T., Sady, M., Bernaś, E., Pogoń, K. Quality of apple-whey and apple beverages over 12-month storage period. Journal of Food and Nutrition Research, 2014, 53(2), 117–126.

Kalman, D. S., Feldman, S., Krieger, D. R., and Bloomer, R. J. Comparison of coconut water and a carbohydrate-electrolyte sport drink on measures of hydration and physical performance in exercise-trained men. Journal of the International Society of Sports Nutrition, 2012, 9(1), 1.

Kalman, D.S., Feldman, S., Krieger, D.R., and Bloomer, R.J. Comparison of coconut water and a carbohydrate-electrolyte sport drink on measures of hydration and physical performance in exercise-trained men. Journal of the International Society of Sports Nutrition, 2012, 9(1), 1.

Kausar, H., Saeed S., Ahmad M.M., and Salam A. Studies on the development and storage stability of cucumber-melon functional drink. Journal of Agricultural Research. 2012, 50, 239–248.

Jiang, Y and Nie, W.J. Chemical properties in fruits of mulberry species from the Xinjiang province of China. Food Chemistry, 2015, 174, 460–466.

Kawashima, L.M. and Valente Soares, L.M. Mineral profile of raw and cooked leafy vegetables consumed in Southern Brazil. Journal of Food Composition and Analysis 2003, 16(5), 605–611.

Khandpur, P. and Gogate, P.R. Understanding the effect of novel approaches based on ultrasound on sensory profile of orange juice. Ultrasonics Sonochemistry, 2015, 27, 87–95.

Kim, H.J., Kim, P., and Shin, C.Y. A comprehensive review of the therapeutic and pharmacological effects of ginseng and ginsenosides in central nervous system. Journal of Ginseng Research, 2013, 37(1), 8–29.

Kinsella, J.E. and Damodaran, S. Flavor problems in soy proteins: origin nature, control and binding phenomena. In: Charalambous, G. (Ed.), The analysis and control of less desirable flavors in foods and beverages, Academic Press: New York, 1980, 95–109.

Knorr, D. Effect of high hydrostatic pressure processes on food safety and quality. Food Technology, 1993, 47, 156–161.

Knorr, D. High pressure processing for preservation, modification and transformation of foods. High Pressure Research, 2002, 22(3–4), 595–599.

Knorr, D., Heinz, V., and Buckow, R. High pressure application for food biopolymers. Biochimica et Biophysica Acta, 2006, 1764(3), 619–31.

Kocadağli, T., Özdemir, K.S., and Gökmen, V. Effects of infusion conditions and decaffeination on free amino acid profiles of green and black tea. Food Research International, 2013, 53(2), 720–725.

Kratchanova, M., Pavlova, E., and Panchev, I. The effect of microwave heating of fresh orange peels on the fruit tissue and quality of extracted pectin. Carbohydrate Polymers, 2004, 56, 181–186.

Kumar, R.S. Development, quality evaluation and shelf life studies of probiotic beverages using whey and aloe vera Juice. Journal of Food Processing & Technology, 2015, 06(09).

Kumar, B.V., Vijayendra, S.V.N., and Reddy, O.V.S. Trends in dairy and non-dairy probiotic products–a review. Journal of Food Science and Technology, 2015, 52(10), 6112–6124.

Lal, G.G. Processing of beverages for the health food market consumer. In: Boye, J.I. (Ed.), Nutraceutical and Functional Food Processing Technology, Canada: John Wiley & Sons, Ltd., 2015,

Lau, T.C., Chan, M.W., Tan, H.P., Kwek, C.L. Functional food: a growing trend among the health conscious. Asian Social Science, 2013, 9, 198–208.

Lawless, L.J.R., Threlfall, R.T., Meullenet, J.F., Howard L.R. Applying a mixture design for consumer optimization of black cherry, concord grape and pomegranate juice blends. Journal of Sensory Studies, 2013, 28, 102–112.

Lawless, H.T., Sinopoli, D., Chapman, K.W., A comparison of the labeled affective magnitude scale and the 9-point hedonic scale and examination of categorical behavior. Journal of Sensory Studies, 2009, 25, 54–66.

Lawrence, N.D., Kentish, S. E., O'Connor, A.J., Barber, A.R., and Stevens, G.W. Microfiltration of skim milk using polymeric membranes for casein concentrate manufacture. Separation and Purification Technology, 2008, 60(3), 237–244.

Levaj, B., Vahčić N, Dragović-Uzelac V, Svetličić S, Sabljak V, Herceg K, Stanić D, Marinčić D, Elez I, Kovačević DB, Lončarić S. Influence of processing on yield and quality of cloudy plum juices. Croatian Journal of Food Technology, Biotechnology and Nutrition, 2012, 7(Special Issue), 34–38.

Lee, P.R., Boo, C.X., and Liu, S.Q. Fermentation of coconut water by probiotic strains *Lactobacillus acidophilus* L10 and *Lactobacillus casei* L26. Annals of Microbiology 2013, 63(4), 1441–1450.

Liepa, M., Zagorska, J., and Galoburda, R. High-pressure processing as novel technology in dairy industry: a review. Research for Rural Development, 2016, 1(1), 76–83.

Lomer, M.C.E., Parkes, G.C., and Sanderson, J.D. Review article: lactose intolerance in clinical practice–myths and realities. Alimentary Pharmacology & Therapeutics, 2008, 27, 93–103.

Lopez-Serrano, M. and Barceloä, A.R. Comparative study of the products of the peroxidase-catalyzed and the polyphenoloxidase-catalyzed (+)-catechin oxidation, their possible implications in strawberry (Fragaria x ananassa) browning reactions. Journal of Agricultural and Food Chemistry, 2002, 50, 1218–1224.

Mana Kialengila, D., Wolfs, K., Bugalama, J., Van Schepdael, A., and Adams, E. Full evaporation headspace gas chromatography for sensitive determination of high boiling point volatile organic compounds in low boiling matrices. Journal of Chromatography A, 2013, 1315, 167–175.

Markets and Markets. HPP (High Pressure Processing) market by equipment type (orientation, vessel size), application (meat, seafood, beverage, fruit & vegetable), product type (meat & poultry, seafood, juice, ready meal, fruit & vegetable) & geography—2013, Forecast to 2018. Report Code: FB 2151.

Matta V.M., Moretti R.H., Cabral. L.M.C. Microfiltration and reverse osmosis for clarification and concentration of acerola juice. Journal of Food Engineering, 2004, 61 (3), 477–482.

Mark-Herbert, C. Innovation of a new product category—functional foods. Technovation, 2004, 24, 713–719.

Massaro, M., Scoditti, E., Carluccio, M.A., and De Caterina, R. Nutraceuticals and prevention of atherosclerosis: focus on ω-3 polyunsaturated fatty acids and Mediterranean diet polyphenols. Cardiovascular Therapeutics, 2010, 28, 13–19.

McKay, D.L. and Blumberg, J.B. The role of tea in human health: an update. Journal of the American College of Nutrition, 2002, 21, 1–13.

Mirsaeedghazi, H., Emam-djomeh, Z., and Mohammad, S. Clarification of pomegranate juice by microfiltration with PVDF membranes, Desalination, 2010, 264, 243–248.

Mitchell, J.B., Grandjean, P.W., Pizza, F.X., Starling, R.D., and Holtz, R.W. The effect of volume ingested on rehydration and gastric emptying following exercise-induced dehydration. Medicine & Science in Sports & Exercise, 1994, 26, 1135–1143.

Morell, P. and Fiszman S. Revisiting the role of protein-induced satiation and satiety. Food Hydrocolloids, 2017, 68, 199–210.

Naghma, K. and Hasan, M. Tea polyphenols for health promotion. Life Sciences, 2007, 81, 519–533.

Naik, L., Sharma, R., Rajput, Y.S., and Manju, G. Application of high pressure processing technology for dairy food preservation—future perspective: a review. Journal of Animal Production Advances, 2013, 3(8), 232–241.

Nanasombat, S., Thonglong, J., and Jitlakha, J. Formulation and characterization of novel functional beverages with antioxidant and anti-acetylcholinesterase activities. Functional Foods in Health and Disease, 2015, 5(1), 1–16.

Nicolas, L., Marquilly C., O'Mahony M. The 9-point hedonic scale: are words and numbers compatible? Food Quality and Preference, 2009, 21, 1008–1015.

Niva, M. Consumers, functional foods and everyday knowledge. Conference of Nutritionists Meet Food Scientists and Technologists, 2000.

Oey, I., Van der Plancken, I., Loey, A., and Hendrickx, M. Does high pressure processing influence nutritional aspects of plant based food systems? Trends in Food Science & Technology, 2008, 19, 300–308.

Ogundele, O.M., Awolu, O.O., Badejo, A.A., Nwachukwu, I.D., and Fagbemi, T.N. Development of functional beverages from blends of *Hibiscus sabdariffa* extract and selected fruit juices for optimal antioxidant properties. Food Science & Nutrition, 2016, 4(5), 679–685.

Ohama, H., Ikeda, H., and Moriyama, H. Health foods and foods with health claims in Japan. Toxicology, 2006, 22, 95–111.

Oliveira, M., Ramos, S., Delerue-Matos, C., and Morais, S. Espresso beverages of pure origin coffee: mineral characterization, contribution for mineral intake and geographical discrimination. Food Chemistry, 2015, 177, 330–338.

Omoni, A.O. and Aluko, R.E. Soybean foods and their benefits: potential mechanisms of action. Nutrition Reviews, 2005, 63, 272–283.

Ordóñez-Santos, L.E., Martinez-Girón, and Arias-Jaramillo, M.E. Effect of ultrasound treatment on visual colour, vitamin C, total phenols and carotenoids content in Cape gooseberry juice. Food Chemistry, 2017, 233, 96–100.

Ottaway, P.B. Fortification of beverages with vitamins and minerals. In: Paquin, P. (Ed.), Functional and Speciality Beverage Technology, Cambridge, United Kingdom: Woodhead Publishing, 2009, 71–91.

Ozen, A.E., Pons, A., and Tur, J.A. Worldwide consumption of functional foods: a systematic review. Nutrition Reviews, 2012, 70, 472–481.

Ozer, B.H., and Kirmaci, H.A. Functional milks and dairy beverages. International Journal of Dairy Technology, 2010, 63, 1–15.

Oziemblowski, M. and Kopec W. Pulsed electric fields (PEF) as an unconventional method of food preservation. Polish Journal of Food and Nutrition Sciences, 2005, 14(55), 31–35.

Pagliarini, E, Tomaselli N, Brenna O.V. Study on sensory and composition changes in Italian Amarone valpolicella red wine during aging. Journal of Sensory Studies, 2004, 19, 422–432.

Paniwynk, L. Applications of ultrasound in processing of liquid foods: a review. Ultrasonics Sonochemistry, 2017, 38, 794–806.

Pereira, C.G., Barreira, L., da Rosa Neng, N., Nogueira, J.M.F., Marques, C., Santos, T.F., Varela, J., Custódio, L. Searching for new sources of innovative products for the food industry within halophyte aromatic plants: In vitro antioxidant activity and phenolic and mineral contents of infusions and decoctions of *Crithmum maritimum* L. Food and Chemical Toxicology, 2017, 107, 581–589.

Pinto, E., Almeida, A.A., Aguiar, A.A.R.M., and Ferreira, I.M.P.L.V.O. Changes in macro-minerals, trace elements and pigments content during lettuce (*Lactuca sativa* L.) growth: influence of soil composition. Food Chemistry, 2014, 152, 603–611.

Prado, F.C., Parada, J.L., Pandey, A., and Soccol, C.R. Trends in non-dairy probiotic beverages. Food Research International, 2008, 41, 111–123.

Pravst, I. Functional foods in Europe: a focus on health claims. In: Valdez, B. (Ed.), Scientific, Health and Social Aspects of the Food Industry. Rijeka, Croatia: In Tech, 2012, 165–208.

Puri, G., Sadana, B., and Singla, N. Development and sensory evaluation of beverages having high antioxidant activity. International Journal of Current Microbiology and Applied Sciences, 2017, 6(11), 2253–2259.

Ramachandra, C.T. and Rao, P.S. Shelf-life and colour change kinetics of aloe vera gel powder under accelerated storage in three different packaging materials. Journal of Food Science and Technology, 2013, 50(4), 747–754.

Ravikumar, C. Review on herbal teas. Journal of Pharmaceutical Sciences and Research, 2014, 6(5), 236–238.

Rifnaz, M.B.M., Jayasinghe-Mudalige, U.K., Guruge, T.P.S.R., Udugama, J.M.M., Herath, H.M.L.K., and Edirisinghe, J.C. Perceived health status of consumers and incorporation of functional ingredients into their diet. Procedia Food Science, 2016, 6, 56–59.

Ripps, H., and Shen, W. Review—taurine: a "very essential amino" acid. Molecular Vision, 2012, 18, 2673–2686.

Rodriguez, E.B., Flavier, M. E., Rodriguez-Amaya, D. B., Amaya-Farfań, J. Phytochemicals and functional foods. Current situation and prospect for developing countries. Seguranca Alimentar e Nutricional, 2006, 13, 1–22.

Rodriguez-Fragoso, L., Reyes-Esparza, J., Burchiel, S., Herrera-Ruiz, D., and Torres, E. Risks and benefits of commonly used herbal medicines in Mexico. Toxicology and Applied Pharmacology, 2008, 227(1), 125–135.

Rodriguez-Roque, M.J., Rojas-Grau, M.A., Elez-Martinez, P., and Martin-Belloso, O. Soymilk phenolic compounds, isoflavones and antioxidant activity as affected by in vitro gastrointestinal digestion. Food Chemistry, 2013, 136, 206–225.

Rossi, M., Giussani, E., Morelli, R., Lo Scalzo, R., Nani, R. C., and Torregiani, D. Effect of fruit blanching on phenolics and radical scavenging activity of high bush blueberry juice. Food Research International, 2003, 36, 999−1005.

Rupasinghe, H.P.V and Yu, L.J. Emerging preservation methods for fruit juices and beverages, Food Additive, Yehia El-Samragy (Ed.), InTech, 205.

Rysstad, G., and Johnstone, K. Extended shelf-life beverages. In: Paquin, P. (Ed.), Functional and Speciality Beverage Technology, 2009, 107–132.

Saarela, M. Probiotics as ingredients in functional beverages. In: Paquin, P. (Ed.), Functional and Speciality Beverage Technology, 2009, 55–70.

Sala-Vila, A., Estruch, R., and Ros, E. New insights into the role of nutrition in CVD prevention. Current Cardiology Reports, 2015, 17(5), 26.

Saucedo-Pompa, S., Martínez-Ávila, G.C.G., Rojas-Molina, R. and Sánchez-Alejo, E.J. Natural beverages and sensory quality based on phenolic contents. In : Emad Shalaby (Ed.), Antioxidants in Foods and Its Applications. Rijeka, Croatia: IntechOpen, 2018.

Salvia-Trujillo, L., Morales-de la Peña, M., Rojas-Graü, A., Welti-Chanes, J., and Martín-Belloso, O. Mineral and fatty acid profile of high intensity pulsed electric fields or thermally treated fruit juice-milk beverages stored under refrigeration. Food Control, 2017, 80, 236–243.

Sampedro, L., Moreira, X., Martíns, P., and Zas, R. Growth and nutritional response of Pinus pinaster after a large pine weevil (Hylobiusabietis) attack. Trees, 2009, 23(6), 1189–1197.

Sanchis-Gomar, F., Pareja-Galeano, H., Cervellin, G., Lippi, G., and Earnest, C.P. Energy drink overconsumption in adolescents: implications for arrhythmias and other cardiovascular events. Canadian Journal of Cardiology, 2015, 31(5), 572–575.

Servili, M., Rizzello, C. G., Taticchi, A., Esposto, S., Urbani, S., Mazzacane, F., Di Maio, I., Selvaggini, R., Gobbeti, M., Di Cagno, R. Functional milk beverage fortified with phenolic compounds extracted from olive vegetation water and fermented with functional lactic acid bacteria. International Journal of Food Microbiology, 2011, 147(1), 45–52.

Sfakianakis, P. and Tzia, C. Conventional and innovative processing of milk for yogurt manufacture; Development of texture and flavor: a review. Foods, 2014, 3(1), 176–193.

Sharma, S., Vaidya, D., Rana, N. Honey as natural sweetener in lemon ready-to-serve drink. International Journal of Bio-resource and Stress Management, 2016, 7(2), 320–325.

Siguemoto, É. S., Gut, J. A. W., Martinez, A., and Rodrigo, D. Inactivation kinetics of *Escherichia coli* O157:H7 and *Listeria monocytogenes* in apple juice by microwave and conventional thermal processing. Innovative Food Science & Emerging Technologies, 2018, 45, 84–91.

Srivastava, J. K., Shankar, E., and Gupta, S. Chamomile: a herbal medicine of the past with bright future. Molecular Medicine Reports, 2010, 3(6), 895–901. (accessed March 20, 2019)

Stratakos, A. C., Delgado-Pando, G., Linton, M., Patterson, M.F., and Koidis, A. Industrial scale microwave processing of tomato juice using a novel continuous microwave oven. Food Chemistry, 2016, 190, 622–628.

Stroup, B.M., Murali, S.G., Nair, N., Sawin, E.A., Rohr, F., Levy, H.L., and Ney, D.M. Dietary amino acid intakes associated with a low-phenylalanine diet combined with amino acid medical foods and glycomacropeptide medical foods and neuropsychological outcomes in subjects with phenylketonuria. Data in Brief, 2017, 13, 377–384.

Sun-Waterhouse, D. The development of fruit-based functional foods targeting the health and wellness market: a review. International Journal of Food Science & Technology, 2011, 46, 899–920.

Sushilkumar, SM, Sawate, A.R., Patil, B.M., Kshirsagar, R.B., Kulkarni S.P. Study on effect of custard apple leaf extract on physico-chemical properties of aonla juice. Asian Journal of Dairy and Food Research, 2016, 35(1), 81–84

Suzuki, Y., Miyoshi, N., and Isemura, M. Health-promoting effects of green tea. Proceedings of the Japan Academy. Series B, Physical and Biological Sciences, 2012, 88(3), 88–101.

Švarc-Gajić, J., Cvetanović, A., Segura-Carretero, A., Mašković, P., and Jakšić, A. Functional coffee substitute prepared from ginger by subcritical water. Journal of Supercritical Fluids 2017, 128, 32–38.

Sworn, G. Introduction to food hydrocolloids. In: Williams P.A. (Ed.), Handbook of Industrial Water Soluble Polymers, Cambridge: Blackwell Publishing, 2007, 18, 24.

Szilagyi, A. Adaptation to lactose in lactase non persistent people: effects on intolerance and the relationship between dairy food consumption and evaluation of diseases. Nutrients, 2015, 7(8), 6751–6779.

Tripathi, A.K., and Misra, A.K. Soybean—a consummate functional food: a review. Journal of Food Science and Technology, 2005, 2, 42–46.

Turk, M., Perino, S., Cendres, A., Petitcolas, E., Soubrat, T., and Chemat, F. Alternative process for strawberry juice processing: microwave hydrodiffusion and gravity. LWT—Food Science and Technology, 2017, 84, 626–633.

Udabage, P., Augustin, M., Versteeg, C., Puvanenthiran, A., Yoo, J., Allen, N., and Kelly, A. Properties of low-fat stirred yoghurts made from high-pressure-processed skim milk. Innovative Food Science & Emerging Technologies, 2010, 11(1), 32–38.

Vaillant, F., Cisse, M., Chaverri, M., Perez, A., Dornier, M., Viquez, F., and Dhuique-Mayer, C. Clarification and concentration of melon juice using membrane processes. Innovative Food Science and Emerging Technologies, 2005, 6(2), 213–220.

Vieira da Silva, B., Barreira, J.C.M., and Oliveira, M.B.P.P. Natural phytochemicals and probiotics as bioactive ingredients for functional foods: Extraction, biochemistry and protected-delivery technologies. Trends in Food Science & Technology, 2016, 50, 144–158.

Venkatesh, M. and Raghavan, G.S.V. An overview of microwave processing and dielectric properties of agri-food materials. Biosystems Engineering, 2004, 88(1), 1–18.

Verma, A. and Singh, A. Optimization of ingredients for a herbal beverage with medicinal attributes. Asian Journal of Dairy & Food Research, 2013, 32(4), 318–322.

Wang, L. and Weller, C.L. Recent advances in extraction of nutraceuticals from plants. Trends in Food Science & Technology, 2006, 17, 300–335.

Wang, B., Kammer, L.M., Ding, Z., Lassiter, D.G., Hwang, J., Nelson, J.L., and Ivy, J.L. Amino acid mixture acutely improves the glucose tolerance of healthy overweight adults. Nutrition Research, 2012, 32(1), 30–38.

Warriner, K., Movahedi, S., and Waites, W.M. Laser-based packaging sterilisation in aseptic packaging. In: Richardson, P. (Ed.), Improving the Thermal Processing of Foods, Cambridge: Woodhead Publishing, 2004.

Wolf, P. H., Siverns, S., and Monti, S. UF membranes for RO desalination pretreatment. Desalination, 2005, 182(1e2), 293–300.

Xing Chen, L., Xing, Y., Cao, L., Xu, Q., Li, S., Wang, R., Jiang, Z., Che, Z., Lin, H. Effects of six commercial *Saccharomyces cerevisiae* strains on phenolic attributes, antioxidant activity, and aroma of kiwifruit (*Actinidia deliciosa* cv.) wine., BioMed Research International, 2017, 1–10. Article ID 2934743.

Yang, Y. Scientific substantiation of functional food health claims in China. Journal of Nutrition, 2008, 138, 1199S–205S.

Yuan, L., Xiong, L., Deng, T., Wu, Y., Li, J., Liu, S., Huang, J., Liu, Z. Comparative profiling of gene expression in *Camellia sinensis* L. cultivar AnJiBaiCha leaves during periodic albinism. Gene, 2015, 561(1), 23–29.

Yuan, T., Li, L., Zhang, Y., and Seeram, N.P. Pasteurized and sterilized maple sap as functional beverages: Chemical composition and antioxidant activities. Journal of Functional Foods, 2013, 5(4), 1582–1590.

Zhao, Y., Flugstad, B., Kolbe, E., Park, J.W., and Wells, J.H. Using capacitive (radio frequency) dielectric heating in food processing and preservation—a review. Journal of Food Process Engineering, 2000, 23(1), 25–55.

INTERNET SOURCES

Anonymous (2012). How the food industry defines botanicals (June 05, 2012). https://www.foodprocessing.com/articles/2012/defining-botanicals/. (accessed March 20, 2019)

Beer, E. (2017). Storming growth for energy, water and tea in functional beverages. https://www.nutraingredients.com/Article/2017/03/15/Storming-growth-for-energy-water-and-tea-in-functional-beverages. (accessed March 13, 2019)

Cowland, D. (2013). Health and wellness beverages outperform wider drinks industry. http://blog.euromonitor.com/2013/02/health-and-wellness-beverages-outperform-wider-drinks-industry.html. (accessed March 13, 2019)

Diabetes.co.uk. (2017). Sports drinks. http://www.diabetes.co.uk/sports-drinks.html. (accessed March 20, 2019)

Enos, D. (2011). "Superfruits" may bring some health benefits (November 10, 2011). Retrieved from https://www.livescience.com/35968-superfruit-pomegranate-acai-gogi-health-benefits.html. (accessed March 20, 2019)

Euromonitor (2015). Global state of the industry: a review of the top tea trends and markets around the world. Retrieved from https://www.slideshare.net/Euromonitor/euromonitor-world-tea-expo-final. (accessed March 20, 2019)

Euromonitor (2017a). RTD tea in Malaysia. http://www.euromonitor.com/rtd-tea-in-malaysia/report. (accessed March 20, 2019)

Euromonitor (2017b). RTD tea in Australia. http://www.euromonitor.com/rtd-tea-in-australia/report. (accessed March 20, 2019)

FAO/WHO (2006). Probiotics in food: health and nutritional properties and guidelines for evaluation. Retrieved from http://www.fao.org/3/a-a0512e.pdf. (accessed March 13, 2019)

Food Insight (2009). Functional foods fact sheet: soy. (October 14, 2009). http://www.foodinsight.org/Functional_Foods_Fact_Sheet_Soy. (accessed March 13, 2019)

Fortitech (2011). New innovation in functional beverages. Retrieved from https://www.fortitech-premixes.com/wp.../Functional_Beverages_FINAL_ENG.pdf. (accessed March 14, 2019)

Global Industry Analysts, Inc. (2016). Innovative flavors and organic & functional juices to sustain growth in the global fruit and vegetable juices market. http://www.strategyr.com/MarketResearch/Fruit_and_Vegetable_Juices_Market_Trends.asp. (accessed March 14, 2019)

Kubitz, L. (April 07, 2016). Super berry: is acai a "Superfood"? http://www.foodinsight.org/acai-superfood-antioxidants-flavonoids-blueberries. (accessed March 20, 2019)

Statisca (2017). Sales volume of soy milk worldwide in 2015 and 2018 (in billion liters). https://www.statista.com/statistics/645662/soy-milk-sales-volume-worldwide/. (accessed March 20, 2019).

PART II
Nutraceuticals

CHAPTER 6

Herbal Antiobesity Products and Their Function in the Gut-Brain Axis

NURIA-ELIZABETH ROCHA-GUZMÁN[1*], PAOLA FLORES-RODRÍGUEZ[2],
GUADALUPE MONTIEL-RAMÍREZ[1], MARTHA-ROCÍO MORENO-JIMÉNEZ[1],
JOSÉ ALBERTO GALLEGOS-INFANTE[1],
RUBÉN-FRANCISCO GONZÁLEZ-LAREDO[1], and
CARLOS-ALONSO SALAS-RAMÍREZ[1]

[1]*Department of Chemical and Biochemical Engineering of the National Technology of Mexico/Technological Institute of Durango (TecNM/IT Durango)*

[2]*Centro Interdisciplinario de Investigación para el Desarrollo Integral Regional (CIIDIR-IPN Unidad Durango). Sigma 119, Fracc. 20 de Noviembre II, 34234, Durango, Dgo. México*

*Corresponding author. E-mail address: nrocha@itdurango.edu.mx.

ABSTRACT

Obesity is associated with different health complications and a reduced life expectancy. Different nutraceuticals are offered in the market, but present serious concerns are regarding the safety of their use, due to the side effects they may induce in the body. Recently, sibutramine and rimonabant have been withdrawn from the European market, because they are associated with anorexia problems. This has given rise to several investigations, being focused on the study of the gastroenteropancreatic system. Reciprocally, a series of endocrine peptides and hormones that send signals to the brain have been identified in the intestine. These biomolecules coordinate various functions involved in the regulation of food intake and energy balance. Several research groups have explored the effect of continued consumption of different extracts in induced obesity models without establishing their impact on the neuroendocrine regulation of appetite and homeostatic energy. Currently, the effect of the consumption of infusions and nutraceuticals from herbal products as a therapy for weight control is being studied. Therefore in this chapter, it is proposed to document the effect of these natural products on the gut–brain axis to relate weight loss to the influence of hormones and peptides in the metabolic activity of neurons.

6.1 INTRODUCTION

Unquestionably, obesity is a pathology that has been increasing in recent years and that by itself brings complications in the biopsychosocial field. Its increase is such that according to data reported by the world obesity federation, it is possible to visualize most countries with a prevalence of obesity greater than 25.0%. Mexico, for its part, has been listed as the country with the highest prevalence worldwide, according to the results of the National Health and Nutrition Survey, which points to a prevalence of 71.2% in the population over the next 20 years.

The pathology indicates a significant alteration on the body, highlighting a strong involvement of the gut–brain axis, defined this as the bidirectional neural processing of information between the central and digestive nervous systems. This chapter integrates the implicit participation—for operational purposes—of study of the neural axis, the neuroendocrine axis, and the inflammatory axis, which maintain direct activities in the regulation of hunger and satiety, as well as the influence on the consumption of various xenobiotics.

Reviewing the current treatment lines of obesity, it is possible to find on the market six drugs approved by the Food and Drug Administration (FDA), the same that by data reported in the literature have serious side effects, as well as the disadvantage of generating adverse effects in the short term.

Thus the importance of offering alternative treatments for the control and treatment of obesity justifies the study of different plants reported empirically in the control of said pathology. These plants have been consumed either as food, infusion, or extracts, and at the present time the use of various emerging technologies has guided the creation of nutraceutical products derived from the bioactive compounds of plants and/or foods. For instance, the compounds with more relevance are polyphenols, which have been proven with healthy effects by their chemical structure on different pathologies but compromising their preventive therapeutic value.

6.2 OBESITY

6.2.1 DEFINITION

In the current decade and throughout history, organizations that contemplate the health sciences worldwide continue in the search to combat diseases considered pandemics, such as malaria, malnutrition, and today a multifactorial disease like obesity. There are several definitions adopted by the different organizations, thus according to the World Health Organization, obesity is defined as abnormal or excessive fat accumulation that presents a health risk. Meanwhile, the Obesity Medicine Association defines it as a chronic, relapsing, multifactorial, and neurobehavioral disease wherein an increase in body fat promotes adipose tissue dysfunction and physical forces for abnormal fat mass, resulting in adverse metabolic, biomechanical, and psychosocial health consequences. Thus unquestionably, the common characteristic of both definitions is the increase in fat mass stored in adipose tissue, a condition related to different comorbidities that gives rise to various public health problems.

It is recognized, therefore, an obesogenic environment in which it is difficult to avoid excessive weight gain, since physiologically the body maintains a tendency to minimize effort and store excess food as fat mass, especially triglycerides (World Obesity, 2020).

6.2.2 ETIOLOGY DISEASE

The etiology of the disease focuses on the excess of calories that are consumed through foods that are rich in fats, starches, and simple sugars contents with the consequent gain of fat mass by the body; a condition that usually exceeds the energy expenditure during physiological metabolism. Thus by combining risk factors such as inadequate eating habits with limited physical activity results in the development of obesity.

Among the multiple hypotheses that try to explain the epidemic, we can mention the influence of biological, psychological, and socioeconomic factors.

6.2.2.1 BIOLOGICAL FACTORS

There are isolated cases in which the influence factor that corresponds to the environment is not directly responsible for the development of obesity but is caused by genetic abnormalities, endocrine alterations, and metabolic diseases.

6.2.2.1.1 Genetic Abnormalities

Several studies have suggested that the genetic load contributes from 50% to 70% of reported cases of obesity, since there are thousands of different genes associated with the pathology (Lamiquiz-Moneo et al., 2019). Likewise, it has been documented that some genetic mutations or alterations directly cause obesity. Thus mutations in a single gene can have an important effect on the accumulation of total body fat. In addition, reference has been made to nonsyndromic monogenic obesity, which determines extreme forms of obesity and opens the door for understanding the mechanisms involved in the eating behavior. Table 6.1 shows some of the main genes, which when presenting a mutation are directly related to the cause of severe obesity.

TABLE 6.1 Main Genes Related to the Development of Monogenic Obesity

No.	Gene	Name
1	CRHR1	Corticotropin-releasing hormone receptor 1
2	CRHR2	Corticotropin-releasing hormone receptor 2
3	LEP	Leptin
4	LEPR	Leptin receptor
5	GPR24	Melanin-concentrating hormone
6	PCSK1	Prohormone convertase 1
7	POMC	Proopiomelanocortin
8	MC3R	Melanocortin receptor 3
9	MC4R	Melanocortin receptor 4
10	NTRK2	TrkB brain neurotrophic factor receptor
11	SIM1	Single-minded homologue 1

Source: Adapted from González Jiménez, 2011.

6.2.2.1.2 Metabolic Diseases

There are several metabolic conditions in which the development of obesity may exist as one of its consequences. Among the most common metabolic disorders are Cushing Syndrome, hypogonadism, insulinoma, hypothyroidism, and hypothalamic lesions (Table 6.2).

TABLE 6.2 Metabolic Disorders Related to the Development of Obesity

Disease	Definition	Causes	Reference
Cushing syndrome	Autonomous cortisol secretion. Disorder with physical and mental changes	(↑) Blood cortisol for a prolonged period	Fassnacht et al. (2016)
Hypogonadism	A syndrome that results from failure of the testis to produce physiological levels of testosterone and an abnormal number of spermatozoa due to disruption of one or more levels of the hypothalamic-pituitary-testicular axis	(↓) Hormone secretion or other physiological activity of the gonads	Sigalos et al. (2018)
Insulinoma	Functional neoplasm of the pancreas characterized by a proliferation of beta cells in pancreatic islets	(↑) Insulin causing hypoglycemia	Diéguez-Felechosa et al. (2009)
Hypothyroidism	Thyroid-stimulating hormone concentration and free thyroxine concentrations below the reference range	(↓) Activity of the thyroid gland	Chaker et al. (2017)
Hypothalamic lesions	Inability to transduce afferent hormonal signals of adiposity, mimicking a starvation state of the central nervous system	(↓) Efferent sympathetic activity (↓) Waste of energy (↑) Vagal activity (↑) Insulin secretion and adipogenesis	Lustig (2011)

6.2.2.2 PSYCHOLOGICAL FACTORS

Many are the psychological factors linked to obesity; however, for an appropriate approach, those with more relevance for different authors are presented below. First, the personality of an individual is considered a predisposing factor to the development of obesity, since it has been reported that highly impulsive individuals show little commitment and adherence to balanced diets and regular physical activity programs (Sutin et al., 2011).

Equally important, a trigger for obesity is the stress that is present in today's society and the demand for an increasingly competitive world. That is to say, it is known that a situation of acute stress tends to inhibit appetite, but when stress is chronic it triggers alterations in physiological processes, highlighting those in which the adrenal glands release cortisol that increases appetite. Another manifestation associated with chronic stress is insomnia with the consequent lower physical activity, which predisposes to weight gain (Bennett et al., 2013; Sinha and Jastreboff, 2013). There are also emotional factors that tend to cause the individual to eat food as a way to suppress or attenuate negative emotions, such as fear, boredom, anger, and loneliness (Schneider et al., 2010; Bennett et al., 2013).

Depression is another emotional aspect, which is linked to mental health that can significantly predispose to obesity. Depression has a psychopathological origin, which must be treated in a multidisciplinary way, contributing with it in a positive response to the prevention of obesity; according to statistics, it is indicated that the previous diagnosis of depression is associated with a greater probability of being overweight or obese in approximately 7% (Dave et al., 2011).

6.2.2.3 SOCIOECONOMIC FACTORS

Worldwide, there is evidence of an association between low-socioeconomic status and obesity in adults and children. Thus it has been cited that parental obesity is a factor that influences the increase in childhood obesity and its comorbidities, since there is a correlation in certain dietary patterns that can trigger obesity in the younger members of families. In this sense, the involvement of parental obesity in the severity of childhood or juvenile obesity and its associated comorbidities continues to be controversial, and its possible influence on the success of the treatment is insufficiently characterized (Martínez-Villanueva et al., 2019). Consequently, it is necessary to point out that this is not supported by population epidemiological studies; however, the phenomenon has been observed in some specialized medical centers (Tornero, 2019).

Analyzed from another perspective, some research works suggest that there are social inequalities that influence the rates of obesity shown in different countries. Thus, as in general, the highest rates of obesity are observed in those countries that have a low-socioeconomic position or are suffering from immigration phenomena. A country where we can focus on trying to explain the causality of obesity is the United States, where socio-economic inequalities seem not to be correlated with the increase in obesity rates. In this sense, this phenomenon has begun to be related with the access to education (OECD, 2017). Thus different indices have been proposed to measure inequality among populations with lower and higher education levels, which would help to determine their correlation with obesity rates in different countries (Figure 6.1).

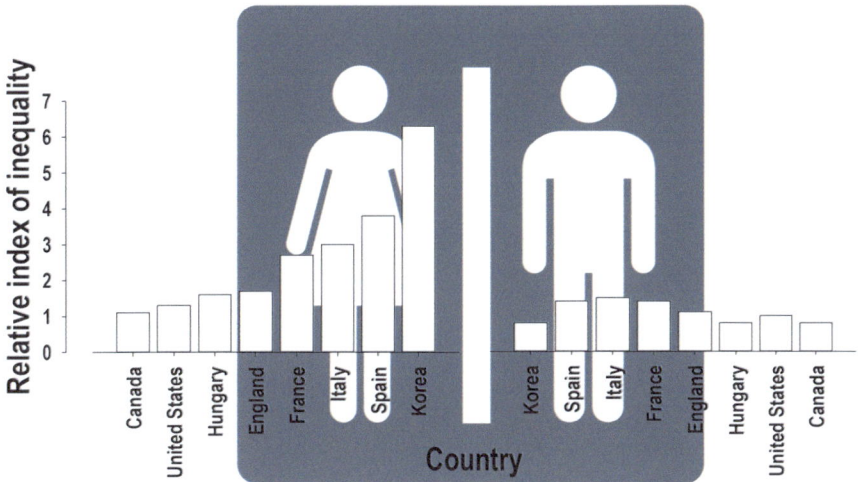

FIGURE 6.1 Education-related inequality in overweight defined as body mass index ≥ 25 kg m^{-2}. On the Y-axis, the relative index of inequality measures by education level. Adapted from OECD (2017). https://www.oecd.org/els/health-systems/Obesity-Update-2017.pdf

It is, therefore, necessary to note that it is not possible to describe obesity as a pathology developed through poor eating habits as well as the adoption of a sedentary life. Likewise, the etiology of obesity can be seen as a multi-factorial problem, in this sense the search for its treatment must be equally multidisciplinary.

6.2.3 EPIDEMIOLOGY

Obesity has been considered as a major health problem worldwide, not because of the chronic degenerative diseases to which it is directly related but because of its high incidence now that is occurring. During the latest update report on obesity presented by the OECD, it has shown alarming rates of overweight and obesity that have grown rapidly in England, Mexico, and the United States since the 1990s, whereas the corresponding increase has been slower in the other seven OECD countries (Figure 6.2).

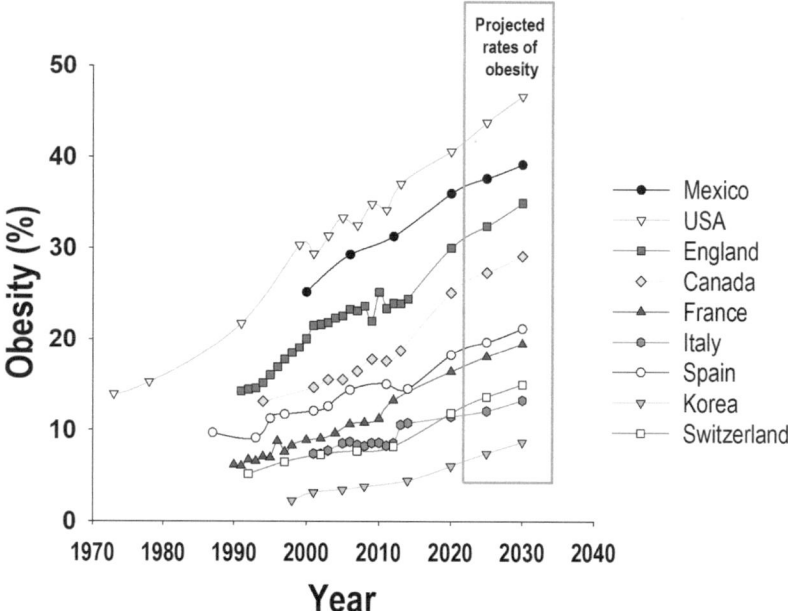

FIGURE 6.2 Projected rates of obesity defined as body mass index ≥ 30 kg m^{-2}. Adapted from OECD (2017). https://www.oecd.org/els/health-systems/Obesity-Update-2017.pdf

Among the statistics provided by the study carried out by the National Institute of Public Health and the Ministry of Health of Mexico is from the National Survey of Health and Nutrition in which important data on the prevalence of obesity indicated that in Mexico is well-established by age groups. Accordingly, information at the national level, the combined prevalence of overweight and obesity in the school-age population in 2016 was 33.2% and prevalence of obesity alone was 15.3%. The distribution by place of residence showed a higher combined prevalence of overweight and obesity in urban locations (34.9%) compared to rural locations (29.0%). The combined prevalence of overweight and obesity in the adolescent population was 36.3% and in adults aged 20 years or older it was 71.2% (ENSANUT-halfway, 2016).

6.2.4 PATHOPHYSIOLOGY

The fat buildups in the adipose tissue in subcutaneous and visceral deposits but maintains the particularity of expanding to other deposits throughout the body; the commonplaces of accumulation are kidneys, blood vessels, and heart. The function of adipokine secretion by the adipose organ can affect the physiology of the different organs and the systemic metabolism, or in other words, the implication of obesity in the different pathologies is developed through this process. The composition of adipose tissue consists of preadipocytes, adipocytes, endothelial cells, pericytes, fibroblasts, and immune cells (Suganami et al., 2012).

The adipocyte is defined as the main cell of adipose tissue. One of its primary functions is the storage of energy from the diet mainly in the form of triglycerides, to be used as an energy source when there is a lack of it or in some process of pathophysiological stress in which the demands are high; therefore, it can sense energy demands. There are two processes in which the adipocyte develops, being these by hypertrophy and hyperplasia. The first is defined as an increase in size and the second one described as the increase from a cell that goes through a series of basic steps to differentiate at its last preadipocyte–adipocyte stage.

Previously, there was no knowledge on the role of the adipocyte as an endocrine cell, and today its participation in either energy, physiological, and metabolic balances is well recognized (Blüher and Mantzoros, 2015), in this sense at least 600 bioactive factors are considered adipokines (Dahlman et al., 2012).

When the organism is exposed to a high fat intake, the preadipocyte proliferates viscerally without verifying if there has been a sign of hypertrophy in the adipocytes. On the contrary, it is observed that the adipocyte undergoing a process of hypertrophy presents dysfunctions in its activity as hypoxia, affecting insulin sensitivity, increased intracellular stress, autophagy, and apoptosis, as well as tissue inflammation (Klöting and Blüher, 2014).

It is through obesity when dysfunction occurs in adipose tissue. It results in an imbalance that triggers a characteristic lipoinflammation scenario with high serum leptin and low adiponectin levels (Blüher and Mantzoros, 2015; Wang and Scherer, 2016). In this condition, a state of phenotypic change is presented, which is defined as a transformation in the polarization state of macrophages. Particularly, the change is observed with the transformation of the secretory macrophages of anti-inflammatory adipocytokines (M2) to those secretory macrophages of proinflammatory adipocytokines (M1) with the consequent inflammatory response. In summary, inflammation is finally a self-defense response, whose purpose is to isolate and recover the associated lesion.

Thus in obese organisms, adipose tissue shows a massive infiltration of M1 macrophages in response to increased secretion of the macrophage chemoattractant protein (MCP-1), who plays a crucial role in the inflammatory response in obesity. A key intermediary is a nuclear factor that enhances the kappa light chains of activated B cells (NF-kB), one of the major inducers of the expression of this adipokine (Prieur et al., 2011). The activation of these inflammatory processes induces a state of hypoxia in the adipose tissue and a consequent cell death of peripheral adipocytes. Besides, there is an infiltration of immune cells and the transformation of macrophages M2 to M1, with the deregulation of homeostasis associated with the increase in the secretion of proinflammatory adipocytokines mainly interleukin-6 (IL-6), C-reactive protein (PCR), tumor necrosis factor alpha (TNF-alpha). Among additional responses to this deregulation of homeostasis is the decrease in anti-inflammatory adipocytokines such as adiponectin that suppresses the synthesis of several proinflammatory cytokines, that is, TNF-alpha and interferon gamma (IFN-γ) (Izaola et al., 2015). Being through these biochemical processes, it is how we can explain the metabolic abnormalities associated with obesity.

Within the characteristics that obesity presents, it is relevant to mention that a characteristic sign is the proportion in the gain of body fat mass, reflected through the total increase in weight. In this sense, those other signs and symptoms that do not describe obesity itself are presented as comorbidities associated with this pathology.

Due to the comorbidities associated with obesity, different studies show that the life expectancy of those individuals who suffer from it is reduced by at least 8 years. Likewise, obesity has been associated with 236 medical conditions, highlighting its severity and a high degree of mortality, different types of cancer and physiological disorders (Table 6.3).

It is relevant to indicate that the design of prospective studies is complicated due to the difficulty in interpreting the pathophysiology. Despite this, there are concrete data in which a correlation between obesity and the development of pathologies is determined through the pathophysiological effects that this condition implies. That is why obesity must have efficient long-term control with the proposal of nonpharmacological therapies. For this, it is necessary to study the gut–brain axis and its direct implication in the activity of appetite regulation and direct link with the genesis of obesity.

TABLE 6.3 Percentage of Association of Obesity and Comorbidities

Associated Pathology	Association Percentage or Relative Risk	Reference
Breast cancer	(↑) 25 units in BMI indicates an increase in the risk of 71%	Aguilar et al. (2012)
Liver cancer	(↑) 50% risk, higher prevalence in men	Campbell et al. (2016)
Multiple myeloma	(↑) 12% risk	Wallin and Larson (2011)
Type II diabetes	(↑ >) 2.9 times higher in subjects with obesity	De Lorenzo et al. (2019)
Sleep apnea	(↑) 14%–50% risk	Peppard et al. (2013)

6.3 GUT–BRAIN AXIS

There are various connections established through the circulatory system, the immune system, and the vagus nerve. The gut–brain axis involves a network of complex organs, such as the enteric nervous system, the neuroimmune and neuroendocrine system, and the central and autonomous nervous system, including the gut microbiota (Han et al., 2018). The implication of this axis maintains an important homeostasis in different physiological functions (Mayer et al., 2015; Westfall et al., 2017).

During the process of digestion, absorption, metabolism, and excretion of food consumed in the diet, the participation of most of the organs and systems in the body is involved. For this reason, the appetite is strongly regulated under normal conditions. However, under conditions such as anorexia and obesity this regulation is altered, so it is important to understand the processes that control it. In this sense, the gut–brain axis, which refers to the bidirectional neural processing of information between the central and digestive nervous systems, plays an essential role in the feeding process through motility regulation, secretion, and gastrointestinal blood flow, as well as appetite and energy balances (Andermann and Lowell, 2017). Thus the gut–brain axis also modulates mucosal immunity, thereby providing a protective effect against various xenobiotics from the diet that can be pathological in nature. Based on this, it is considered important to study the involvement of the gut–brain axis in the maintenance of homeostasis (Khlevner et al., 2018).

6.3.1 NEURONAL AXIS

The human brain can become the most complex of all biological systems. It is here that afferent signals are interpreted from the different organs that make up the organism's systems; as well as, efferent responses are generated for homeostasis control. There is a complex connection and interpretation process between the anatomofunctional cells of the brain, the neurons, and their interaction by a mechanism called synapse described by the doctor and humanist Santiago Ramón y Cajal during the 20th century. It is through these connections that is carried out the control of the different processes and the signaling between the neuronal cells.

As mentioned earlier, the brain as a part of the central nervous system plays a fundamental role in the regulation of energy balance, thereby involving the center of appetite regulation.

The appetite regulation center is located in the nuclear region of the brain (i.e., the hypothalamus). Within this structure is the arcuate nucleus, which contains two different types of neurons that oppositely regulate appetite. One promotes appetite through agouti-related neuropeptide expression (AgRP) (Andermann and Lowell, 2017) and another expresses proopiomelanocortin (POMC), a protein that comprises several neuropeptides of multiple nature, such as the melanocyte-stimulating hormone α (α-MSH). The α-MSH suppresses appetite, and together with different peripheral signals, such as hormones and peptides secreted by specialized cells along the gastrointestinal tract (i.e., serotonin (5-HT), cholecystokinin (CCK), peptide tyrosine-tyrosine (PYY), among other), induces neuronal regulation (Waterson and Horvath, 2015). Figure 6.3 shows the anatomical structures described previously, as well as the different nuclei of the hypothalamus that are involved in appetite regulation.

6.3.2 NEUROENDOCRINE AXIS

The activity of the arcuate nucleus or infundibular integrates signals that provide information about the status of energy balance, consisting of two groups of neurons that perform opposing activities (Figure 6.4). The first group of neurons, known as POMC, sends signals to other parts of the brain to reduce appetite. Here the synthesis of peptides derived from POMC is performed, which includes the α and β forms of the melanocyte-stimulating hormone that are activated by leptin and glucose, exerting anorexigenic activity. The melanocyte-stimulating hormone also exerts activity through melanocortin-4 receptors found in the different nuclei of the hypothalamus to reduce food intake (Shen et al., 2017). Concerning the second group of neurons that form the infundibular nucleus, they are the neuropeptide Y (NPY)/AgRP neurons, named for two proteins they produce, the NPY and the AgRP in the basal middle hypothalamus; they are prototypical neurons that establish a functional link between neuronal activity and feeding behavior (Yang et al. 2015).

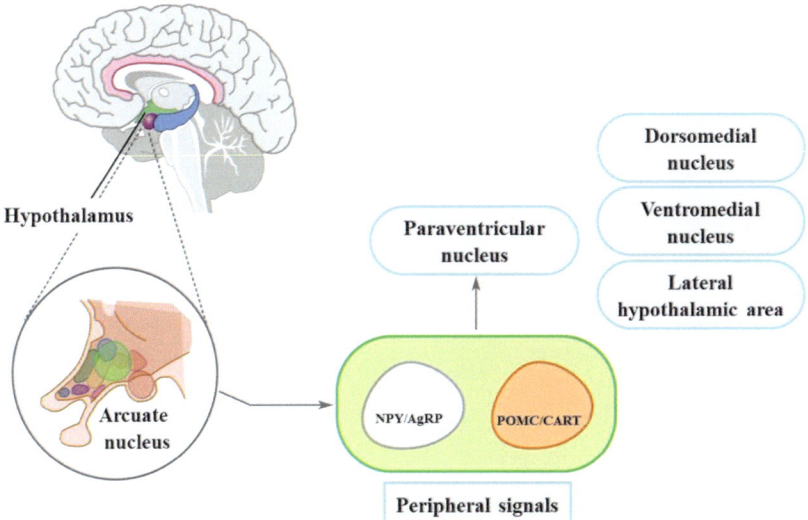

FIGURE 6.3 Regions of appetite regulation in the central nervous system. NPY—neuropeptide Y; AgRP—agouti-related peptide; POMC—proopiomelanocortin; CART—cocaine and amphetamine-regulated transcript.

It is important to mention that neuroendocrinal axis involves the peptide hormones, such as cholecystokinin, serotonin, and the tyrosine–tyrosine peptide, which come from the small intestine. They have as target cells the neuronal axis for an energetic control, giving special importance to the topic that is addressed to the neurons expressing AgRP, and whose activity signals is a great appetite stimulation.

AgRP neurons produce agouti-related peptides in the hypothalamus and participate in energy metabolism through responses to the different circulating factors that occur during the preprandial period. Thus neurons promote food intake and may be involved in the development of obesity. Other forms of action recently found, point to these neurons responsible for metabolic processes such as white fat tissue browning. AgRP neurons also regulate other metabolic processes such as thermogenesis (Ruan et al., 2014;

Waterson and Horvath, 2015; Burke et al., 2017) and glucose metabolism of brown adipose tissue (Steculorum et al., 2016). It is important to mention that this chapter analyzes the participation of AgRP neurons in the regulation of appetite, making it relevant to describe the neuroendocrine axis which is directly related to the satiety signals that are expressed in the intestine in response to the nature of the food intake.

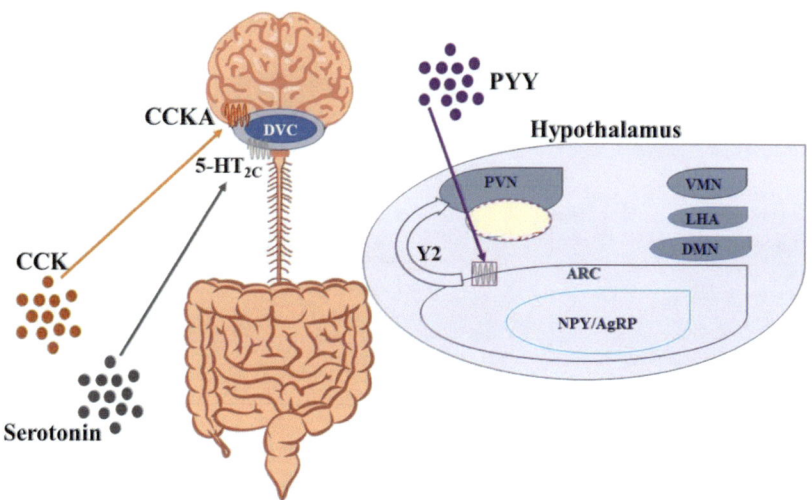

FIGURE 6.4 Integration of signals that provide information on the energy balance in the arcuate nucleus. PYY—peptide tyrosine-tyrosine; PVN—paraventricular nucleus; VMN—ventromedial nucleus; LHA—lateral hypothalamic area; DMN—dorsomedial nucleus; ARC—arcuate nucleus; NPY—neuropeptide Y; AgRP—agouti-related peptide; Y_2R—Y_2 receptor; CCKA—cholecystokinin A receptor; CCK—cholecystokinin; $5\text{-}HT_{2c}$—5-hydroxytryptamine 2c receptor; DVC—dorsal vagal complex.

As mentioned in this chapter, it is through the participation of the neuronal axis that the control of food intake in the diet is carried out. Thus this control will be mediated by a series of multifactorial aspects, including environmental, social, neuroendocrinological, and neurobiological aspects. Since the regulation of eating behavior is accomplished by the hypothalamus, a communication through peripheral signals both orexigenic and anorexigenic is necessary. These signals that travel to the nuclear center of the brain involve hormonal signals such as serotonin, ghrelin, and leptin, as well as circulating signals in the blood such as glucose levels and gastrointestinal peptides (CCK and PYY); these together integrate the neuroendocrine axis (Zanchi et al., 2017). Table 6.4 shows the peptides and hormones with direct influence on appetite regulation and energy expenditure.

TABLE 6.4 Neuroendocrine Mechanisms That Govern Eating Behavior

Signal	Source	Effect	Target	Reference
Melanocortin system	Hypothalamus	Suppress appetite Body weight regulation	Melanocortin receptors	do Carmo et al. (2017)
NPY/AgRP neurons	Hypothalamus	Promote food intake	MC4 receptors (MC4R)	Krashes et al. (2013)
Serotonin	Peripheral tissues and brain	Reduction of food intake Promotion weight loss Control of energy expenditure	5-HT2C receptor 5-HT1B receptor	Palamiuc et al. (2016)
PYY	Small intestine	Reduction of food intake Promotion weight loss Increase energy expenditure	Y2 receptors in neurons	Jones et al. (2019)
CCK	Gastrointestinal system Central nervous system	Reduction of food intake	CCK^{NTS} neurons	Roman et al. (2017)
Ghrelin	Stomach	Stimulates food intake Increase fat mass	Growth Hormone Secretagogue Receptor (GHSR)	Korbonits et al. (2004)
Leptin	Peripheral adipose tissue	Stimulates activity of POMC neurons Inhibit activity of NPY/AgRP neurons	Leptin receptor (Ob-Rb) in the arcuate nucleus of the hypothalamus (ARC)	Shen et al. (2017)
Insulin	Pancreas	Suppression of food intake	Insulin receptors	Shen et al. (2017)

6.4 TREATMENT

The increase in obese population worldwide has promoted research on the mechanisms of appetite regulation, which involve the centers that participate at the hypothalamic level, as well as the hormonal signals coming from the intestine and the periphery. These investigations propose treatments that regulate energy expenditure, as well as satiety control. Therefore considering that the control of satiety and appetite involves different complex signaling pathways, together hormonal, neural, metabolic, and even emotional, this chapter is aimed at reviewing different treatments for their control.

6.4.1 PHARMACOTHERAPY

In order to offer a comprehensive management for the treatment of obesity, in which in many cases the nutritional therapy, physical activity, as well the psychological aspect were offered with no indicated favorable results, drugs have been established that contribute to the weight loss contribution.

Worldwide, FDA has authorized five drugs for the treatment of obesity, these are orlistat, lorcaserin, naltrexone/bupropion, phentermine/topiramate, and liraglutide (Cohen and Gadde, 2019). However, in Mexico, the drugs authorized by the Federal Commission for the Protection Against Sanitary Risks (COFEPRIS in Spanish) are orlistat and liraglutide. These drugs action is in two approaches, the first modifying the metabolism of macronutrients and the second modulating neuroendocrine processes of appetite and satiety regulation (Table 6.5).

Pharmacological treatments with effects on the central nervous system have been linked to adverse responses, so the sale of some of them has been prohibited. Citing an example, sibutramine, having a structure similar to that of amphetamine, is a selective inhibitor of monoamine recapture. This drug has an established efficacy in sustained weight reduction and a favorable safety profile; however, its action on the sympathetic nervous system has been related to alterations in blood pressure and heart frequency (Florentin et al., 2008). Another example is rimonabant, whose activity is related to the blockade of cannabinoid type 1 receptors. This drug has been used in the last decade in clinical practice for weight loss and maintenance. Some adverse effects attributed to the drug are associated with psychiatric effects such as anxiety, depression, and suicide ideas (Blasio et al., 2013).

TABLE 6.5 FDA-Approved Medications for the Treatment of Obesity

Drug	Mechanism of Action	Effect	Adverse Response	Reference
Lorcaserin	Serotonergic agonist (↑ binding affinity and activation of 5-hydroxytryptamine 2c)	(↓) Appetite	Headache, nausea, dry mouth, fatigue, constipation, hematologic changes.	Thomsen et al. (2008), Fleming et al. (2013)
Orlistat	Inhibition of pancreatic lipase	(↓) Absorption of lipids from the diet	↓ Absorption of fat-soluble vitamins, flatulence, fecal incontinence	Shirai et al. (2019), Halpern and Halpern (2015)
Naltrexone/bupropion	Opioid antagonist (↑affinity for μ-opioid receptor) / Inhibit reuptake of the catecholamines (dopamine and norepinephrine)	(↓) Appetite/antidepressant	↑ Risk of cardiovascular events, nausea, constipation	Billes et al. (2014)
Phentermine/topiramate	Sympathomimetic amine with pharmacologic activity similar to that of amphetamine	(↓) Appetite (↓) Body weight (↑) Energy expenditure	Paresthesia, headache, dysgeusia, constipation, dry mouth, upper respiratory tract infection, nasopharyngitis	Fleming et al. (2013)
Liraglutide	Glucagon-like peptide 1 (GLP-1) receptor agonist	(↓) Appetite (↓) Body weight	Nausea, hypoglycaemia, diarrhea, constipation, vomiting, headache, fatigue	Khoo et al. (2017), Krentz et al. (2016)

6.4.2 NONPHARMACOLOGICAL TREATMENT

As discussed in this chapter, drug treatments aimed at reducing body weight provide benefits, both to reduce obesity and the risk of complications; however, some drugs have been withdrawn from the market due to their serious adverse effects. Hence when considering obesity as a pandemic in the last two decades as the fifth leading cause of death worldwide, it is important to find alternatives for improving the quality of life of the affected population without the risk of experiencing adverse effects.

It has been constantly mentioned that changes in lifestyle through diet and exercise may be an appropriate alternative in the etiology and treatment of the disease. Despite this, the results have not been sufficient because changing lifestyle depends on the ability of patients to modify their behavior patterns, particularly with regard to diet and exercise.

Studies conducted by Alonso-Castro et al. (2019), in some states of the Mexican Republic show that a high percentage of overweight individuals diagnosed with hypertension and diabetes tends to self-medicate. In particular, 50% of the study population usually consume at least one product per day, seven days a week, and for a period of six months. Of this population, 67% use herbal products, recommended by a friend or relative. Some of the herbal or botanical products (H/bp) that have been traditionally consumed for weight control have not shown to promote adverse effects in the body. Thus various research groups have postulated the use of natural products derived from plants (Gamboa-Gómez et al., 2015). In the following section, it is summarized the documented phytoconstituents with effect on obesity and its complications, even if the underlying mechanism of action is still not entirely clear.

6.4.2.1 HERBAL PRODUCTS IN ANTIOBESITY TREATMENT

Medicinal herbs are an important source of phytoconstituents with the potential to suppress appetite and promote weight loss. It is based on the hypothesis that coming from natural sources, whose millenary consumption has not shown adverse effects to their consumption, they can be an adequate source for the control and management of obesity and its complications.

In this sense, the analysis of the effect on the consumption of H/bp has established its impact on the neuroendocrine regulation of appetite and homeostatic energy as necessary. Various investigations with H/bp are being carried out to propose therapies for weight control, depending on which properties they are attributed, that is, thermogenic effects, lipase inhibitors, appetite suppressants, regulators of lipid metabolism, and adipocyte differentiation. Therefore it is important to establish the action of phytoconstituents present in these natural sources in the gut–brain axis to relate them to weight loss through the modulation of the action of hormones and peptides [i.e., serotonin (5-HT), cholecystokinin (CCK), peptide tyrosine-tyrosine (PYY), among other)] in the NPY/AgRP neuron inhibition. Also, by highlighting the neuroendocrine-metabolic point of view and integrating their responses to the insulin-like growth hormone/growth factor axis, which is responsible for the long-term regulation of weight and obesity control.

The H/bp has been traditionally used to treat health disorders. These products are not regulated by the Food and Drug Administration as these are considered as dietary supplements within the food category. H/bp are currently available in many presentations such as tablets, liquids, powders, tea bags, among others and were obtained through direct purchase in supermarkets, pharmacies, or the Internet. The most popular forms of consumption of H/bp is in tablets or infusions. In these dietary supplements, the presence of saponins, polyphenols, alkaloids, among other phytoconstituents with the potential effect on weight control therapies has been documented.

It is important to mention that traditionally several sources have been recognized for their effect on appetite suppression. However, the effects of these H/bp have not been fully related to the signals involved in the gut–brain axis. Table 6.6 summarizes some of the most used herbal products, as well as some of the effects that they can promote. An important source is *Hoodia gordonii*, which is traditionally used by the Khoi-San tribe for this purpose. Studies conducted by Jain and Singh (2013, 2016) indicate that in effect extracts rich in steroids glycosides obtained with organic solvents decrease appetite. Based on Jain and Singh's study and available information, the research workers hypothesize that this response is enhanced when *H. gordonii* extract is supplemented with L-carnitine. Among the responses associated with its consumption is the decrease in the levels of NPY and IGF-1. Likewise, when its consumption is supplemented with L-carnitine, significant changes in the levels of ghrelin, leptin-corticosterone, and thyroid hormones have been observed. Another H/bp used for weight loss are those developed with *Garcinia cambogia*, which can be found in supermarkets as a dietary supplement rich in hydroxycitric acid. Its mode of action is associated with an inhibition in serotonin reuptake. It is also related to cholinesterase inhibition, fatty acid oxidation, and reduction of lipogenesis (Nguyen et al., 2019; Attia et al., 2019).

TABLE 6.6 Herbal Supplements Used for Weight Loss

Herbal Supplement	Phytoconstituent	Action Mode	Reference
Garcinia cambogia	Hydroxycitric acid (HCA)	Selective serotonin reuptake inhibitor and cholinesterase inhibitor (↑) Fatty acid oxidation (↓) Lipogenesis	Nguyen et al. (2019), Attia et al. (2019)
Moringa oleifera	Polyphenol extract	Downregulation of mRNA of leptin and resistin and upregulation of adiponectin gene expression	Metwally et al. (2017)
Camellia sinensis	MeOH and BuOH extracts Chakasaponin II (acylated polyhydroxyoleanane-type triterpene oligoglycoside) and flavonol glycosides	(↑) Release of 5-HT (↓) Appetite signals (↓) mRNA levels of neuropeptide Y (NPY)	Hamao et al. (2011)
Hoodia gordonii	Organic solvent extract rich in steroid glycosides	Hoodia gordonii (alone) (↓) Appetite (↓) NPY and IGF-1 (↑) Leptin and CCK (marginal) Anorectic activity Hoodia gordonii + L-carnitine Significant changes in ghrelin, leptin, corticosterone and thyroid hormones	Jain and Singh (2013), Jain and Singh (2016)
Capsicum oleoresin	Nonivamide (N-nonanoyl 4-hydroxy-3-methoxy-benzylamide)	(↑) Release of 5-HT	Hochklogler et al. (2017)
Coleus forskohlii	Forskolin (diterpene) 7β-Acetoxy-8, 13-epoxy-1α, 6β, 9 α-trihydroxy-labd-14-ene-11-one	Inhibition pancreatic lipase activity (↑) The rate of lipolysis	Badmaev et al. (2015), Loftus et al. (2015)

TABLE 6.6 (Continued)

Herbal Supplement	Phytoconstituent	Action Mode	Reference
Citrus aurantium	p-synephrine (protoalkaloid nonstimulant thermogenic agent with similarity to ephedrine)	∝ and ß-adrenergic receptor agonist	Stohs (2017), Haaz et al. (2006), Rios-Hoyo and Gutiérrez-Salmeán (2016), Nakajima et al. (2014)
Yerba maté	Aqueous solution [chlorogenic acid, caffeine, theobromine, quercetin and rutin]	(↑) of leptin (↓) Appetite	Lima et al. (2013), Yimam et al. (2016)
Curcumae longa	Beverage [rich in curcuma]	(↓)Desire to eat (↑) PYY in plasm	Zanzer et al. (2017)

A popular source, whose consumption is recognized for weight loss, is *Camellia sinensis*. Particularly, it has been demonstrated by Hamao et al. (2011) that methanolic extracts obtained from flower buds from this source have high concentrations of flavonol glycosides. Also, in BuOH extracts, the main constituent is Chakasaponin II (acylated poly-hydroxy-oleanane-type triterpene oligo-glycoside). Research carried out by these authors in obese animal models have demonstrated that Chakasaponin II enhances the release of serotonin, suppresses mRNA levels of NPY and consequently suppresses appetite signals.

Some reports documented by Metwally et al. (2017) indicate that the use of hydroethanolic extracts of *Moringa oleifera* in animal models promote weight loss. In these reports, the consumption of the extract of *M. oleifera* at a concentration of 600 mg kg^{-1} of body weight for 12 weeks has shown downregulation mRNA of leptin and resistin and upregulation of adiponectin gene expressions.

A product that is part of traditional Mexican food is chili pepper (*Capsicum* spp). Its consumption has been associated with beneficial effects on lipid metabolism and thermogenesis. These effects have been mainly associated with capsaicin (8-methyl-*N*-vanillyl-6-nonenamide). However, the development of this phytochemical in high doses can promote irritation in the gastrointestinal system. These responses have caused that the development of products for weight control based on this phytochemical is not allowed in Europe, therefore alternative analog structures have been explored. In this regard, a structural analog of capsaicin found in low concentrations in *Capsicum oleoresin* is Nonivamide (*N*-nonanoyl-4-hydroxy-3-methoxy-benzylamide). This constituent has been authorized by the European Union because its pungent potential is low and that it induces the peripheral release of serotonin (Hochklogler et al., 2017).

Other types of extracts aimed at reducing the risk factors of overweight and its complications include diterpene-rich formulations such as *Coleus forskohlii*, whose active substance (forskolin) improves insulin resistance, inhibits pancreatic lipase activity, and increases lipolysis via accumulation of cyclic adenosine monophosphate by mechanisms independent of hormonal stimulation (Badmaev et al., 2015; Loftus et al., 2015). However, it is important to indicate that forskolin induces alterations in the gastrointestinal system.

p-Synephrine, a sympathomimetic agent, which is a protoalkaloid with thermogenic properties, shows a structure similar to ephedrine. A rich source of this constituent is *Citrus aurantium*, whose consumption has been related to an increase in energy expenditure and a decrease in food intake. Its effect is associated with the demonstrated ability as agonists of a and β-adrenergic receptors, resulting in a decrease of gastric motility. This response is analogous to those that exhibit compounds such as cholecystokinin and other intestinal peptides (Stohs, 2017; Haaz et al., 2006; Ríos-Hoyo and Gutiérrez-Salmeán, 2016; Nakajima et al., 2014).

As previously mentioned, one of the ways to consume herbal products is as infusions. Some sources used for the management of obesity are yerba mate and *Curcumae longa*, these sources are characterized by their abundance in phenolic compounds and methylxanthines. Yerba mate can control several parameters related to obesity and metabolic syndrome, particularly increasing leptin levels and suppressing appetite (Lima et al., 2013; Yimam et al., 2016). On the other hand, turmeric-rich beverages decrease the desire to eat and increases the plasma levels of PYY (Zanzer et al., 2017).

To recapitulate, herbal products show an important effect on signals involved in appetite control. However, we must also consider that even when they are sources that have traditionally been consumed, it is imperative to rule out adverse effects that can be promoted. Particularly, the consumption of *Garcinia cambogia* has been associated with cases of psychosis. Other studies have shown that *Coleus forskohlii* promotes alterations in the gastrointestinal system or that *Citrus aurantium* induces alterations in the cardiovascular system.

6.5 CONCLUSIONS

In summary, few studies with herbal supplements have been focused on the control of obesity via modulation of the neuronal and neuroendocrine axis. In this sense, it is important to expand the study of herbal products aimed at exploring their effect on the signals involved in the gut–brain axis, with the purpose of having alternatives for long-term weight control.

KEYWORDS

- **herbal products**
- **phytoconstituents**
- **antiobesity**
- **gut–brain axis**

REFERENCES

Aguilar J.L., Martínez I.A., Villarreal-Garza C.M., Lara G.A., Lara-Medina F., Alvarado-Miranda A., De La Garza J.G., Mohar A., Meneses A., Herrera-Gomez A., Olvera-Caraza D., Granados-Garcia M., Arrieta O. (2012). Impact of obesity and overweight in the prognosis of women diagnosed with non metastativ breast cancer in a Mexican cohort. Journal of Clinical Oncology. 30:15 suppl, 1607–1607.

Alonso-Castro A.J., Ruiz-Padilla A.J., Ramírez-Morales M.A., Alcocer-García S.G., Ruiz-Noa Y., Ibarra-Reynoso L. del R., Solorio-Alvarado C.R., Zapata-Morales J.R., Mendoza-Macías C-L., Deveze-Álvarez M.A., Alba-Bentancourt C. (2019). Self-treatment with herbal products for weight-loss among overweight and obese subjects from central Mexico. Journal of Ethnopharmacology. 234:21–26.

Andermann M.L. and Lowell B.B. (2017). Toward a wiring diagram understanding of appetite control. Neuron. 95:757–778.

Attia R.T., Abdel-Mottaleb Y., Abdallah D.M., El-Abhar H.S., El-Maraghy N.N. (2019). Raspberry ketone and *Garcinia cambogia* rebalanced disrupted insulin resistance and leptin signaling in rats fed high fat fructose diet. Biomedicine & Pharmacotherapy. 110:500–509.

Badmaev V., Hatakeyama Y., Yamazaki N., Noro A., Mohamed F., Ho C-T., Pan M-H. (2015). Reprint of "Preclinical and clinical effects of *Coleus forkohlii*, *Salacia reticulata* and *Sesamum indicum* modifying pancreatic lipase inhibition in vitro and reducing total body fat". Journal of Functional Foods. 15: 44–51.

Bennett J., Greene G., Schwartz-Barcott D. (2013). Perceptions of emotional eating behavior. A quality study of college students. Appetite. 60:187–192.

Billes S.K., Sinnayah P., Cowley M.A. (2014). Naltrexone/bupropion for obesity: an investigational combination pharmacotherapy for weight loss. Pharmacology Research. 84:1–11.

Blasio A., Iemolo A., Sabino V., Petrosino S., Steardo L., Rice K.C., Orlando P., Iannotti F.A., Di Marzo V., Zorrilla E.P., Cottone P. (2013). Rimonabant precipitates anxiety in rats withdrawn from palatable food: role of the central amygdala. Neuropsychopharmacology. 38: 2498–2507.

Blüher M. and Mantzoros C.S. (2015). From leptin to other adipokines in health and disease: Facts and expectations at the beginning of the 21st century. Metabolism Clinical and Experimental. 64:131–145.

Burke L.K., Darwish T., Cavanaugh A.R., Virtue S., Roth E., Morro J., Liu S-M., Xia J., Dalley J.W., Burling K., Chua S., Vidal-Puid T., Schwartz G.J., Blouet C. (2017). mTORC1 in AGRP neurons integrates exteroceptive and interoceptive food-related cues in the modulation of adaptive energy expenditure in mice. eLife. 6:e22848(1-22).

Campbell P.T., Newton C.C., Freedman N.D., Koshiol J., Alavanja M.C., Beane-Freeman L.E., Buring J.E., Chan A.T., Chong D.Q., Datta M., Gaudet M.M., Gaziano J.M., Giovannucci E.L., Graubard B.I., Hollenbeck J.L., King L., Lee I.-M., Linet M.S., Palmer J.R., Petrick J.L., Pounter J.N., Purdue M.P., Robien K., Rosenberg L., Sahasrabuddhe V.V., Schairer C., Sesso H.D., Sigurdson A.J., Stevens V.L., Wactawski-Wende J., Zeleniuch-Jacquitte A., Renehan A.G., McGlynn K.A. (2016). Body mass index, waist circumference, diabetes, and risk of liver cancer for U.S. adults. Cancer Research. 75(20):6076–6083.

Chaker L., Bianco A.C., Jonklaas J., Peeters R.P. (2017). Hypothyroidism. The Lancet. 390 (10101): 1550–1562. doi:http://dx.doi.org/10.1016/S0140-6736(17)30703-1

Cohen J.B. and Gadde K.M. (2019). Weight loss medications in the treatment of obesity and hypertension. Current Hypertension Reports. 21(16):1–9.

Dahlman I., Elsen M., Tennagels N., Korn M., Brockmann B., Sell H., Eckel J., Arner P. (2012). Functional annotation of the human fat cell secretome. Archives of Physiology and Biochemistry. 118(3):84–91.

Dave D.M. and Colman G. (2011). Isolating the effect of major depression on obesity: role of selection bias. J. Ment Health Econ. 14(4):165–186.

De Lorenzo A., Gratteri S., Gualtieri P., Cammarano A., Bertucci P., Di Renzo L. (2019). Why primary obesity is a disease? Journal of Translation Medicine. 17:169 (1–13).

Diéguez-Felechosa M., Riestra-Fernández M., Menéndez-Torre E. (2009). Insulinoma. Criterios diagnósticos y tratamiento. Avances en Diabetología. 25:293–299.

do Carmo J.M., da Silva A.A., Wang Z., Fang T., Aberdein N., Perez de Lara C.E., Hall J.E. (2017). Role of the brain melanocortins in blood pressure regulation. Biochimica et Biophysica Acta (BBA)—Molecular Basis of Disease. 1863:2508–2514.

Fassnacht M., Art W., Bancos I., Dralle H., Newell-Price J., Sahdev A., Taarin A., Terzolo M., Tsagarakis S., Dekkers O.M. (2016). Management of adrenal incidentalomas: European Society of Endocrinology Clinical Practice Guideline in collaboration with the European Network for the Study of Adrenal Tumors. European Journal of Endocrinology. 174(2):G1–G34.

Fleming J.W., McClendon K.S., Riche D.M. (2013). New obesity agents: Lorcaserin and Phentermine/Topiramate. Annals of Pharmacotherapy. 47:1007–1016.

Florentin M., Liberopoulos E.N., Elisaf M.S. (2008). Sibutramine-associated adverse effects: a practical guide for its safe use. Obesity Reviews. 9:378–387.

Gamboa-Gómez C.I., Rocha-Guzmán N.E., Gallegos-Infante J.A., Moreno-Jiménez M.R., Vázquez-Cabral B.D., González-Laredo R.F. (2015). Plants with potential use on obesity and its complications. EXCLI Journal 14:809–831.

González-Jiménez E. (2011). Genes y obesidad: una relación de causa-consecuencia. Endocrinología y Nutrición. 58(9):492–496.

Haaz S., Fontaine K.R., Cutter G., Limdi N., Perumean-Chaney S., Allison B.D. (2006). *Citrus aurantium* and synephrine alkaloids in the treatment of overweight and obesity: an update. Obesity Reviews. 7: 79–88.

Halpern B. and Halpern A. (2015). Safety assessment of FDA-approved (orlistat and lorcaserin) anti-obesity medications. Expert Opinion on Drug Safety. 14(2):305–315.

Hamao M., Matsuda H., Nakamura S., Nakashima S., Semura S., Maekubo S., Wakasugi S., Yoshikawa M. (2011). Anti-obesity effects of the methanolic extract and chakasaponins from flower buds of Camellia sinensis in mice. Bioorganic & Medicinal Chemistry. 19:6033–6041.

Han W., Tellez L.A., Perkins M.H., Perez I.O., Qu T., Ferreira J., Ferreira T.L., Quinn D., Liu Z-W., Gao X-B., Kaelberer M.M., Bohórquez D.V., Shammah-Lagnado S.J., de Lartigue G., de Araujo I. (2018). A neural circuit for gut-induced reward. Cell. 175:665–678.

Hochkogler C.M., Lieder B., Rust P., Berry D., Meier S.M., Pignitter M., Riva A., Leitinger A., Bruk A., Wagner S., Hans J., Widder S., Ley J.P., Krammer G.E., Somoza V. (2017). A 12-week intervention with nonivamide, a TRPV1 agonist, prevents a dietary-induced body fat gain and increases peripheral serotonin in moderately overweight subjects. Molecular Nutrition Food Research. 61(5):1600731 (1–13).

Izaola O., de Luis D., Sajoux I., Domingo J.C., Vidal M. (2015). Inflamación y obesidad (lipoinflamación). Nutrición Hospitalaria. 31(6):2352–2358.

Jain S. and Singh S.N. (2013). Metabolic effect of short term administration of *Hoodia gordonii*, an herbal appetite suppressant. South African Journal of Botany. 86:51–55.

Jain S. and Singh S.N. (2016). Effect of L-carnitine and *Hoodia gordonii* supplementation on metabolic markers and physical performance under short term calorie restriction in rats. Defence Science Journal. 66(1): 11–18.

Jones E.S., Nunn N., Chambers A.P. Ostergaard S., Wulff B.S., Luckman S.M. (2019). Modified peptide YY molecule attenuates the activity of NPY/AgRP neurons and reduces food intake in male mice. Endocrinology. 160(11):2737–2747.

Khlevner J., Park Y., Margolis K.G. (2018). Brain–gut axis. Gastroenterology Clinics of North America. 47:727–739.

Khoo J., Hsiang J., Law N-M., Ang T-L. (2017). Comparative effects of liraglutide 3mg vs structures lifestyle modification on body weight, liver fat and liver function in obese patients with non-alcoholic fatty liver disease: a pilot randomized trial. Diabetes, Obesity and Metabolism. 19:1814–1817.

Klöting N. and Blüher M. (2014). Adipocyte dysfunction, inflammation and metabolic syndrome. Reviews in Endocrine and Metabolic Disorders 15(4):277–287

Korbonits M., Goldstone A.P., Gueorguiev M., Grossman A.B. (2004). Ghrelin—a hormone with multiple functions. Frontiers in Neuroendocrinology. 25(1):27–68.

Krashes M.J., Shah B.P., Koda S., Lowell B.B. (2013). Rapid versus delayed stimulation of feeding by the endogenously released AgRP neuron mediators GABA, NPY, and AgRP. Cell Metabolism. 18:588–595.

Krentz A.J., Fujioka K., Hompesch M. (2016). Evolution of pharmacological obesity treatments: focus on adverse side-effect profiles. Diabetes, Obesity and Metabolism. 18:558–570.

Lamiquiz-Moneo I., Mateo-Gallego R., Bea A.M., Dehesa-García B., Pérez-Calahorra S., Marco-Benedí V., Baila-Rueda L., Laclaustra M., Civeira G., Cenarro A. (2019). Genetic predictors of weight loss in overweight and obese subjects. Scientific Reports. 9:10770. 1–9.

Lima N da S., Franco J.G., Peixoto-Silva N., Maia L.A., Kaezer A., Felzenszwalb I., de Oliveira E., de Moura E.G., Lisboa P.C. (2013). *Ilex paraguariensis* (yerba mate) improves endocrine and metabolic disorders in obese rats primed by early weaning. European Journal of Nutrition. DOI 10.1007/s00394-013-0500-3.

Loftus H.L., Astell K.J., Mathai M.L., Su X.Q. (2015). *Coleus forskohlii* extract supplementation in conjunction with a hypocaloric diet reduces the risk factors of metabolic syndrome in overweight and obese subjects: a randomized controlled trial. Nutrients. 7:9508–9522.

Lustig R.H. (2011). Hypothalamic obesity after craniopharyngioma: mechanisms, diagnosis, and treatment. Frontiers in Endocrinology. 2:1–8.

Martínez-Villanueva J., González-Leal R., Argente J., Martos-Moreno G.A. (2019). La obesidad parental se asocia con la gravedad de la obesidad infantil y de sus comorbilidades. Anales de Pediatría. 90(4): 224–231.

Mayer E.A., Tillisch K., Gupta A. (2015). Gut/brain axis and the microbiota. Journal of Clinical Investigation. 125(3):926–938.

Metwally F.M., Rashad H.M., Ahmed H.H., Mahmoud A.A., Raouf E.R.A., Abdalla A.M. (2017). Molecular mechanisms of the anti-obesity potential effect of *Moringa oleifera* in the experimental model. Asian Pacific Journal of Tropical Biomedicine. 7(3):214–221.

Nakajima V.M., Macedo G.A., Macedo J.A. (2014). Citrus bioactive phenolics: Role in the obesity treatment. LWT—Food Science and Technology. 59:1205–1212.

Nguyen D.C., Timmer T.K., Davison B.C., McGrane I.R. (2019). Possible *Garnicia cambogia*-induced mania with psychosis: a case report. Journal of Pharmacy Practice. 32(1): 99–102.

Organisation for Economic Co-operation and Development (OECD). Obesity Update 2017. https://www.oecd.org/health/health-systems/Obesity-Update-2017.pdf. (accessed Jan 08, 2020).

Palamiuc L., Noble T., Witham E., Ratanpal H., Vaughan M., Srinivasan S. (2016). A tachykinin-like neuroendocrine signaling axis couples central serotonin action and nutrient sensing with peripheral lipid metabolism. Nature Communications. 8:14237.1–14.

Peppard P.E., Young T., Barnet J.H., Palta M., Hagen E.W., Hla K.M. (2013). Increased prevalence of sleep-disordered breathing in adults. American Journal of Epidemiology. 177 (9):1006–1014.

Prieur X., Mok C.Y.L., Velagapudi V.R., Núñez V., Fuentes L., Montaner D., Ishikawa K., Camacho A., Barbarroja N., O´Rahilly S., Sethi J.K., Dopazo J., Oresic M., Ricote M., Vidal-Puig A. (2011). Differential lipid partitioning between adipocytes and tissue macrophages modulates macrophage lipotoxicity and M2/M1 polarization in obese mice. Diabetes. 60:797–809.

Ríos-Hoyo A. and Gutiérrez-Salmeán G. (2016). New dietary supplements for obesity: what we currently know. Current Obesity Reports. 5:262–270.

Roman C.W., Sloat S.R., Palmiter R.D. (2017). A tale of two circuits: CCK^{NTS} neuron stimulation controls appetite and induces opposing motivational states by projections to distinct brain regions. Neuroscience. 358: 316–324.

Ruan H-B., Dietrich M.O., Liu Z-W., Zimmer M.R., Li M-D., Singh J.P., Zhang K., Yin R., Wu J., Horvath T.L., Yang X. (2014). O-GlcNAc transferase enables AgRP neurons to suppress browning of white fat. Cell. 159:306–317.

Schneider K.L., Appelhans B.M., Whited M.C., Oleski J., Patogo S.L. (2010). Trait anxiety, but not trait anger, predisposes obese individuals to emotional eating. Appetite. 55:701–706.

Shen W-j., Yao T., Kong X., Williams K.W., Liu T. (2017). Melanocortin neurons: multiple routes to regulation of metabolism. BBA-Molecular Basis of Disease. 1863:2477–2485.

Shirai K., Fujita T., Tanaka M., Fujii Y. (2019). Efficacy and safety of lipase Orlistat in Japanese with excessive visceral fat accumulation: 24-week, double-blind, randomized, placebo-controlled study. Advances in Therapy. 36:86–100.

Sigalos J.T., Pastuszak A.W., Khera M. (2018). Hypogonadism: therapeutic risks, benefits and outcomes. Medical Clinics of North America. 102:361–372.

Sinha R., and Jastreboff A.M. (2013). Stress as a common risk factor for obesity and addiction. Biological Psychiatry. 73:827–835.

Steculorum S.M., Ruud J., Karakasilioti I., Backes H., Ruud L.E., Timper K., Hess M.E., Tsaousidou E., Mauer J., Vogt M.C., Perger L., Bremser S., Klein A.C., Morgan D.A., Frommolt P., Brinkkötter P.T., Hammerschmidt P., Benzing T., Rahmouni K., Wunderlich F.T., Kloppenburg P., Brüning J.C. (2016). AgRP neurons control systemic insulin sensitivity via myostatin expression in brown adipose tissue. Cell. 165:125–138.

Stohs S.J. (2017). Safety, efficacy and mechanistic studies regarding *Citrus aurantium* (Bitter orange) extract and S-synephrine. Phytotherapy Research. 31: 1463–1474.

Suganami R., Tanaka M., Ogawa Y. (2012). Adipose tissue inflammation and ectopic lipid accumulation. Endocrine Journal. 59(10):849–857.

Sutin A.R., Ferrucci L., Zonderman A.B., Terracciano A. (2011). Personality and obesity across the adult life span. Journal of Personality and Social Psychology. 101(3):579–592.

Thomsen W.J., Grottick A.J., Menzaghi F., Reyes-Saldana H., Espitia S., Yuskin D., Whelan K., Martin M., Morgan M., Chen W., Al-Shamma H., Smith B., Chalmers D., Behan D. (2008). Lorcaserin, a novel selective human 5-hydroxytryptamine$_{2c}$ agonist: in vitro and in vivo pharmacological characterization. Journal of Pharmacology and Experimental Therapeutics. 325(2):577–587.

Tornero P. S. (2019). Importancia de los factores socioeconómicos en estudios de obesidad. Anales de Pediatría. 91:422–423.

Wallin A. and Larsson S.C. (2011). Body mass index and risk of multiple myeloma: A meta-analysis of prospective studies. European Journal of Cancer. 47:1606–1615.
Wang Z.V. and Scherer P.E. (2016). Adiponectin, the past two decades. Journal of Molecular Cell Biology. 8(2):93–100.
Waterson M.J. and Horvath R.L. (2015). Neuronal regulation of energy homeostasis: beyond the hypothalamus and feeding. Cell Metabolism. 22:962–970.
Westfall S., Lomis N., Kahouli I., Dia S.Y., Singh S.P., Prakash S. (2017). Microbiome, probiotics and neurodegenerative diseases: deciphering the gut brain axis. Cellular and Molecular Life Sciences. 74:3769–3787.
World Obesity. Causes of obesity. https://www.worldobesity.org/about/about-obesity/causes-of-obesity (accessed Jan 08, 2020).
Yang D., Liu T., Williams K.W. (2015). Motivation to Eat – AgRP neurons and homeostatic need. Cell Metabolism. 22:62–63.
Yimam M., Jiao P., Hong M., Brownell L., Lee Y-C., Hyun E-J., Kim H-J., Kim T-W., Nam J-B., Kim M-R., Kia Q. (2016). Appetite suppression and antiobesity of a botanical composition composed of *Morus alba*, Yerba mate and *Magnolia officinalis*. Journal of Obesity. Article ID 4670818, 12 pages. http://dx.doi.org/10.1155/2016/4670818
Zanchi D., Depoorter A., Egloff L., Haller S., Mählmann L., Lang U.E., Drewe J., Beglinger C., Schmidt A., Borgwardt S. (2017). The impact of gut hormones on the neural circuit of appetite and satiety: a systematic review. Neuroscience and Biobehavioral Reviews. 80:457–475.
Zanzer Y.C., Plaza M., Dougas A., Turner C. Björck I., Östmann E. (2017). Polyphenol-rich spice-based beverages modulated postprandial early glycaemia, appetite and PYY after breakfast challenge in healthy subjects: a randomized, single blind, crossover study. Journal of Functional Foods. 35:574–583.

CHAPTER 7

Plants as a Potential Source of Acetylcholinesterase Inhibitors for Nutraceutical Therapy Disease

YESENIA ESTRADA-NIETO[1], DANIEL GARCÍA-GARCÍA[1],
ALEJANDRA I. VARGAS-SEGURA[2], ROBERTO ARREDONDO-VALDÉS[1],
RADIK A. ZAYNULLIN[3], RAIKHANA V. KUNAKOVA[3],
MÓNICA CHÁVEZ-GONZÁLEZ[1], RODOLFO RAMOS-GONZÁLEZ[3],
JOSÉ L. MARTÍNEZ–HERNÁNDEZ[1], MAYELA GOVEA-SALAS[1],
ANNA ILYINA[1], and E. PATRICIA SEGURA-CENICEROS[1*]

[1]*Nanobioscience Group, Chemistry School, Autonomous University of Coahuila, Blvd. V. Carranza e Ing. J. Cardenas V., Saltillo 25280, Mexico*

[2]*Faculty of Dentistry, Autonomous University of Coahuila, Saltillo 25280, Mexico*

[3]*Ufa State Petroleum Technological University, 1 Cosmonauts St., Bashkortostan, Ufa 450062, Russia*

[4]*CONACYT, Autonomous University of Coahuila, Saltillo 25280, México*

*Corresponding author. E-mail: psegura@uadec.edu.mx.

ABSTRACT

Acetylcholinesterase (AChE) inhibitors are an important therapeutic strategy in Alzheimer's disease (AD). AChE hydrolyzes a neurotransmitter acetylcholine in choline and acetate, which reduces the amount of acetylcholine. The inhibition of AChE has been studied for a long time not only for AD but also in anxiety treatment. There are some plants with an inhibitory effect on AChE that can be used as an adjuvant in AD in dietary supplements. Extracts from *Cucurbita pepo* L., *Solanum melongena* L., and *Spinacia oleracea*

L. contain a considerable number of compounds with inhibition effect on AChE. Evernic acid, usnic acid, atranorin, and vulpinic acid are plant growth inhibitors from *Umbilicaria esculenta*. Thus, plants seem to be a potential alternative to become a valuable nutraceutical source for the preparation of functional food. The purpose of this chapter consists on describe a great variety of plants with the inhibitory effect on AChE and to provide a broad view of their applications. The fact that naturally occurring compounds from plants are a potential source of new inhibitors has led to the discovery of an essential number of secondary metabolites and plant extracts with the ability to inhibit the enzyme AChE.

7.1 INTRODUCTION

Alzheimer's disease (AD) is a neurological disorder, which is characterized by memory loss, cognitive dysfunction, behavioral disorders, and deficits in daily activities (Konrath et al., 2012). Currently, drugs based on the use of acetylcholinesterase inhibitors (AChEIs) allow to control the levels of the neurotransmitter acetylcholine (ACh), which is involved in one of the memory mechanisms, whose effect ends due to the hydrolytic action of the enzyme acetylcholinesterase (AChE). This effect is important in older adults with AD who show a decrease in the number of ACh-producing cells (Jamerson et al., 2008). The use of AChEI in order to improve cholinergic function in the brain is the main strategy in the treatment of said disease (Barbosa et al., 2006; Ortiz et al., 2013). The degeneration and loss of the cholinergic innervation of the basal brain are considered as one of the causes of cognitive impairment and memory loss as symptoms of the disease. Currently, the strategy in the treatment of AD is based on cholinergic neurotransmission, where this "cholinergic hypothesis" postulates that the symptoms presented by patients are the result of poor cholinergic transmission and deficit in the level of neurotransmitters mainly ACh. Several drugs—AChE inhibitors, such as tacrine, donepzil, rivastigmine, and galantamine, are available for the treatment of AD; however, its success is limited by its dose-dependent liver toxicity and its adverse side effects such as anorexia, diarrhea, fatigue, nausea, muscle cramps, as well as gastrointestinal, cardiorespiratory, genitourinary, and sleep disorders (Jamerson et al., 2008; Obulesu and Lakshmi, 2014). Thus, plants seem to be a potential alternative to become a valuable nutraceutical source for preparation of functional food and new medical alternatives for the treatment of the disease.

7.1.1 ENZYME ACETYLCHOLINESTERASE

AChE (EC 3.1.1.7) has been one of the most studied enzymes in terms of its physiological effect, mechanism of action, nature of its active center, as well as its distribution and location in different tissues, it belongs to a family of enzymes known as cholinesterase (Sánchez and Salceda, 2008), which are defined as a group of serine esterase capable of hydrolyzing choline esters, such as ACh. Cholinesterases have a very wide distribution, they have been found from single-celled organisms, plants, invertebrates, and in vertebrates (Sánchez and Salceda, 2008; Oran et al., 2015; López et al., 2018) it appears from very early stages of embryonic development before synaptogenesis, which suggests that these enzymes may have different functions (Muñetón, 2009).

The enzyme AChE is a polymorphic enzyme that is composed of several catalytic subunits of an approximate of 80 kDa, which are assembled to form an oligomeric structure, whose active site includes a hydroxyl group that performs a nucleophilic attack on the carbonyl carbon of ACh, developing a tetrahedral transition and the product is the acetylation of serine and the release of acetate and choline (Johnson and Moore, 2006; Sánchez and Salceda, 2008; Mosquera et al., 2004). The active site of this enzyme is formed by two subsites: (1) steric and (2) anionic, both are necessary for the catalytic mechanism. The steric subsite contains the catalytic triad consisting of Ser 200, His 440, and Glu 327, as seen in Figure 7.1 (Muñetón, 2009; Mosquera et al., 2004) in the latter the ACh is hydrolyzed, generating acetate and choline (López et al., 2018).

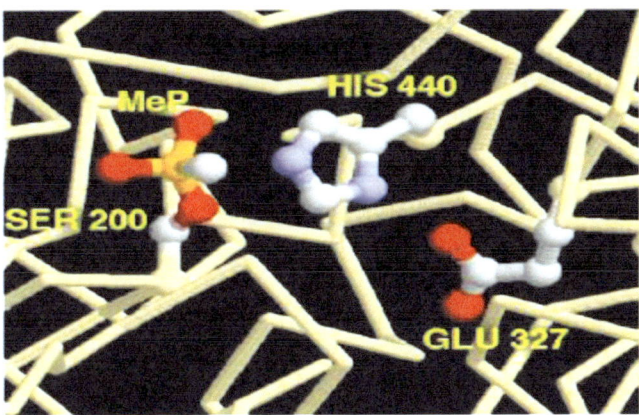

FIGURE 7.1 Structure of the active site of the AChE enzyme (Mosquera et al., 2004).

AChE is a serine hydrolase that breaks choline esters, the main substrate being the neurotransmitter ACh. It has a high catalytic activity, and each molecule is capable of degrading about 25,000 ACh molecules per second. It is considered an almost perfect enzyme whose working limit depends on the diffusion of the substrate (1). AChE is responsible for regulating the transmission of the nerve impulse in the cholinergic synapse by hydrolysis of the neurotransmitter ACh in acetic acid and choline (Oran et al., 2015; López et al., 2018; Johnson and Moore, 2006; Merchan et al., 2008).

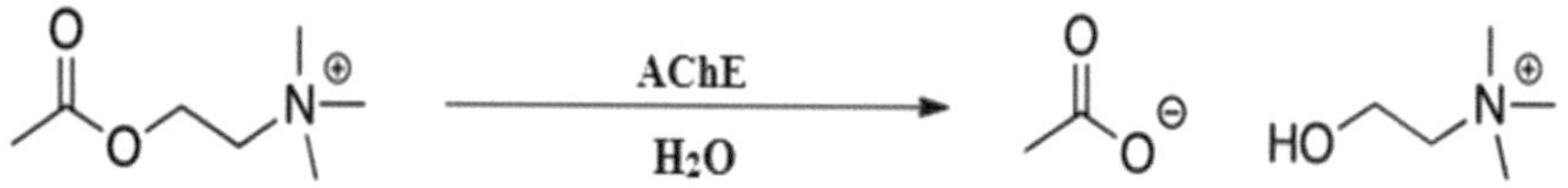

FIGURE 7.2 Reaction catalyzed by AChE on ACh obtaining acetate and choline products (Shaikh et al., 2014).

ACh is a neurotransmitter present in the central and peripheral nervous system of vertebrates and insects; it is involved in motor mediation, in autonomic functions, and in the mechanism of cognitive events, such as memory function (Shuvaey et al., 2001). This neurotransmitter is synthesized by acetylcholinetransferase from acetyl coenzyme A (Lu et al., 2011) and degraded by AChE into neuromuscular and neuroeffective junctions (Castellani et al., 2010). Trials of AChE activity can be used for the verification of treatment effectiveness, for example, for AD therapy. Novel drugs for AD are tested by in vitro methods when AChE is implicated in the treatment process (Oran et al., 2015). A trial of nerve agents and selected pesticides by devices with AChE is another application of this enzyme (Sánchez and Salceda, 2008). Experimental protocols for AChE activity trials have been proposed. The main role of AChE is the termination of nerve impulse transmission to the cholinergic synapse by the rapid hydrolysis of ACh. AChE is found in the serum fluid between each cholinergic synapse and it is in that place where the enzyme regulates the nerve impulse, eliminating the neurotransmitter and thus ending the impulse. Although there is a balance between the speed with which the AChE reaction is carried out on the ACh and the amount of enzyme and substrate, there are some anomalies that may occur, due to the absence of ACh or some other disorder related to the same. A low concentration of ACh can cause various diseases, one of the most common and reported is AD. Other pathologies that are related to the effect of AChE are myasthenia gravis, schizophrenia, attention deficit or cognitive deficit disorders, glaucoma, and Parkinson's disease (Sánchez and Salceda,

2008; Brambati et al., 2009; Neumann et al., 2006; López, 2003; Toledo et al., 2017; Bateman, 2017)

7.1.2 ALZHEIMER'S DISEASE

AD known as senile dementia, because it occurs in elderly patients, is the most common cause of irreversible dementia, accounting for up to 70% of all dementia cases that involve progressive deterioration of intellectual function: memory, language, reasoning, decision making, visuospatial function, attention, and orientation. Cognitive disorders can be accompanied by changes in personality, emotional regulation, and social behaviors which influences the work, social activities, and relationships and impair a person's ability to perform daily activities (e.g., driving, shopping, housekeeping, cooking, managing finances, and personal care) (Korolev, 2014). According to the National Institute of Neurology and Neurosurgery in Mexico, there are more than 350,000 people who are affected by this disease and more than 5 million people suffer from this disease in the United States alone, with an increase of 13.8 million in the year 2050 (Korolev, 2014). In addition, the world's population is aging rapidly, and the number of people with dementia is expected to grow from 35 million today to 65 million by 2030.

The costs of care for patients with AD and other dementias in 2015 was approximately 226 billion dollars, which is expected to increase beyond a trillion dollars in annual cost in 2050, which is a major economic challenge unless disease-modifying treatments are developed (Sperling et al., 2011; Merchan et al., 2008). Approximately 10% of patients are older than 65 years and 47% are older than 85 years.

AD is the most common cause of dementia, being a brain disorder that causes problems related to memory, thinking, and behavior (Neumann et al., 2006; Merchan et al., 2008). Usually, the patient gets progressively worse, showing perceptual, language, and emotional problems as the disease progresses (Houghton et al., 2006). It is a neurodegenerative disorder, so it continues to affect the person as time passes, there is a deficit of neurotransmitters (ACh), which causes a low cholinergic transmission (Rubinztein, 2006). It was also found that, in the hippocampus and cerebral cortex of patients, there is up to 90% decreased of the enzyme choline acetyltransferase, which catalyzes the formation of ACh, so this considerable decrease, produces a deficit of ACh (Pinar et al., 2017). Normally, upon receiving the nerve impulse signal, ACh is released into the serum liquid where it meets AChE that fulfills its normal function by degrading one part of the released

ACh while the other ends the impulse reaching the postsynapse. However, in this disease, there is not enough concentration of ACh, so AChE degrades ACh and since there is a deficit, it is not possible to end the nerve impulse. Alzheimer's patients have a progressive loss of cholinergic synapse, in brain regions related to major mental functions mainly in the hypothalamus. In these patients, the critical element for the development of the disease consists in the decrease of ACh, being this an important neurotransmitter.

The acquisition and maintenance of memory is a multicausal function dependent on individual, collective, nutritional, historical, traumatic, and environmental factors, which are strongly influenced when the processes are cumulative, degenerative, or simply by the age of the individuals. The set of symptoms and signs that develop with the progressive loss of memory are classified as neurodegenerative syndromes that are pathological processes of the central nervous system that appear over time and are characterized by neurological disorders that affect damaging brain cells, thus deteriorating memory and movement. Among them, especially Alzheimer's (AD), Parkinson's, and Huntington's diseases (Chopra et al., 2011; Salles et al., 2003). AD has become a very serious social problem for millions of families and for national health systems around the world, it is a major cause of death in developed countries, behind cardiovascular diseases and cancer. However, what makes this dementia have such a strong impact on the health system and society is, without a doubt, its irreversible character, the lack of curative treatment and the burden it represents for families of those affected (Houghton et al., 2006). The disease usually has an average duration of approximately 10–12 years, although this may vary from one patient to another. The most important finding associated with memory loss and motor skills is the decrease in the neurotransmitter ACh, due to increased levels of the enzyme AChE (Chopra et al., 2011; Salles et al., 2003; Mukherjee et al., 2007).

There is no cure for AD, and drug therapy for the disease is still in its infancy. Approved medications for the treatment of probable AD help control the symptoms of AD but do not slow down the progression or reverse the course of the disease itself. Treatments for this condition are scarce and most are linked to the inhibition of AChE to reduce the destruction of ACh and ensure higher levels for the benefit of brain health (Mukherjee et al., 2007). AChEIs help to improve memory function and attention in AD patients by interfering with the breakdown of ACh, thereby increasing the levels of the neurotransmitter at the synapse. Many compounds with this property have been investigated, but there is not yet a drug of choice for this purpose (Francis et al., 1999). One of the therapeutic strategies for the treatment of AD is to counteract the deterioration of cholinergic activity in the brain using AChEIs (Marco and Carreiras, 2006), which, although they do not cure or prevent the disease, stabilize the amount of ACh at the synapse of the cerebral cortex by preventing its hydrolysis (Baquero, 2007). AChEIs were the first medication approved for the treatment of AD. Postmortem studies of the brains of patients with AD revealed lower levels of the neurotransmitter (ACh) and acetyltransferase. The shortage of ACh in the brain shows a strong correlation with impaired cognitive function in these patients (Martin et al., 2014). The development of drugs for the treatment of cognitive deficits of AD has focused on agents that counteract the loss of cholinergic activities. It is in this field, where research has achieved the greatest therapeutic advances in the treatment of AD. AChE inhibition allows regulation of this system allowing nerve impulse to be carried out.

There are currently some AChE inhibitors approved by the Food and Drug Administration (Ringman et al., 2005; Birks, 2006) that have been used as medicines for the treatment of symptomatic patients of this disease, such as donepezil (for all stages of AD), rivastagmine and galantamine (for mild to moderate AD), and tacrine among others (Shaikh et al., 2014; Neumann et al., 2006; López, 2003; Sperling et al., 2011; Mukherjee et al., 2007; Birks and Flocker, 2003; Loy and Schneider, 2006; Keith, 2010). There is evidence of the clinical efficacy of AChEI in the treatment of AD, double-blind, placebo-controlled clinical trials; using donepezil, galantamine, and rivastigmine (Keith, 2010; Chattipakorn et al., 2006; Weinmann et al., 2010; Ming et al., 2012), showed that this medication can help preserve functions of daily life and reduce behavioral problems by producing sustained benefits for a period exceeding 2 years or more (Weinmann et al., 2010; Ming et al., 2012; Ortiz et al., 2013; Feitosa et al., 2011); however, these compounds can cause adverse side effects (Feitosa et al., 2011), among the most common being anorexia, diarrhea, fatigue, nausea, muscle cramps, as well as gastrointestinal disturbances, respiratory cardio, genitourinary, and sleep disorders (Korolev, 2014; Herrero et al., 2013). Given the limited effectiveness and possible adverse effects of current medication, there is a growing interest in using herbal remedies as a complementary therapy for AD, although medicinal plants have been used for thousands of years for medicinal purposes; nowadays, science has turned its attention to them, interested in their possible effectiveness in favoring certain diseases, Table 7.1 lists some medicinal plants used as an alternative to treat this disease (Akhondzadeh and Abassi, 2006; Howes and Perry, 2011; Jiménez, 2009; Kim et al., 2010), these do not act as AChEIs, their mechanism of action is to stimulate nerve receptors, blood vessel dilation achieving with this a better blood flow in the brain cells, these observations are presented with the reservation of not having a scientific veracity, what could have no healing property and fall into pseudoscience; however, from the point of view of herbalism they are recommended as an alternative treatment. Similarly, there are medicinal plants with an inhibitory effect on AChE as a treatment for AD, mainly belonging to the following families: *Apiaceae, Lamiaceae, Magnoliaceae,* and *Tiliaceae* (Perry and Howes, 2010).

TABLE 7.1 List of Medicinal Plants Without Scientific Evidence of the Presence of AChEIs Used in Alternative Medicine (Herbal Medicine) for the Treatment of Alzheimer's (Weinmann et al., 2010; Ming et al., 2012)

Plant (Scientific Name)	Metabolites Present	Origin	Uses
Ginkgo biloba	Flavones, lactones, terpenic, and phytosterols	Japan	It improves blood flow and brain activity, promotes good memory, and mental agility
Uncaria tomentosa	Oxindole alkaloids, glycosides, polyphenols, terpenes, and steroids	Peru	Cancer, Alzheimer's disease, and dilates blood vessels
Panax ginseng	Citric acid, fumaric acid, paraxanes, beta-carotenes, ginsenosides, and vitamins	Siberia, China, and Korea	Diabetes, colds, cancer, and Alzheimer's disease,
Curcuma longa	Phenolic derivatives, peptides, proteins, and methionine residues with antioxidant properties	India, China, and Middle East	Anti-inflammatory, Alzheimer's disease
Smilax aspera	Saponins, acids (ascorbic, stearic, linoleic, oleic, palmitic), tannins, glycosides, phytosterols, and minerals	Amazonia	Acne, Alzheimer's disease, arthritis, asthma, kidney stones, cancer, and among others

7.1.3 ACETYLCHOLINESTERASE INHIBITORS

The use of medicinal plants (PM) dates to the beginnings of human civilization; ancient cultures such as China, India, and North Africa have provided evidence on the use of this resource for healing purposes. According to the World Health Organization (WHO), a medicinal plant contains substances that can be used for therapeutic purposes and/or can serve as active ingredients or as precursors for the semi-synthesis of new drugs, estimates that more than 80% of the world's population routinely use them to meet their main health needs (Adams et al., 2007; Yang et al., 2011). Thus, Eastern cultures, such as India or China, have been incorporating experiences in this line of study for years

(Akhondzadeh and Abassi, 2006; Howes and Perry, 2011). Chinese herbs are particularly promising and have undergone multiple clinical trials. However, most of these studies primarily used formulas (that is, combinations of several plants) instead of assessing the efficacy of the plants individually. From a scientific point of view, it is important to understand the pharmacological properties of the herb in isolation, as well as its active chemical compounds. In the African herbal pharmacopoeia (Jiménez, 2009), numerous plants have been used for the treatment of various diseases, including inflammation and neuropharmacological disorders. Thus, complementary medicine is becoming increasingly popular in many developed countries; 48% of Australia's population, 70% in Canada, 42% in the USA, 38% in Belgium, and 75% in France have used it mainly in homeopathy and acupuncture (Lai et al., 2006). The world market for herbal remedies used in traditional medicine is now estimated at 60,000 million US dollars (Llorens et al., 2014).

The use of medicinal plants to cure some health discomforts is a very common practice in many countries. According to the WHO, 80% of people in less developed regions use traditional medicine using plants for health care. This popular knowledge is based on effectiveness, that is, it is accepted and adopts what is useful; however, a problem with phytotherapy is the difficulty of controlling the dose and quality of the product, leading to health risks and damages. Many of the traditional remedies are manufactured from wild populations whose chemical content varies due to genetic or environmental reasons. In addition, there is not enough information on the abundance and distribution of all medicinal plants, much less on the range of species variability (Kim et al., 2010).

Studies in recent years have shown that some plant species may have molecules that have acetylcholinesterase inhibitory activity (AChEI), which have served as a basis for the development of medications that can be used to prevent or delay the development of AD or decrease the severity of symptoms (Marco and Carreiras, 2006; Jiménez, 2009) particularly by means of the drug galantamine isolated from several species of the *Amaryllidaceae* family (Jiménez, 2009; Lai et al., 2006; Kim et al., 2010; Kwon et al., 2015). This is due to the fact that AD is associated with the decrease in the levels of the neurotransmitter ACh, as a consequence of the increase in AChE; thus, treatments focus on drugs that increase ACh levels by inhibition of the enzyme AChE. Studies in different parts of the world have described some secondary plant metabolites as a possible source to inhibit enzyme activity (Jiménez, 2009; Lai et al., 2006; Loraine and Mendoza, 2010). Barbosa et al. (2006); Lai et al. (2006) carried out a selection of plants used in various studies throughout history in the treatment of AD, among which are the families of plants *Amaryllidaceae, Apiaceae, Asteraceae, Fabaceae, and Fumariaceae.*

Trevisan (Kim et al., 2010) conducted a study with plants native to Brazil, finding excellent results in species of *Amburana cearensis, Lippia sidoides, Paullinia cupana, Plathymiscium floribundum,* and *Solanum asperum* as AChEIs. Similarly, Feitosa et al. (Kim et al., 2010) evaluated extracts of medicinal plants belonging to the *Convolvulaceae, Crassulaceae, Euphorbiaceae, Leguminosae, Malvaceae, Moraceae, Nyctaginaceae,* and *Rutaceae* families, finding that *Ipomoea asarifolia, Jatropha curcas, Jatropha gossypiifolia, Kalanchoe brasiliensis,* and *Senna alata* are the most active species as AChEIs. Akhondzadeh *et al.* (2006); Wilkinson, (2007)[34] have observed that the treatment of AD with aqueous extracts of the *Melissa officinalis* plant could have a moderate positive effect on memory. Adams *et al.* (2007) [49] used plants from the *Zygophyllaceae* family to help memory. While the *Vervena officinalis* plant, used in traditional medicine as a nutritional supplement for the treatment of dysentery, amenorrhea, and depression can have cytoprotective activities of the cells of the central nervous system (Howes and Perry, 2011). The above being just some of the many investigations in this field of study. As can be seen in several studies, results of studies focused on the effect of extracts from various plants are reported showing the presence of compounds that have inhibitory activity on AChE is an enzyme that interferes in the control of AD (Kim et al., 2010; Lai et al., 2006).

Table 7.2 shows a list of medicinal plants that have been studied as AChEIs. These plants were selected because they were used in alternative medicine (herbal medicine) for the treatment of certain diseases. The sale of herbs and herbal supplements has increased by 101% (Jayasri et al., 2009). The most popular products included ginseng (*Ginkgo biloba*), garlic (*Allium sativum*), *Echinacea spp.* and St. John's wort (*Hypericum perforatum*), traditional medicine and complementary alternative medicine are gaining more and more respect from national governments and health providers

7.2 CONCLUSIONS

AD is a very serious social problem for millions of families and for national health systems around the world, treatments for this condition are scarce and most are related to AChE inhibition. Although there is some research in this regard there is not yet a drug of choice since inhibitors used without synthetics and these cause adverse side effects. Therefore, one of the therapeutic strategies for the treatment of said AD could be the use of medicinal plants with molecules that have AChE inhibitory activity, as a complementary therapy for the disease. Although these inhibitors do not cure or prevent the disease, they can be used to prevent or delay the development of AD or decrease the severity of symptoms

TABLE 7.2 List of Plants With Scientific Evidence of Potential Presence of AChEIs

Scientific Name	Family	Use in Herbal Medicine	Observations and References
Pimpinella anisum	*Apiaceae*	Aromatic, antispasmodic, digestive, and diuretic	Babosa et al. (2006)
Salvia officinalis	*Lamiaceae*	Against depression, vertigo, Alzheimer's	
Melissa officinalis	*Lamiaceae*	Antiseptic, to improve memory, against stress	Akhondzadeh et al. (2006)
Larrea tridentada	*Zygophyllaceae*	Analgesic, against swelling, headache, colds, and antifungal	Adams et al. (2007)
Valeriana edulis	*Valerianaceae*	Diuretic, soothing, to improve memory	
Magnolia officinalis	*Magnoliaceae*	Against cough, heart problems, and Alzheimer's	Barbosa et al. (2006)
Tilia mexicana	*Tiliaceae*	Antispasmodic, sleeping and relaxing	
Mentha piperita var officinalis	*Lamiaceae*	Antispasmodic, antiseptic, and analgesic	
Verbena officinialis	*Verbenaceae*	Antispasmodic, antiseptic, analgesic, and relaxing	Lai et al. (2006)

ACKNOWLEDGMENTS

The authors would like to thank CONACYT for the financial support that is provided for the master's studies of the YGEN student (922133).

KEYWORDS

- medicinal plants
- Alzheimer's disease
- acetylcholinesterase
- inhibitors

REFERENCES

Adams M, Gmünder F, Hamburger M. Plants traditionally used in age related brain disorders: a survey of ethnobotanical literature. J. Ethnopharmacol. 2007; 111: 363–381.

Akhondzadeh S, Abassi, S. Herbal medicine in the treatment of Alzheimer's disease. Am. J. Alzheimer's Dis. Other Demen. 2006; 21: 113–118.

Baquero TM. Tratamiento de la enfermedad de Alzheimer. Medicine 2007; 9(77): 4936–4943.

Barbosa F, Medeiros P, Diniz M, Batista L, Athayde F, Silva M. Natural products inhibitors of the enzyme acetylcholinesterase. Rev. Bras. Farmacogn. 2006; 16(2): 258–285.

Bateman RJ. The DIAN-TU next generation Alzheimer's prevention trial: adaptive design and disease progression model. Alzheimer's Dement. 2017; 13(1): 8–19.

Birks J, Flocker L. Selegiline for Alzheimer's disease. Cochrane Database Syst Rev. 2003: CD000442.

Birks JS. Cholinesterase inhibitors for Alzheimer's disease. Cochrane Database Syst. Rev. 2006: CD005593.

Brambati S, Degroot C, Kullmann B, Strafella A, Lafontaine AL, Chouinard S. Regional brain stem atrophy in idiopathic Parkinson's disease detected by anatomical MRI. PLos One 2009; 4(12): 1–4.

Castellani RJ, Rolston RK, Smith MA. Alzheimer disease. Dis. Mon. 2010; 56(9): 484–546.

Chattipakorn S, Pongpanparadorn A, Pratchayasakul W, Pongchaidacha A, Ingkaninan K, Chattipakorn N. Tabernaemontana divaricate extract inhibits neuronal acetylcholinesterase activity in rats. J. Ethnopharmacol. 2006; 110(1): 61–68.

Chopra K, Misra S, Kuhad A. Current perspectives on pharmacotherapy on Alzheimer's disease. Expert Opin. Pharmacother. 2011;12(3): 335-350.

Feitosa C, Freitas R, Luz N, Bezerra M, Trevisan M. Acetylcholinesterase inhibition by some promising Brazilian medicinal plants. Braz. J. Biol. 2011; 71(3): 783–789.

Francis PT, Palmer AM, Snape M, Wilcock GK. The cholinergic hypothesis of Alzheimer's disease: a review progress. Neurol. Neurosurg. Psychiatry 1999; 66: 137–147.

Herrero MT, Terradillos MJ, Ramírez MV, Capdevila LM, López A, Riera K. Especias, hierbas medicinales y plantas. Usos en medicina: Revisión de la bibliografía científica (Medline). Medicina Balear. 2013; 28(2): 35–42

Houghton PJ, Ren Y, Howes MJ. Acetylcolinesterase inhibitors from plants and fungi. Nat. Prod. Rep. 2006; 23(2): 181–199.

Howes MJ, Perry E. The role of phytochemicals on the treatment and prevention of dementia. Drugs Agind. 2011; 28: 439–468.

Jamerson K, Webber M, Bakris G, Dahlöf B, Pitt B, Dhi V. Benazepril plus amlodipine or hydrochlorothiazide for hypertension in high-risk patients. N. Eng. J. Med. 2008; 359(23): 2417–2428.

Jayasri MA, Radha A, Mathew TL. α-amylase and α-glucosidase inhibitory activity of costus pictus D. Don in the management of diabetes. J. Herbal Med. Toxicol. 2009; 3(1): 91–94.

Jiménez V. Inhibidores de Acetilcolinesterasa en el tratamiento de la enfermedad Alzheimer. Rev. Méd. de Costa Rica Centroamérica 2009; 66(588): 203–206.

Johnson G, Moore S. The peripheral anionic site of acetylcholinesterase: structure, functions and potential role in rational drug design. Curr. Pharm. Des. 2006; 12(2): 217–225.

Keith AW, Alzheimer's disease: the pros and cons of pharmaceutical, nutritional, botanical and stimulatory therapies with a discussion of treatment strategies from the perspective of patients and practitioners. Altern. Med. Rev. 2010; 15(3): 223–244.

Kim J, Lee H, Lee K. Naturally occurring phytochemicals for the prevention of Alzheimer's disease. J. Neurochem. 2010; 112: 1415–1430.

Kim J, Lee HJ, Lee KW. Naturally occurring phytochemicals for the prevention of Alzheimer's disease. J. Neurochem. 2010; 112: 1415–1430.

Konrath S, Fuhrel F, Lou A, Brown S. Motives for volunteering are associated with mortality risk in older adults. Health Psychol. 2012; 31(1): 87–96.

Korolev O. Alzheimer's disease: a clinical and basic science review. Med. Stud. Res. J. 2014; 4: 24–33.

Kwon O, Lee S, Ban S, Im J, Lee D, Lee E, Kim J, Lim S, Kang I, Kim K, Yoon S. Effects of the combination herbal extract on working memory and white matter integrity in healthy individuals with subjective memory complaints: a randomized, double-blind, placebo-controlled clinical trial. Korean J. Biol. Psychiatry 2015; 22(2): 63–77.

Lai S, Yu M, Yuen W, Chang R. Novel neuroprotective effects of the aqueous extracts from *Verbena officinalis* Linn. Neuropharmacology. 2006; 50(6): 641–650.

Llorens M, Jurado J, Hernández F, Ávila J. GSK3β, apivotalkinasein Alzheimer disease. Front. Mol. Neurosci. 2014; 7(46): 1–11.

López OL. Clasificación del deterioro cognitivo leve en un estudio poblacional. Rev. Neurol. 2003; 37(2): 140–144.

López R, Valencia R, Sánchez J, Pérez B, Salinas A, Serrano H, García M, Muñoz H, Hernández A, Vidal C, Gómez J. La estructura y función de las colinesterasas: blanco de los plaguicidas. Rev. Int. Contam. Ambie. 2018; 34: 69–80.

Loraine S, Mendoza J. Las plantas medicinales en la lucha contra el cáncer, relevancia para México. Rev. Mex. Cienc. Farm. 2010; 41(4): 18–27.

Loy C, Schneider L. Galantamine for Alzheimer's disease and mild cognitive impairment. Cochrane Database Syst. Rev. 2006; CD001747.

Lu SH, Wu JW, Liu HL, Zhao JH, Liu KT, Chuang CK. The discovery of potential acetylcholinesterase inhibitors: a combination of pharmacophore modeling, virtual screening, and molecular docking studies. J. Biomed. Sci. 2011; 18: 8.

Marco L, Carreiras C. Galanthamine, a natural product for the treatment of Alzheimer's disease. Recent Pat. CNS Drug Discov. 2006; 1: 105–111.

Martin ML, Jurado J, Hernández F, Ávila J. GSK-3β a pivotal kinase in Alzheimer disease. Front. Mol. Neurosci. 2014; 7(46): 1–11.

Merchan DR, Vargas LY, Kouznetov VV. Nuevos agentes inhibidores de la acetilcolinesterasa con fragmentos estructurales de lignanos. Revista Salud UIS 2008; 40(2): 166–168.

Ming Z, Zhen D, Zhong Y, Shi X, Ya L. Effects of Ginkgo biloba extract in improving episodic memory of patients with mild cognitive impairment: a randomized controlled trial. J. Chin. Integr. Med. 2012; 10(6): 628–634.

Mosquera O, Niño J, Correa Y. Detección in vitro de inhibidores de la acetilcolinesterasa en extractos de cuarenta plantas de la flora colombiana mediante el método cromatográfico de Ellman. Scientia et Technica 2004; 10(26): 155–160.

Mukherjee PK, Kumar V, Mal M, Houghton PJ. Acetylcholinesterase inhibitors from plants. Phytomedicine. 2007; 14(4): 289–300.

Muñetón P, Plantas medicinales: un complemento para la salud de los mexicanos. Entrevista con el Dr. Erick Estrada Lugo. Revista Digital Universitaria 2009; 10(9): 1067–1076.

Neumann M, Sampathu D, Kwong L, Truax A, Micsenyi M, Chou T. Ubiquitinated TDP-43 in frontotemporal lobar degeneration and amyotrophic lateral sclerosis. Science 2006; 314: 130–133.

Obulesu M, Lakshmi M. Apoptosis in Alzheimer's disease: an understanding of the physiology, pathology and therapeutic avenues. Neurochem. Res. 2014; 39(12): 2301–2312.

Oran K, Sunho L, Soonhyun B, Jooyeon J, Doo, Eun HL. Effects of the combination herbal extract on working memory and white matter integrity in healthy individuals with subjective memory complaints: a randomized, double-blind, placebo-controlled clinical trial. Korean J. Biol. Psychiatry 2015; 22(2): 63–77.

Ortiz D, Valdez A, López L, Gaitán I, Paz M, Cruz S, Álvarez L y Cáceres A. Actividad inhibitoria de la acetilcolinesterasa por extractos de 18 especies vegetales nativas de Guatemala usadas en el tratamiento de afecciones nerviosas. Revista Científica 2013; 23(1): 17–25.

Ortiz D, Valdez A, López L, Gaitán I, Paz M, Cruz S. Actividad inhibitoria de la acetilcolinesterasa por extractos de 18 especies vegetales nativas de Guatemala usadas en el tratamiento de afecciones nerviosas. Revista Científica-IIQB 2013; 23(1): 17–25.

Perry E, Howes MJR. Medicinal plants and dementia therapy: herbal hopes for brain aging. CNS Neurosci. Ther. 2010; 17: 1–16.

Pierce AL, Bullain SS, Kawas CH. Late-onset Alzheimer disease. Neurol. Clin. 2017; 35(2): 283–293.

Pinar A, Mahir C, Hakan Y, Remise G, Esma A, Sena O. The efficacy of donepezil administration on acetylcholinesterase activity and altered redox homeostasis in Alzheimer's disease. Biomed. Pharmacother. 2017; 90: 786–795.

Ringman JM, Frautschy SA, Cole GM, Masterman DL, Cummings JL. A potential role of the curry spice curcumin in Alzheimer's disease. Curr. Alzheimer Res. 2005; 2: 131–136.

Rubinztein DC. The roles of intracellular protein degradation pathways in neurodegeneration. Nature 2006; 443: 780–786.

Salles MT, Viana FV, Van de Meent M, Rhee IK, Verpoorte R. Screening for acetylcholinesterase inhibitors from plants to treat Alzheimer disease. Quím. Nova 2003; 26: 117.

Sánchez G y Salceda R. Enzimas polifuncionales: el caso de la Acetilcolinesterasa. Rev. Educ. Bioquimica 2008; 27(2): 44–51.

Sánchez G, Salceda R. Enzimas polifuncionales: el caso de la acetilcolinesterasa. Revista de Educación Bioquímica 2008; 27(2): 44–51.

Shaikh S, Verma A, Siddiqui S, Ahmad SS, Rizvi SMD, Shakil S, Biswas D, Singh D, Siddiqui MH, Shakil S, Tabrez S, Kamal MA. Current acetylcholinesterase-inhibitors: a neuroinformatics perspective. CNS Neurol. Disord. Drug Targets 2014; 13(3): 391–401.

Shuvaey W, Laffont I, Serott JM, Fujii J, Taniquchi N, Siest G. Increased protein glycation in cerebrospinal fluid of Alzheimer's disease. Neurobiol. Aging 2001; 22(3): 397–402.

Sperling RA, Aisen PS, Beckett LA, Bennett DA, Craft S, Fagan AM. Toward defining the preclinical stages of Alzheimer's disease: recommendations from the National Institute on Aging-Alzheimer's Association workgroups on diagnostic guidelines for Alzheimer's disease. Alzheimer's Dement. 2011; 7(3): 280–292.

Toledo JB, Arnold M, Kastenmüller G, Chang R, Baillie RA, Han X. Metabolic network failures in Alzheimer's disease—a biochemical road map. Alzheimer's Dement. 2017; 13(9): 965–984.

Weinmann S, Roll S, Schwarzbach C, Vauth C, Willich S. Effects of Ginkgo biloba in dementia: systematic review and meta-analysis. BMC Geriatr. 2010; 10: 14.

Wilkinson D. Pharmacotherapy of Alzheimer's disease. Psychiatry. 2007; 7: 9–14.

Yang YH, Wu SL, Chou MC, Lai CL, Chen SH, Liu CK. Plasma concentration of donepezil to the therapeutic response of Alzheimer's disease in Taiwanese. J. Alzheimer's Dis. 2011; 23(3): 391–397.

CHAPTER 8

Characterization of *Heliopsis longipes* and the Potential of Its Ethanolic Extract as an Adjuvant in the Treatment of Pharyngitis

MARIELA CORREA-DELGADO, VÍCTOR MANUEL OROZCO-GONZÁLEZ, ANA PAOLA ALDAMA-NÚÑEZ, VICTOR OLVERA-GARCÍA, and VÍCTOR MANUEL RODRÍGUEZ-GARCÍA

Tecnologico de Monterrey, Escuelade Ingeniería y Ciencias, San Pablo, Querétaro 76130, Mexico

*Corresponding author. E-mail: vmrodrigg@tec.mx.

ABSTRACT

The ethanol-soluble compounds present in the roots of chilcuague (*Heliopsis longipes*) were obtained with three different extraction methods and analyzed using GC-MS. Affinin was isolated using preparative high-pressure liquid chromatography and then analyzed with ultraperformance liquid chromatography-mass spectrometer. Affinin quantification was carried out using high-performance liquid chromatography. Total phenolic compounds were evaluated following the Folin–Ciocalteu method, whereas antioxidant activity was analyzed performing 2,2 Diphenyl-1-picrylhydrazyl assays. Additionally, the bactericidal effect of the extracts was evaluated on Gram-positive and Gram-negative bacteria, as well as on bacteria obtained from buccal microbiota. Finally, genomic DNA was extracted from *H. longipes* leaves in order to characterize the barcoding sequences *trn*H-*psb*A, *rbc*L, and *mat*K for the proper identification of the plant species. Based on these analyses, the use of *H. longipes* ethanolic extract was proposed as a coactive agent in the treatment of pharyngitis.

8.1 INTRODUCTION

Pharyngitis, commonly called sore throat, is the inflammation in the pharynx and the surrounding lymphoid tissue. Figuring within the top 10 causes of health care visits, it can be caused either by bacteria, fungi, or viruses (Aalbers et al., 2011; Farah and Visintini, 2018); however, the majority of the cases related to this disease are associated with viral infections. Nonviral common pathogens include groups A, C, and G *Streptococcus, Neisseria gonorrhoeae, Corynebacterium diphtheriae, Haemophilus influenzae, Corynebacterium haemolyticum, Mycoplasma pneumoniae,* and *Chlamydia pneumoniae* (Woodley and Whelan, 1993). Recently, *Fusobacterium necrophorum* was recognized as a pathogen that could lead to pharyngitis, and then result in Lemierre syndrome, a destructive, suppurative condition with a mortality rate of 5% (Aalbers et al., 2011). Patients with Lemierre syndrome develop internal jugular thrombophlebitis; this infection can invade the lungs, joints, and even the brain. Physicians advise early diagnostic of patients with pharyngitis symptoms, paying attention to persisting or worsening of these symptoms (Aalbers et al., 2011).

Group A *Streptococcus* (GAS) is the most common bacteria associated with this disease in both children and adults, making 40% to 60% of the cases (Jiang et al., 2016). These bacteria can cause not only pharyngitis but lead to other conditions such as rheumatic fever (RF) and rheumatic heart disease (RHD), turning the treatment of this bacteria into something crucial, especially in low- and medium-income countries, where RHD is the principal cause of cardiovascular mortality (Fischer Walker et al., 2006). Patients with streptococcal pharyngitis manifest irritation of the throat, fever, tonsillar exudates, headache, dry mouth, swallowing difficulties, and cervical lymphadenopathy, whereas cough, coryza, and diarrhea are present more frequently in viral pharyngitis (Choby, 2009; Muller et al., 2016).

The symptoms of streptococcal pharyngitis may resemble other diseases, making it difficult to establish an accurate diagnosis based only on physical examinations (Choby, 2009). The conventional diagnosis method in clinical microbiology laboratories involves the performance of throat culture; this is mainly done when the patient has RF antecedents, streptococcal pharyngitis exposure, or when they present complications. If the clinical history suggests it, physicians may try to find *N. gonorrhoeae* or diphtheria, applying specific culture techniques (Woodley and Whelan, 1993). Streptococcal pharyngitis can be recognized in a more effective manner using rapid streptococcal tests, which consists of the detection of a streptococcal antigen through immunoassays. Although this method is highly specific, the sensibility of this test is considered low in comparison to microbial cultures, especially when physicians are not properly trained to perform the test (Toepfner et al., 2013); consequently, a negative result must not be taken conclusively, and it is necessary to confirm with other methods to obtain a reliable diagnosis (Van Limbergen et al., 2006). More recently, a molecular biology approach for the detection of streptococcal pharyngitis was developed, yielding a sensitivity of 96% and a specificity of 98.6%. Authors concluded that PCR detection performance was comparable to conventional throat culture, with the advantage of obtaining promptly results with a low risk of cross-contamination; however, although this technique may be helpful for clinical laboratories, in physician offices medical personnel may find easier to make use of more conventional detection methods (Slinger et al., 2011).

Even though the prevention of RF and RHD is still a concern in underdeveloped countries, the situation differs a lot in countries with advanced economies, where RF and RHD are so unusual; the government encourages clinicians not to prescribe antibiotics to treat pharyngitis, advocating that the harm related to the use of antibiotics overcomes any possible positive outcome (Tanz et al., 2019). Equally important, some physicians and researchers are well aware that a wrong diagnosis and incorrect prescription of antibiotics may lead to antibiotic resistance in bacteria (Van Brusselen et al., 2014). Furthermore, in most cases, pharyngitis is originated from viruses and the symptoms are relieved in three to five days, discouraging physicians from prescribing antibiotics unless the symptoms endure or aggravate (Aalbers et al., 2011; Van Brusselen et al., 2014).

Given the controversy surrounding the prescription of antibiotics and the necessity of making an accurate diagnosis, physicians must make use of methodologies and guidelines to ensure the distinction between cases where the use of antibiotics is mandatory, and cases where this may not be the best option (Aalbers et al., 2011; Choby, 2009). The Centor Score, which involves the analysis of patients through physical examination, is used to decide whether a patient should be treated symptomatically, with the use of antibiotics, or if it is necessary to perform immunoassays or throat culture analysis (Kalra et al., 2016).

According to the Centor Score, physicians must avoid the prescription of antibiotics in patients with low GAS infection risk, managing the symptoms with palliative drugs while observing the development of the disease (Choby, 2009). The symptomatic treatment of pharyngitis involves the use of anti-inflammatories, antipyretics, analgesics, disinfection solutions, and lozenges or sprays containing anesthetics (Lozano, 2003; Muller et al., 2016). In a study performed by the Family Physicians Inquiring Network, nonsteroidal anti-inflammatory drugs, paracetamol, steroids, and oropharyngeal sprays with benzocaine and lidocaine were compared, all of them demonstrating efficiency in alleviating pharyngeal pain (Frye et al., 2011).

Several drugs are used in the formulation of oropharyngeal sprays to treat pharyngitis, namely ectoine, with anti-inflammatory properties (Muller et al., 2016), flurbiprofen, also with anti-inflammatory properties (Looze et al., 2018), chlorhexidine, with antibacterial and anti-infective properties (Golac-Guzina et al., 2019; pubchem.ncbi.nlm.nih.gov, 2019d), lysozyme, with antibiotic properties (Catic et al., 2016; Golac-Guzina et al., 2019), and benzydamine, with anti-inflammatory, analgesic, and antipyretic properties (Kim et al., 2019; pubchem.ncbi.nlm.nih.gov, 2019c).

8.1.1 ALTERNATIVE AND COMPLEMENTARY MEDICINE FOR THE TREATMENT OF PHARYNGITIS.

Alongside clinical examinations and medical treatment, pharyngitis treatment can be accompanied by the use of alternative and complementary medicine. The Food and Drug Administration (FDA) definition of complementary and alternative medicine includes several practices that attempt to treat diseases through therapies that differ from standard (allopathic) medicine (FDA, 2007). Standard medicine refers to medical practices that are executed by professionals with a medical degree, physical therapists, psychologists, and registered nurses (NCCIH, 2019).

As complementary and alternative medicine continue growing, regulatory measures to ensure the quality, efficiency, and to prevent them from exerting any damage to consumers. These legislative controls may vary from country to country, as there are different manners in which each country defines medicine plants and the products that are derived from them. Additionally, some underdeveloped countries may have vast empirical knowledge in the use of medicinal plants, notwithstanding, they may have few or null progress introducing the usage of plants into drug legislation (WHO, 2001). Moreover, although the use of herbal medicine is mostly recognized as secure, its efficiency is not always guaranteed, as the requisites to commercialize an herbal product are not always as strict as in regular medicine (Moreira et al., 2014). For this reason, it is always important to conduct scientific research before launching a herbal product in order to legitimate not only its safety but also its functionality.

Herbal medicines are dietary supplements used by people to improve their health (MedlinePlus, 2019). Herbs have been used for therapeutic purposes since prehistory, as a result of the properties given by their active compounds. These properties, also called biological activities or bioactivities, are a focus of attention in scientific research, as many pharmaceuticals are developed based on plants, for example, aspirin, quinine, and morphine, derived from willow bark, cinchona bark, and opium poppy, respectively (Vickers et al., 2001).

The medicinal properties of plants, conferred by their metabolites, had been widely used to treat bacterial infections through human history (Wijesundara and Rupasinghe, 2019); several studies had demonstrated the capacity of herbal medicine to treat pharyngitis not only symptomatically, but also etiologically. In a study carried out by Wijesundara and Rupasinghe (2019), different herbal teas used in Canadian traditional medicine were tested to determine their potential to inhibit the growth and formation of biofilm against GAS, finding out licorice root, barberry root, oregano flowering shoot, and thyme flowering were the most efficient. The formation of biofilm can make bacteria more resistant to antibiotics; thus, the study conducted by Wijesundara suggests that these herbal teas may be used against antibiotic-resistant GAS (Wijesundara and Rupasinghe, 2019). Furthermore, aqueous *Costus speciosus* rhizome extracts had been tested in pediatric and adult pharyngitis patients based on the anti-inflammatory, antipyretic and analgesic properties that this plant has demonstrated. They found out that 60% of the patients showed symptom improvement, with a remission rate of 93% by day 5 (Bakhsh et al., 2015).

Similar to the aim of our research, one oropharyngeal spray formulated with *Salvia officinalis* ethanolic extract demonstrated efficiency relieving pharyngitis pain intensity significantly in the first 2 h when compared against placebo (Hubbert et al., 2006). Nevertheless, another study compared the efficacy of *S. officinalis* infusions against a benzydamine-based spray in the treatment of postoperative sore throat, finding out that *S. officinalis* infusions were less effective managing pain than benzydamine sprays (Lalicevic and Djordjevic, 2004).

8.1.2 CITRUS VOLATILE COMPOUNDS

Essential oils are chemical compounds present in plants that are produced as secondary metabolites. Within the plant, these are produced as a protection system that protects them from infections and avoid animal consumption.

They are also useful to attract insects that may help in their reproduction by pollination (Bakkali et al., 2008). Some records suggest their use since 10,000 BC; nonetheless, it was not until the 10th century that the process to extract them through steam distillation was described (Baser and Buchbauer, 2015). Essential oils have a broad and well-studied set of biological properties (El Asbahani et al., 2015); they are used against infections as a result of their antimicrobial activity (Chouhan et al., 2017), even dealing with super-resistant bacteria, such as methicillin-resistant *Staphylococcus aureus,* keeping them from producing biofilms and expressing virulence factors (Kim et al., 2015). Their use as food preservatives has also been reported, and this is becoming an essential field of research since people are increasing their negative perception of synthetic food additives, making industries concerned in finding suitable alternatives (Hyldgaard et al., 2012).

Plant species within the *Citrus* genus are rich in secondary metabolites that may have properties of interest in different areas, especially in the food, cosmetic, pharmaceutical, and perfume industry. Their essential oils, obtained by distillation, are composed of 85%–90% by volatile compounds, about 200 different chemicals (González-Mas et al., 2019). The compounds found in citrus plants essential oils are known to have great potential as bacterial inhibitors against Gram-positive and Gram-negative bacteria. The main advantage of using these oils instead of chemical-based bactericides is that their components are classified as safe by the FDA, and their odor and flavor are well suited to be used in food (Fisher and Phillips, 2008). Although the process in which citrus essential oils operate against microorganisms is not fully understood, several mechanisms of action have been proposed (Fisher and Phillips, 2008).

8.1.3 MEXICAN TRADITIONAL MEDICINE

In the late 16th century, as a result of the Spanish colonization, the therapeutic practices from indigenous groups consisted mainly of the employment of natural resources for healing purposes. In contrast, the ritual practices were suppressed because the colonizers considered them to be an obstacle to evangelization. Nevertheless, Mexican traditional medicine has evolved by the interaction with other therapeutic practices and exerting an influence on human development and cultural patrimony (Jiménez-Silva, 2017).

Scientific research of the phytochemicals found in Mexican endemic plants and their use for therapeutic purposes have prospered, especially in the discovery of anticancer and antidiabetic alternatives (Texeira-Duarte and

Rai, 2015). In a study conducted at Tecnologico de Monterrey, researchers analyzed the antibacterial in vitro properties of 343 plants used in Mexican traditional medicine, tested in different bacterial strains. Besides, they gathered data on their toxicological activities as well, intending to ensure the safety of the consumers of these plants and products that derive from them (Sharma et al., 2017). The authors identified 225 compounds with antibacterial properties among 75 different plant species, being Asteraceae, Fabaceae, Lamiaceae, and Euphorbiaceae plant families that ingathered the highest number of plant species with antibacterial properties. However, they found out that 40.57% of the examined plants had at least one report of their toxic effects (Sharma et al., 2017).

8.1.4 HELIOPSIS longipes (A. GRAY) S.F. BLAKE

8.1.4.1 HELIOPSIS LONGIPES TRADITIONAL USES

The first records of the use of *H. longipes* medicinal properties date from 1615. Francisco Hernández described the medicinal properties of plants in New Spain, a plant known as "chilmecuan" employed for the relief of toothache, as well as for the alleviation of headache and ear pain when applied along with a preparation of natural resins. The roots of this species have demonstrated to exhibit analgesic properties (Hernández, 1888). Later in 1948, Little (1948) pointed out that "chilmecuan," also known as "chilcuague," "golden root," "chilcuan," or "chimecátl," actually corresponded to *H. longipes.* Moreover, Acree Jr and Haller (1945) mention the use of chilmecuan roots in the preparation of natural insecticides. Similarly, *H. longipes* roots are known to have been used by prehispanic civilization to enhance the taste of food (Rios and del Carmen, 2007).

Cilia-Lopez et al. (2008) gathered the available knowledge regarding the traditional uses of *H. longipes* through a direct questionnaire with users, merchants, and producers of this species. According to their studies, the ground roots are most frequently employed as a condiment in sauces to increase pungency. Besides, the roots are well known for its anti-inflammatory properties and are employed as a homemade antiparasitics. Other medicinal applications of *H. longipes* include its use in the treatment of oral herpes, oral infections, muscular pain, arthritis, and rheumatism (Cilia-Lopez et al., 2008).

In 1947, the Bureau of Entomology and Plant Quarantine of the United States studied the effect of *H. longipes* dust and spray against house flies, mosquitoes, lepidopterous larvae, and bugs. Their studies reflect that the preparation of *H. longipes* roots results in toxic to houseflies (McGovran et al., 1947). Moreover, the preparation of *H. longipes* roots in milk is employed as an insecticide against flies that drink it, according to the studies reported to Little. Additionally, if grounded root tissue is introduced in wounds, it prevents the spread of larvae (Little, 1948).

8.1.4.2 BOTANICAL GENERALITIES

8.1.4.2.1 General description

H. longipes is a perennial plant endemic to Mexico that grows along the mountainous region known as Sierra Gorda and Sierra de Alvarez (Velez-Haro et al., 2018). Its distribution is limited to the south of San Luis Potosi and the northern part of Guanajuato. Plant specimens can also be found in the north of Queretaro (Little, 1948). *H. longipes* is classified within Magnoliophyta and part of the Asteraceae family, tribe Heliantheae (tropicos.org, 2019).

According to the description of *H. longipes* established by Little (1948), the plant varies in size, ranging from 20 to 70 cm. It presents a few small, ovate leaves along the stem, which are characterized by the presence of serrate leaf margins and short petioles. *H. longipes* grows as a large shrub and presents yellow flowers with ligulated petals (Figure 8.1) (Cullen et al., 2011).

8.1.4.2.2 Morphology

Roots are thin and abundant, similar to the roots of *Tanacetum cinerariifolium*. When consumed, these provide a bitter and pungent flavor. Their size ranges from 15 to 30 cm of 2 mm wide. The cortex is brownish, and the axis is yellow and lignified (Castro-González, 2009). This organ is where the affinin is found (Castro-Ruiz et al., 2017). Stems are relatively large, fluted, slightly bend, and presents no ramifications in the base. The peduncles are up to 25 cm in length and have a unique terminal yellow capsule (Cilia-Lopez et al., 2008; Little, 1948). Leaves are deltoid, oval, with serrated margins (Castro-González, 2009). Flowers are pistillate, ligulated yellow flowers with lingual petals. These vary in size from 1.9 to 2.4 cm length (Castro-González, 2009; Cullen et al., 2011). Fruits are fluted triangular or quadrangular achenes, brownish to blackish, absent or rarely present pappus with 2–4 little and membranous edges (Cilia-López et al., 2013).

FIGURE 8.1 *H. longipes* details: roots, inflorescences, and habit.

8.1.5 AFFININ

The main biologically active compound found in *H. longipes* is affinin (IUPAC name *N*-isobutyl-2E,6Z,8E-decatrienamide), also known as spilanthol (Barbosa et al., 2016); it can be found in the roots, where it represents about 0.78% of its dry weight. However, various alkamides have been described and characterized in the roots of this plant (Barbosa et al., 2016). Affinin belongs to the group of compounds denoted as *N*-alkylamides, which are the bioactive compounds found most frequently in *Spilanthes* species, and other members of the Piperaceae, Rutaceae, Solanaceae, and Asteraceae families (Barbosa et al., 2016). Alkamides are found as secondary metabolites in plants belonging to these taxa and have been associated with various biological activities, including insecticidal, antimicrobial, anti-inflammatory, analgesic, and immune-modulatory properties (Boonen et al., 2010).

The structure of alkamides includes a polar aromatic amine residue and an aliphatic C8-C18 saturated fatty acid chain (Figure 8.2). Both ends of the affinin molecule are attached to a stable amide bond (Boonen et al., 2012). The amphiphilic character of affinin makes it susceptible to be extracted with ethanol, methanol, dichloromethane, hexane, and supercritical CO_2 (Barbosa et al., 2016). Affinin is commonly found as *N*-isobutyl-2E,6Z,8E-decatrienamide; however, cis-trans isomerism (i.e., *N*-isobutyl-2E,6E,8E-decatrienamide) in affinin is directly related to the total activity of the molecule. Jacobson (1954) demonstrated that UV treatment of the affinin molecule yielded trans-affinin, which is inactive against houseflies.

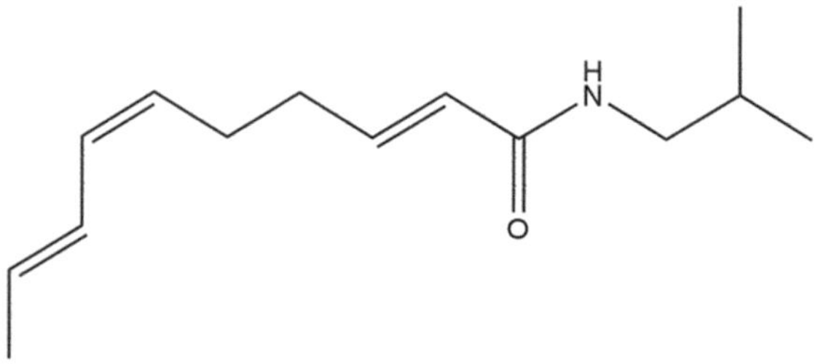

FIGURE 8.2 Affinin chemical structure.

No research has been done on the biosynthesis or function of affinin in *H. longipes;* however, studies made on *Acmella radicans,* other member of the Asteraceae family, have conclusively demonstrated that valine serves as a precursor in the synthesis of affinin while phenylalanine might be further processed in the plant to yield the amine substituents in alkamides (Cortez-Espinosa et al., 2011). Additionally, it has been previously suggested that affinin may have a similar effect as the one produced by auxins on adventitious root formation according to experiments carried out on *Arabidopsis thaliana* (Campos-Cuevas et al., 2008). Affinin has been chemically described by (Barbosa et al., 2016). The chemical characteristics of this compound are listed in Table 8.1.

TABLE 8.1 Chemical Characteristics of Affinin

Chemical Characteristic	Value
Molar mass	221.339 g/mol
Melting point	23 °C
Boiling point	165 °C
Refractive index at 298 °C	1.5135
Max UV absorption	228.5 nm
Ion mass spectrum	Molecular ion m/z = 222
	Fragment m/z = 149 (loss of isobutyl amine group)
	Fragment m/z = 99 (presence of isobutylamine)
Aqueous solubility	18.63 mg/mL

Reviewed by Barbosa et al. (2016).

Regarding the biological activity associated with affinin, it is known that when these plants are chewed, increased amounts of saliva are segregated, and the tongue experiences a feeling of numbness. In general, alkamides found in *Echinacea purpurea,* including affinin, have exhibited rapid oral absorption (C_{max}=10–40 min) and a plasma half-life of 1.5–5.0 h (Boonen et al., 2012).

Besides the oral use of *H. longipes* extracts, some preparations are applied topically as adjuvants in the treatment of fungal infections. Transdermal permeability has been determined for different formulations containing affinin (Barbosa et al., 2016). It has been demonstrated that the ability of affinin to permeate the skin is related to the solvents used as well as the receptor fluids present in the skin. Permeation was higher in preparations containing 10% propylene glycol (PG) than 65% ethanol. Additionally, PG demonstrated to have greater permeability as a receptor fluid than ethanol. In the same study, the theoretically calculated value for the partition coefficient (kp) of affinin was determined to be 2.12×10^{-2} cm/h (Boonen et al., 2012)

Other biological properties associated with this compound include antimicrobial, antifungal, anti-inflammatory, bacteriostatic, larvicidal, analgesic–antinociceptive, and anxiolytic activities, which make it a compound with promising pharmacological applications. These properties are further discussed in the following sections.

8.1.6 BIOLOGICAL ACTIVITY

8.1.6.1 ANTIMICROBIAL PROPERTIES

Affinin has been previously tested against different microbial species and has proved to be a competent antimicrobial agent. The ethanolic extract has shown antimicrobial effect against *E. coli*, *Pseudomonas solanacearum,* and *Bacillus subtilis* at concentrations of 25, 150, and 50 mg/L, respectively (Molina-Torres et al., 1999). Furthermore, the ethanolic solution of affinin has shown bactericidal activity against different bacterial species and fungi responsible for causing plant diseases such as *Erwinia carotovora, S. cerevisiae, Sclerotium rolfsii, Sclerotium cepivorum, Phytophtora infestans*, and *Rhizoctonia solani* (Molina-Torres et al., 2004). Gutierrez-Lugo et al. (1996) reported a significant antimicrobial activity of *H. longipes* root extract against *Staphylococcus aureu*s and, in a lower degree, *Trichophyton mentagrophyte.* For *S. aureus*, an inhibition zone of 26 mm at a concentration of 1000 µg per disc was registered (disc diameter: 13 mm), and 22 mm at a concentration of 500 µg per disc was observed. Meanwhile, *Trichophyton mentagrophyte* exhibited an inhibition zone of 20 mm at a concentration of 1000 µg per disc and 14 mm at a concentration of 500 µg per disc. From their study, Gutierrez-Lugo et al. (1996) concluded that *H. longipes* extract has the potential to be used as an anti-infective agent. Tables 8.2 and 8.3 summarize the bactericidal and fungicidal action of affinin isolated from *H. longipes* extract.

TABLE 8.2 Bactericidal Activity of Affinin

Species	Affinin Concentration (µg/mL)	% Inhibition	Reference
E. coli	75	100	Molina-Torres et al., 2004
B. subtilis	150	100	Molina-Torres et al., 2004
E. carotovora	50 (partially reduced amide)	100	Molina-Torres et al., 2004
P. solanacearum	150	100	Molina-Torres et al., 1999

TABLE 8.3 Fungicidal Aactivity of Affinin

Species	Affinin Concentration (µg/mL)	% Inhibition	Reference
Fusarium sp.	150	38	Molina-Torres et al., 2004
P. infestans	75	100	Molina-Torres et al., 2004
R. solani AG-3	150	100	Molina-Torres et al., 2004
R. solani AG-5	75	91	Molina-Torres et al., 2004
S. cepivorum	75	100	Molina-Torres et al., 2004
S. rolfsii	50	100	Molina-Torres et al., 2004
Aspergillus parasiticus	200	70.56	Velez-Haro et al., 2018

Further studies have been carried out to assess the inhibitory activity of phytopathogenic fungi, particularly *Fusarium oxysporum* f. sp. *lycopersici*. This fungal species affects mainly tomato cultures, and, due to its structure, it can prevail for up to six years in cultivation soil (Ward, 1986). The ethanolic extract from *H. longipes* at affinin concentrations of 75, 150, 300, 600, and 1200 µg/mL was added to the growth medium on a disc, and the diameter of inhibition was measured every 24 h until complete growth inhibition was observed in the control. The values of median lethal dose and lethal dose 90 for affinin obtained from *H. longipes* extract on *Fusarium oxysporum* f. sp. *lycopersici* were 164.2 µg/mL and 348.6 µg/mL, respectively (González Morales et al., 2011).

8.1.6.2 ANALGESIC ACTIVITY

The evaluation of the analgesic effect in *H. longipes* extract has been previously carried out by Ogura et al. (1982). The employed method consisted in measuring the inhibition of acetic acid-induced writhing in mice. The authors assessed analgesia produced by two different fractions obtained from *H. longipes* ethanolic extract. A fraction A consisted of the ethyl acetate dry residue (1.1 g) dissolved in 10% ethanol, while fraction B was the lyophilized aqueous portion of the extract (4.2 g). Fractions A and B were administered by gavage to mice at doses ranging from 1 to 50 mg/kg and from 50 to 50 mg/kg, respectively. Similarly, purified affinin was tested at 2.5–5 mg/kg. The median effective dose (IC_{50}) for both fractions and affinin is reported in Table 8.4. While fraction B did not exhibit secondary effects during its administration, mice treated with fraction A presented depression and tremors starting from a dose of 25 mg/kg. Mice death was observed at a dose of 50 mg/kg 30 min after the administration of the treatment (Ogura et al., 1982).

TABLE 8.4 Reported Median Effective Dose for Fractions Collected From *H. longipes*

Treatment	IC_{50} (mg/kg)
Fraction A	19.04
Fraction B	426.98
Affinin	6.98

Reviewed by Ogura et al. (1982).

Similarly, Cilia-López et al. (2010) evaluated the analgesic properties of the ethanolic extract from *H. longipes*. They induced pain to male albino rats (30–33 kg) with either chemical or thermal stimuli. The chemical stimulus consisted of administering a solution of 3% (v/v) acetic acid intraperitoneally for 30 min after the administration of the ethanolic extract (10 mg/kg) and purified affinin at 1 mg/kg. The number of abdomen stretches and contractions was registered. Previously, the thermal stimulus was evaluated according to the hot plate test (Eddy and Leimbach, 1953). The results obtained revealed that pure affinin and the ethanol extract at the concentrations employed exhibited inhibition of the number of stretches in 96% and 87%, respectively, for the chemical test. Meanwhile, in the case of thermal stimulus, both affinin and ethanolic extract displayed analgesic action by increasing the time required for the onset of pain symptoms up to 30 min and through the end of the experiment (60 min) (Cilia-López et al., 2010).

Acetic acid is an irritant substance that induces pain and inflammation in the abdominal area of mice by promoting the activation of nociceptive neurons (Dzoyem et al., 2017). Writhing in mice is produced as a consequence of the increase in the number of prostaglandins E_2 and F_2 in the peritoneal fluid (Cilia-López et al., 2010). These prostaglandins sensitize nociceptive neurons by interacting with specific receptors in the spinal cord (Bär et al., 2004). It has been reported that alkamides, such as those found in *H. longipes* extract, are capable of inhibiting prostaglandin production without inducing cell death (LaLone et al., 2007); therefore, *H. longipes* extract can exert analgesia due to the presence of alkamides that interfere in the overall inflammation process.

Déciga-Campos et al. (2010) developed a similar experiment as that established by Cilia-López et al. (2010) to measure the analgesic activity of *H. longipes* extract. The induction of abdominal writhes was achieved by administering a 0.6% (v/v) acetic acid solution. The antinociceptive effect was measured according to the number of writhes registered every 5 min during 30 min after the administration of the extract. Similarly, capsaicin-induced nociception was tested in order to evaluate neurogenic nociception. For this purpose, 20 μL of capsaicin was injected subcutaneously into the mouse's paw in order to evaluate nociceptive behavior (amount of time spent liking the paw). The logarithmic doses employed for *H. longipes* extract were ranged from 0.01 to 1.75 mg/kg and were administered 15 min before the capsaicin injection. Morphine was used as a positive control. According to their investigation, *H. longipes* acetone extract demonstrated to have a dose-dependent antinociceptive effect. The administration of *H. longipes* reduced the number of writhes registered in acetic-acid-induced nociception and exhibited a median effective dose (ED_{50}) equal to 2.2 ± 0.2 mg/kg i.p. Meanwhile, the maximum inhibition of nociceptive behavior induced by capsaicin (65.8%) was observed when a logarithmic dose of 1.75 mg/kg i.p. was administered to the mouse. In the same study, affinin isolated from *H. longipes* extract was tested for antinociceptive activity. ED_{50} for affinin (3.6 ± 5 mg/kg) was significantly higher than that registered for *H. longipes* extract in acetic-acid-induced writhing. Moreover, the maximum antinociception observed for affinin was 46.67% at a logarithmic dose of 1.875 mg/kg i.p. These results prove that *H. longipes* extract is more effective in inducing antinociception than affinin alone (Déciga-Campos et al., 2010).

Other methods have been used to evidence the analgesic effect of affinin and *H. longipes* extract. Rios et al. (2007) determined the content of gamma-aminobutyric acid (GABA) produced in mice brain slices after the administration of a 10 µg/mL dichloromethane extract. The increase of GABA concentration in the brain produces analgesia by inhibition of neuronal activity in the insular cortex (Jasmin et al., 2003). Therefore, the high-pressure liquid chromatography (HPLC) quantification of GABA in the culture medium after the administration of the extract was used to estimate a substance's potential to be used as an analgesic agent. The dichloromethane extract from *H. longipes* induced GABA to release from mice brain slices of 61 µmol/mL 3 min after its administration. These results suggest that the ethanolic extract from *H. longipes* may also exert its action on the spinal cord in the central nervous system (Cilia-López et al., 2010).

8.1.6.3 ANTI-INFLAMMATORY PROPERTIES

Inflammation is a natural response to pathogens, damaged cells, and toxic substances that can be harmful to the body (Chen et al., 2017). The inflammation process is sequential and involves the synthesis, activation, and regulation of different inflammatory cells, such as macrophages (Fujiwara and Kobayashi, 2005). The activation of macrophages by recognition receptors induces the release of cytokine genes, such as the tumor necrosis factor (TNFα). The TNFα is responsible for the activation of nuclear factor-kappa b, which, in turn, triggers the production of nitric oxide (NO) by inducible NO synthases (Soufli et al., 2016). Previously, Hernandez et al. (2009) studied the anti-inflammatory effect of *H. longipes* ethanolic extract in vitro by evaluating the release of TNFa and NO on mouse macrophages. The assay was carried out using RAW 264.7 murine macrophage cells activated with bacterial lipopolysaccharide and interferon-γ. After macrophage activation, the effect of *H. longipes* extract was evaluated using concentrations of 1–200 µg/mL. The solutions were dissolved in DMEM. The TNFα production was measured according to the methodology developed by Klostergaard (1985). Meanwhile, NO was quantified by adding Griess reagent (1% sulfanilamide and 0.1% naphtylethylenediamide in 2.5% phosphoric acid) to the culture supernatants and determining NO_2 concentration by comparison with a sodium standard curve. The absorbance was read at $\lambda = 540$ nm. The ethanol extract exhibited a half-maximal inhibitory concentration of TNFa production equal to 223.0 µg/mL. According to their observations, NO production was also reduced (IC_{50} =136.9 µg/mL). These results conclusively reflect the potential of *H. longipes* ethanolic extracts to exert anti-inflammatory action (Hernandez et al., 2009a).

The anti-inflammatory effect has also been evaluated in vivo previously by Hernández et al. (2009). Inflammation was induced with two different compounds: arachidonic acid (AA) and phorbol myristate acetate (PMA), on OF-1 mice. Inflammation was induced topically by the administration of 1 μg/mL AA on the right ear after the administration of *H. longipes* ethanolic extract, affinin, or isobutyl-decanamide at doses varying from 0.5 to 3.0 mg/20 μL of ethanol. Nimesulide was employed as a positive control at a dose of 1 mg/*ear* in acetone. Inflammation persisted for 1 h, after which the mice was euthanized, and slices of 6 mm were retrieved from both ears. Similarly, PMA was used to induce right ear edema at a dose of 2 μg/*ear* in 20 μg ethanol. The same doses of *H. longipes* extract, affinin, and isobutyl-decanamide were applied topically before PMA administration. After 4 h, mice was euthanized, and 6 mm slices were removed from both ears. The positive control group was treated with indomethacin at a dose of 3 mg/ear in acetone. In both cases, the left ear was used as control and the vehicle was applied. Left and right collected ear slices from each specimen were weighed and compared. Decreases in weight were related to the anti-inflammatory effect of the evaluated substances (Hernández et al., 2009b).

As given in Tables 8.5 and 8.6, *H. longipes* extract and the bioactive compounds evaluated by Hernández et al. (2009b) demonstrated the potential to be used as anti-inflammatory agents. *H. longipes* ethanolic extract showed a more significant inhibition percentage than positive control nimesulide on AA-induced inflammation, whereas in PMA-induced inflammation, *H. longipes* ethanolic extract and isobutyl-decanamide anti-inflammatory effect were comparable to that exhibited by Indomethacin at the same dose. In general, inhibition of AA-induced inflammation by *H. longipes* products was significantly higher than the maximum percentage inhibition identified for PMA-induced inflammation.

TABLE 8.5 Anti-Inflammatory Effect From *H. longipes* Extract, Affinin, and Isobutyl-Decanamide on AA-Induced Inflammation

Compound	ED_{50} (mg/*ear*)	Maximum Inhibition (%)	Dose (mg/*ear*)
H. longipes extract	0.8	91.3	3.0
Affinin	1.2	72.6	2.0
Isobutyl-decanamide	0.9	81.4	2.0
Positive Control			
Nimesulide	1.0	66.7	1.0

Reviewed by Hernández et al., 2009b.

TABLE 8.6 Anti-Inflammatory Effect From *H. longipes* Extract, Affinin, and Isobutyl-Decanamide on PMA-Induced Inflammation

Compound	ED_{50} (mg/*ear*)	Maximum Inhibition (%)	Dose (mg/ear)
H. longipes extract	2.0	80.3	3.0
Affinin	1.3	72.4	3.0
Isobutyl-decanamide	1.1	82.9	3.0
Positive Control			
Indomethazin	3.0	98.6	3.0

Reviewed by Hernández et al. (2009b).

AA induces inflammation when it is metabolized through the lipoxygenase pathway and leads to the formation of pro-inflammatory agents, such as prostaglandins and leukotrienes (Trostchansky et al., 2018). Hernández et al. (2009b) concluded that the similarity of the alkamides derived from *H. longipes* extract and AA allows these to act as competitive inhibitors of leukotriene enzymes and results in the suppression of inflammation.

8.1.6.4 ANTIHYPERALGESIA

Antihyperalgesia has also been studied previously by Ortiz et al. (2009) in *H. longipes* ethanolic extracts according to the Hargreaves model of thermal hyperalgesia (Hargreaves et al., 1988). The method consists of the measurement of the frequency of withdrawal of the hind paws in mice when subjected to radiant heat stimuli. Carrageenan is administered to the right-hand paw in order to increase thermal hyperalgesia in mice. The paw withdrawal latencies, meaning the time required for the mouse to retire its paw from the heat-source, were quantified 1–6 h after the administration of carrageenan. For the antihyperalgesic effect of *H. longipes* extract, 100 μL of a solution of concentrations ranging from 10 to 300 mg/kg were administered to mice 30 min before the carrageenan injection. The animal posture and ability to walk normally were observed in order to determine nociception. Antihyperalgesia exhibited by *H. longipes* ethanolic extract increased according to the doses employed. The extract proved to be most effective when 300 mg/mL was applied (Ortiz et al., 2009). Hargreaves model has also been employed to determine the synergy between *H. longipes* extract and diclofenac. Acosta-Madrid et al. (2009) studied the interactions between diclofenac and *H. longipes* extract at doses of 6.8 and 108.7 mg/kg, respectively, when administered orally to mouse samples. The formulation was injected at a volume of 100 µL 30 min before the carrageenan injection. *H. longipes* extract and diclofenac (diclofenac-HLE) were evaluated separately at doses of 10–300 mg/kg and 1–30 mg/kg, respectively. The values of ED_{30} determined experimentally in the same study are summarized in Table 8.7.

TABLE 8.7 Experimental ED30 Values for *H. longipes* Extract, Diclofenac, and Diclofenac-HLE

Treatment	Experimental $ED30$ (mg/kg)
H. longipes extract	105.3
Diclofenac	3.4
Diclofenac-HLE	8.6

Reviewed by Acosta-Madrid et al. (2009).

Acosta-Madrid et al. (2009) determined theoretical ED_{30} values for the diclofenac-HLE formulation, which was equal to 54.4 mg/kg. This value is significantly higher than the reported experimental data; thus, it was concluded that the combination of diclofenac and *H. longipes* extract is capable of inducing antihyperalgesia at lower concentrations than *H. longipes* extract used alone.

8.1.6.5 VASODILATION

Vasodilation is the process by which smooth muscle cells from blood vessels relax, increasing the internal diameter of the blood vessel lumen that allows a higher blood flow (Tucker and Mahajan, 2018). Vasodilator substances have various therapeutic applications, mainly in the treatment of hypertension (Osterziel and Julius, 1982), congestive heart failure (Rubin and Swan, 1981), and myocardial ischemia (Thomas, 1983).

Gasotransmitters, such as NO, CO, and H_2S, play an essential role in vasodilation since they induce relaxation in vascular tissues. These molecules are produced by oxygenase or NO-synthases from the vascular wall (Wang et al., 1997), which trigger a cascade of reactions that result in the conversion of soluble guanylyl cyclase to cyclic guanosine monophosphate (cGMP). cGMP induces the release of Ca from smooth muscle cells thus, resulting in vasorelaxation (Loh et al., 2018). Another mechanism of vasodilation includes Prostacyclin (PGI_2) signaling pathway in which PGI_2 along with GS alpha-protein bound to GTP, bind to the prostacyclin receptors on the surface of vascular smooth muscle cells. Consequently, it causes a series of reactions that result in the activation of protein kinase A, which induces the release of NO in the vascular tissue and initiate the NO vasodilation signaling pathway (Loh et al., 2018).

Castro-Ruiz et al. (2017) evaluated the effect of affinin and *H. longipes* ethanolic extract on the vasodilation of aortic rings. Their results (Table 8.8) showed that purified affinin induces vasodilation at lower concentrations than the ethanolic and dichloromethane extracts from *H. longipes*. Additionally, although affinin proved to be less potent than positive control acetylcholine, the alkamide exhibited a higher value for maximum activity.

TABLE 8.8 Vasodilation Effect of *H. longipes* Extracts, Purified Affinin, and Acetylcholine

Treatment	E_{max} (%)	EC_{50} (µg/mL)
Dichloromethane extract	100	76.99
Ethanolic extract	100	140.5
Affinin	100	27.38
Acetylcholine	70.02	1.094

Reviewed by Castro-Ruiz et al. (2017).

Furthermore, in their investigation, Castro-Ruiz et al. (2017) confirmed that *H. longipes* induces vasorelaxation by enhancing the activity of gasotransmitters, K^+ channels, and prostacyclin. The involvement of gasotransmitters in vasorelaxation was verified by employing endothelial NO synthase inhibitors that reduced the overall effect, demonstrating that NO, CO, and H_2S pathways were involved in *H. longipes* ethanolic extracts and affinin-induced vasodilation. Moreover, when K^+ channel inhibitors were used (glibenclamide and tetraethylammonium), higher concentrations of the treatments assessed were needed to observe an effect. Finally, to demonstrate the involvement of prostacyclins in affinin-induced vasorelaxation, methacin was employed at a concentration of 10 µM to inhibit cyclooxygenase. The pretreatment with this inhibitor decreased the overall effect (Castro-Ruiz et al., 2017).

8.1.6.6 TOXICITY

Gutierrez-Lugo et al. (1996) evaluated the cytotoxic activities of *H. longipes* methanol and chloroform extracts obtained by maceration of root ground tissue. The extracts were assayed for seven days, on A-549 (lung carcinoma), MCF-7 (breast carcinoma), and HT-29 (colon adenocarcinoma) cellular lines. Adriamycin was used as a positive control. *H. longipes* extract exhibited no cytotoxicity. In a similar study performed by Déciga-Campos et al. (2012), *H. longipes* ethanolic extracts, with acetone, were administered via oral in a concentration of 0.2 mL/10 g body weight. LD_{50} value was calculated with a geometrical median according to the Lorke method, demonstrating that the *H. longipes* extract (LD_{50} = 62.14 mg/kg) was more toxic than affinin (LD_{50} = 113.13 mg/kg) in mice. Besides, *H. longipes* and affinin showed to have antimutagenic properties.

Furthermore, Escobedo-Martínez et al. (2017) performed an assay of *H. longipes* extraction with hexane where mice were treated with doses of 10, 100, or 1000 mg/kg. These were supervised over 14 days. The medium lethal dose (LD_{50}) study was performed to determine the dose of the extract that killed 50% of the animal population given; the value p.o. was 775 mg/kg, it was demonstrated to not acquire any toxicity, even though the accumulation by the daily administration.

8.2　PILOT STUDY: METHODOLOGY

In the following sections, methods and preliminary results will be presented from a pilot study focused on the characterization of extracted active compounds from *H. longipes* as well as some DNA analysis from this plant species. Plant materials were provided by the same plant producer and all specimens used in this study were cultivated under the same conditions.

8.2.1　*EXTRACTION AND ANALYSIS METHODS FOR H. LONGIPES ACTIVE COMPOUNDS*

Ultrasonication, maceration, and Soxhlet extraction were compared in order to determine which one exhibited the most significant yield and higher content of affinin. For ultrasonication, 50 mL of 96% (v/v) ethanol was added to 5 g of the ground root. The preparation was sonicated in an ultrasonic water bath at 25 °C. Sonication took place at 5-min intervals threefold, allowing the sample to cool down for 1 min at room temperature. The preparation was vacuum filtered and stored at 4 °C. The extraction by maceration was carried out by adding 50 mL of 96% (v/v) ethanol to 5 g of root powder. The preparation was left to macerate for one week at room temperature in dark conditions. Finally, Soxhlet extraction was carried out using 5 g of ground root tissue on 200 mL of 96% (v/v) ethanol. The extraction was left on reflux for 18 continuous hours.

For all the extraction methods, all extracts were dried on a rotary evaporator at 60 °C until a thick residue was formed in all of them. All residues were weighed to determine the extraction yield for each method. These samples were stored at 4 °C for further analyses.

8.2.1.1 GAS CHROMATOGRAPHY

A sample from Soxhlet extraction was analyzed with gas chromatography using an Agilent Technologies 7890A GC System. The experimental conditions for GC-MS were as follows: DB-17HT nonpolar column, size: 30 m; carrier gas: Helium; and flow rate: 1.295 mL/min. The oven program was raised at 50 °C for 1 min and then 6 °C for 40 min and the injection volume was 1 µL per sample. Results were analyzed using the NIST 2005 library.

8.2.1.2 QUANTIFICATION OF TOTAL PHENOLIC COMPOUNDS

Total phenolic compounds from the *H. longipes* root extract were quantified, from the three types of extraction described above, according to the colorimetric Folin–Ciocalteu (Cheung et al., 2003). This method is based on the reaction of phenolic hydroxyl groups in the plant ethanolic extracts with a mixture of sodium molybdate and sodium tungstate (Folin–Ciocalteu reagent). The reaction results in the formation of a blue complex that may be used to measure total polyphenols by UV-vis spectrophotometry at l = 760 nm (Blainski et al., 2013).

8.2.1.3 EVALUATION OF THE ANTIOXIDANT ACTIVITY

The antioxidant activity was determined by the 2,2 diphenyl-1-picrylhydrazyl (DPPH) radical scavenging method, as described by Cardador-Martínez et al. (2011). The assay measures the scavenging activity of plant ethanolic extracts on DPPH reagent. Structurally, DPPH contains an unpaired electron that can be stabilized by antioxidant molecules present on plant ethanolic extracts. The delocalization of the spare electron in DPPH yields a deep violet color, which is reduced when the molecule is stabilized (Kedare and

Singh, 2011). The change in color is, thus, related to the antioxidant activity of the ethanolic extracts and may be measured by visible-light spectrophotometry at 520 nm.

8.2.2 CYTOTOXICITY ASSAY

Two cellular lines, Hep G2 (human hepatocellular carcinoma) and PC-3 (prostate adenocarcinoma), were grown in DMEM medium supplemented with 10% fetal bovine serum on a 24-well plate medium at 37 °C, 5% CO_2 using a humidified incubator. The medium for PC3 contained 1.5 g/mL glucose, while the medium for Hep G2 cells was supplemented with 2.5 g/mL glucose. A hundred microliters out of 180 μM of *H. longipes* Soxhlet extract were added to each well. Cells were grown until 90% confluence during 120 h. Every 24 h, a column from the 24-well plate is fixed with 70% ethanol (v/v). Cells were stained with crystal violet and resuspended in acidified methanol 10% (v/v). This preparation was analyzed with UV–vis spectrophotometry at $\lambda = 570$ nm.

8.2.3 AFFININ ANALYSES

8.2.3.1 ISOLATION BY PREPARATIVE HPLC

Preparative HPLC was carried out in order to obtain a purified sample of affinin and quantify this compound in the ethanolic extracts. The preparative column used was Agilent Eclipse plus C18, 21.2 × 250 mm, 7 μm. The mobile phase was acetonitrile:water (44:56 v:v); the flow was set at 10 mL/min and 200 μL of the sample from the Soxhlet ethanolic extract was injected. The column remained at room temperature, with 31 bar pressure; DAD was configured at 229 nm; eluate was collected from the minute 56 to 70 and then vacuum dried. Finally, the thick residue formed was then weighted.

8.2.3.2 AFFININ IDENTIFICATION BY UPLC-MS ANALYSIS

The ultraperformance liquid chromatography-mass spectrometer (UPLC-MS) analysis was carried out to make sure that the peak collected by preparative HPLC was indeed affinin, using Acquity UPLC® Class I coupled with MALDI SYNAPT® G2-Si high-definition mass spectrometer. The column used was Acquity UPLC® BEH C18, 1.7 µm, 2.1 × 50 mm. Samples were prepared by filtration through a 0.22 µm filter. The mobile phase A was 0.01% formic acid, whereas the mobile phase B was acetonitrile. The flow was set at 0.4 mL/min, and the column equilibrated for 5 min, with 95% A and 5% B. Pressure limits were set at 500 and 1700 psi. The temperature of the column was set at 35 °C, and the volume of injection was 1 µL. The flow gradient was set to start with 95% A and 5% B, then switched in the minute 5 to 5% A and 95% B, and finally, from minute 7 to minute 10, 95% A and 5% B. The mass spectrometer was configured to detect masses in a 100–300 Da range. Lectures were obtained from 0 to 10 min. The fragmentation voltage was set to 20–40 V. The MS-MS fragmentation was set directed to the monoisotopic mass of affinin.

8.2.3.3 HPLC-DAD QUANTIFICATION

Quantification of affinin for the different types of extraction methods was carried out employing HPLC-DAD analysis. The analysis was performed using a C18 (4.6 × 100 mm, 3.5 µm) column. The column temperature was held at 30 °C. The mobile phase consisted of acetonitrile/water (44:56) and the flow rate was 0.5 mL/min in isocratic mode. The injection volume was 5 µL. Detection wavelength was programmed at l = 229 nm. Affinin exhibited a retention time of 18 min. Total affinin was evaluated based on a calibration curve of affinin from 0 to 0.219 mg/mL.

8.2.4 ANTIMICROBIAL ACTIVITY OF H. LONGIPES ETHANOLIC EXTRACTS

The inhibition of Gram-negative and Gram-positive bacteria, as well as bacteria found in buccal microbiota, was carried out following the modified methodology described by Molina-Torres et al. (1999). For this purpose, we used *Escherichia coli (*HB101) as the Gram-negative representative species. *Lactobacillus acidophilus* and *Streptococcus thermophilus* for Gram-positive antimicrobial analysis. The MRS agar was used as a growth medium for Gram-positive bacteria, while LB was used to grow *E. coli* and buccal microbiota. Bacteria were incubated at 37 °C and 800 r/min (in the case of *E. coli*) for 24 h.

For the antimicrobial disk preparation, 40 mL from the ethanolic extracts, obtained with maceration, sonication, and Soxhlet, were vacuum dried out and resuspended in 5 mL of methanol, then 20 μL of the ethanolic extracts were poured into 6 mm Whatman® discs and left resting until they were completely dried out. Petri dishes were inoculated with bacteria in exponential phase shortly before placing the discs above the media. Bacterial cultures were then incubated for 24 h at 37 °C, and the diameter of the inhibition zone was measured subsequently. Ampicillin, chloramphenicol, and gentamicin from Multibac I.D multidisc insert were used as a positive control.

8.2.5 STATISTICAL ANALYSIS

The variance analyses were performed in Minitab® 18 with a = 0.05, and for the comparison of means, the method used was Hsu's multiple comparison with best.

8.2.6 HELIOPSIS LONGIPES DNA FINGERPRINTING

8.2.6.1 DNA EXTRACTION

DNA was extracted from *H. longipes* roots and leaves. The protocol used for DNA extraction from root tissue was a modified version from Khan et al. (2007). Later, DNA extraction from leaf tissue was carried out employing some variations implemented by the authors. The modifications for both protocols included the use of a smaller amount of tissue (20 mg) as well as the addition of phenol:chloroform:isoamyl alcohol (25:24:1) for protein precipitation and sodium acetate at a concentration of 3M. The DNA isolation from leaf tissue was also performed using PureLink® Plant Total DNA Purification Kit. The purity and concentration of the DNA extract were determined by spectrophotometry at $l = 280$ nm. The DNA visualization was done by SAGE for all samples.

8.2.6.2 BARCODING PROFILE

The amplification of the barcoding regions *rbc*L, *trn*H-*psb*A, and *mat*K was carried out from plant leaf tissue DNA extraction. The primers sequences and optimum annealing temperatures for these regions are given in Table 8.9.

DNA was amplified using BioMix™ Bioline 2x, using 5 μL for each 10 μL of reaction, 0.4 μL of each primer at 10 μM concentration, 3.2 mL of template, and 3.2 mL water. Amplification reaction consisted of 2–4 min initial denaturation at 94 °C, 34–35 cycles of 94 °C for 30 s, optimum annealing temperatures given in Table 8.9 for 20–40 s, followed by a final extension at 72 °C for 5–10 min, depending on the amplified region.

TABLE 8.9 Sequences of Barcoding Regions *trn*H-*psb*A, *rbc*L, and *mat*K

Primers	Sequence	Optimum Annealing Temperature (°C)	Author
*trn*H	5′-CGCGCATGGTGGATTCACAATCC-3′	57	(Tate and Simpson, 2003)
*psb*A	5′-GTTATGCATGAACGTAATGCTC-3′		(Sang et al., 1997)
Forward: *rbc*La-F	5′-ATGTCACCACAAACAGAGACTAAAGC-3′	57	(Levin et al., 2003; Sang et al., 1997)
Reverse: *rbc*La-R	5′-GTAAAATCAAGTCCACCRCG-3′		(Kress et al., 2009)
Forward: *mat*K-xf	5′-TAATTTACGATCAATTCATTC-3′	50.5	(Ford et al., 2009; Kress et al., 2009)
Reverse: *mat*K-MALP	5′-ACAAGAAAGTCGAAGTAT-3′		(Dunning and Savolainen, 2010)

8.2.6.3 PCR ENHANCEMENT WITH BOVINE SERUM ALBUMIN (BSA) SOLUTION

Amplification reactions were enhanced with the aid of a BSA solution modified from the protocol established by Samarakoon et al. (2013). One milliliter of this solution contained 1 mg/mL BSA and 850 mL of 10 mM Tris buffer. The final volume was adjusted to 1 mL with DNAse free water. Two microliters of this solution were added to adjust the 10 mL of each reaction final volume.

8.3 PILOT STUDY: PRELIMINARY RESULTS AND DISCUSSION

8.3.1 EXTRACTION METHODS FOR H. LONGIPES ACTIVE COMPOUNDS

The results of total extract yields obtained from ultrasonication, maceration, and Soxhlet extractions are summarized in Table 8.10.

TABLE 8.10 Extract Yield for Ultrasonication, Maceration, and Soxhlet Extractions

Type of Extraction	Yield (Grams of Extract/ 5 g of dw.)	Yield Percentage
Ultrasonication	0.1275	3%
Maceration	1.5961	32%
Soxhlet extraction	0.749	15%

As given in Table 8.10, the maceration method presented a higher yield; however, as will be discussed in the following pages, this extract exhibited lower antioxidant activity and concentration of polyphenols when compared to Soxhlet extracts. Maceration showed the highest extract yield because, during the process, ground plant tissue was directly in contact with the solvent for prolonged periods of time; thus, it is possible that particles from this tissue remained in the solvent even after filtration. Therefore, extract yield may appear higher in the macerated extract.

8.3.2 PHYTOCHEMICAL PROFILING OF H. LONGIPES ETHANOLIC ROOT EXTRACT

8.3.2.1 GAS CHROMATOGRAPHY

The preliminary GC-MS analysis from the extract revealed the chromatogram shown in Figure 8.3. The analysis of *H. longipes* extract exhibited the presence of diverse compounds that are displayed in Table 8.11.

It has been previously mentioned that the main compound found in *H. longipes* roots is affinin, which is mostly responsible for the analgesic properties attributed to *H. longipes* root extract. However, GC-MS analysis revealed that *H. longipes* root extract is also rich in terpenes. Terpenes are biosynthesized from isoprene units (isopentenyl diphosphate and dimethylallyl diphosphate), which are bound from head to tail (Jiang et al., 2016). According to the number of isoprenoid units in their structure, terpenes can be classified as monoterpenes, sesquiterpenes, diterpenes, triterpenes, or tetraterpenes (Cho et al., 2017). Sesquiterpenes (three isoprene units) are the most common form of terpenes found in *H. longipes* extract. These compounds have proven to be beneficial to human health as anti-inflammatory agents and as inhibitors of tumorigenesis. Borneol, for example, has shown free-radical scavenging activity that provides the compound with a neuroprotective capacity in Alzheimer's disease models (Cho et al., 2017).

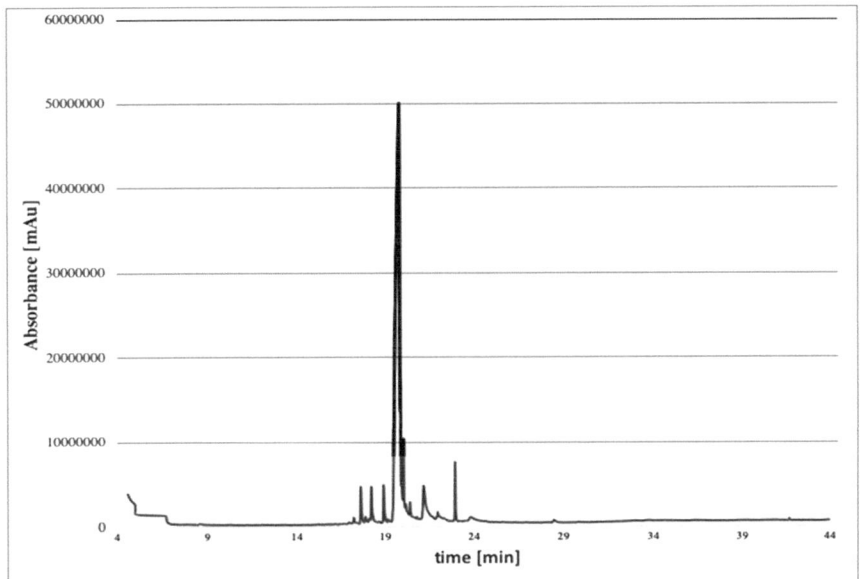

FIGURE 8.3 Chromatogram for GC-MS *H. longipes* ethanolic root extract.

From the compounds listed in Table 8.11, caryophyllene is also known to exert anti-inflammatory and anticancer activities, which suggests its potential to be employed as an adjuvant in cancer treatments. This compound has exhibited cytotoxicity against various cell lines, including HeLa (human cervical adenocarcinoma cells), Hep G2 (human leukemia cancer cells), AGS (human lung cancer cells), and SNU-1 (human gastric cancer cells) (Fidyt et al., 2016).

Similarly, squalene is an oxygen scavenger that may be used in the protection of skin, since it retains unstable dioxygen molecules that arise from UV radiation and that are involved in lipid peroxidation (Kelly, 1999). Likewise, squalene is a compound capable of inhibiting carcinogenesis by reducing in 46% aberrant crypt foci development in rats (Smith, 2000).

Besides the alkamides identified by GC-MS hereby, other minor alkamides have previously been isolated from *H. longipes* roots. These minor alkamides include *N*-isobutyl-8,10-diynoic-3Z-undecenamide, *N*-isobutyl-syn-8,9-dihydroxy-2E, 6Z-decadienamide, and *N*-isobutyl-syn-6,9-dihydroxy-2E, 7E-decadienamide, which are thought to be partly responsible for the overall analgesic activity in the extract (López-Martínez et al., 2011). Other compounds purified from *H. longipes* ethanolic root extract include beta-sitosterol palmitate, and sitosterol which are phytosterols that act as

TABLE 8.11 Compounds Found in *H. longipes* Root Extract

Compound	Common Name	Coincidence Percentage	Retention Time (min)	Molecular Mass (Da)	Reported Product Ions (m/z)	Product Ions References
Alkamides						
N-(2-Isobutyl)-(2E,6Z,8E)-decatrienamide	Affinin, Spilanthol	90	19.259	221.17	67, 81, 98, 99, 121, 123, 126, 141, 149	(Bae et al., 2010; Barbosa et al., 2016; Barbosa et al., 2017)
N-(2-Methylbutyl)-(2E,6Z,8E)-decatrienamide	–	89	20.667	235.36	123, 81, 121, 166	(Bae et al., 2010)
Terpenes						
(1R,7R,10R)-4,10,11,11-Tetramethyltricyclo [5.3.1.0^{1,5}] undec-4-ene	Cyperene	95	8.026	204.35	204, 189, 41	(pubchem.ncbi.nlm.nih.gov, 2019e)
(1R,4E,9S)-4,11,11-trimethyl-8-methylidenebicyclo[7.2.0]undec-4-ene	Caryophyllene	99	8.422	204.35	93, 133, 91	(NIST, 2019d)
1,6,10-Dodecatriene,7,11- dimethyl-3-methylene	β-Farnesene	87	8.906	204.35	41, 69, 93	(NIST, 2019b)
1H-Cycloprop[e]azulene, 1a, 2, 3, 4, 4a, 5, 6, 7 b-octa	Viridiflorene	91	9.269	204.35	109, 43, 69	(NIST, 2019e)
(2S,4aS)-4a,8-dimethyl-2-prop-1-en-2-yl-2,3,4,5,6,7-hexahydro-1H-naphthalene	4,11-Selinadiene	81	9.445	204.35	189, 95, 105	(NIST, 2019a)
Other Compounds						
(6E,10E,14E,18E)-2,6,10,15,19,23-hexamethyltetracosa-2,6,10,14,18, 22-hexaene	Squalene	99	27.743	410.7	69, 81, 41	(pubchem.ncbi.nlm.nih.gov, 2019f)

TABLE 8.11 (Continued)

Compound	Common Name	Coincidence Percentage	Retention Time (min)	Molecular Mass (Da)	Reported Product Ions (m/z)	Product Ions References
Phenol, 2-methyl-5-(1-methylethyl)-	Carvacrol	93	6.584	150.22	135, 150, 91	(NIST, 2019c)
4a-methyl-1,3,4,5,8,8a-hexahydronaphthalen-2-one	-	53	9.445	169.24	41, 39, 68	(pubchem.ncbi.nlm.nih.gov, 2019a)
Cyclohexene,6-butyl-1-nitro	-	59	21.261	183.25	41, 81, 79	(pubchem.ncbi.nlm.nih.gov, 2019e)
4a,8-dimethyl-2-prop-1-en-2-yl-2,3,4,5,6,7-hexahydro-1H-naphthalene	-	81	45.036	204.35	189, 133, 91	(pubchem.ncbi.nlm.nih.gov, 2019b)

hepatoprotective agents and are involved in the reduction of cholesterol levels (Kim et al., 2019).

8.3.2.2 QUANTIFICATION OF TOTAL PHENOLIC COMPOUNDS

Polyphenols are another remarkable group of compounds found in *H. longipes* ethanolic root extract. These prevent the development of cancer, diabetes, neurodegenerative, and cardiac diseases when consumed frequently (Pandey and Rizvi, 2009). The total phenolic concentration for each type of extraction is given in Table 8.12.

TABLE 8.12 Total Polyphenols in *H. longipes* Ethanolic Extract

Type of Extraction	Total Phenolic Concentration (µg GAE/ g dw.)
Maceration	40.5 ± 0.0304
Ultrasonication	31.3 ± 0.0467
Soxhlet	134.3± 0.0071

Total phenolic content in the Soxhlet extraction was significantly higher than in the macerated and sonicated ethanolic extracts ($p = 0.015$). Total phenolic compounds quantified in *H. longipes* were considerably lower than those reported for other *Heliopsis* species. For the methanolic extract of *H. sinaloensis*, the total phenolic account was determined to be 16.064 mg GAE/g dw when the extract was obtained from leaves, and 5.807 mg GAE/g dw when the extract was obtained from stems (Olivas-Quintero et al., 2017).

8.3.2.3 EVALUATION OF THE ANTIOXIDANT ACTIVITY

The antioxidant activity of *H. longipes* ethanolic extracts expressed as the IC_{50} are given in Table 8.13.

The highest radical scavenging activity found in *H. longipes* ethanolic extracts was found in the Soxhlet extraction. It exhibited an $IC_{50} = 0.245$ mg/mL, comparable to data previously reported for the methanolic extracts of *Spilanthes acmella* ($IC_{50} = 0.223$ mg/mL), and *Helicteres vegae* ($IC_{50} = 0.218$ mg/mL) (Nikolova 2011; Olivas-Quintero 2017). However, radical scavenging activity was significantly smaller than that reported for *Heliopsis sinaloensis* ($IC_{50} = 0.051$ mg/mL) (Nikolova, 2011; Olivas-Quintero et al., 2017).

TABLE 8.13 Antioxidant Activity in *H. longipes* Ethanolic Extract

Type of Extraction	Expected IC$_{50}$ (mg/mL)	Maximum DPPH Radical Scavenging %	Concentration for Maximum DPPH Radical Scavenging % (mg/mL)
Maceration	3.642	30.941 ± 0.027	1.804
Sonication	0.362	19.598 ± 0.015	0.768
Soxhlet	0.245	22.880 ± 0.003	0.168

From these preliminary results, it is possible to observe that the conditions employed during Soxhlet extraction improved the extraction of phenolic compounds, as well as other antioxidant molecules. Soxhlet extraction is a common method used in vitro for the extraction of these compounds, and its efficiency may vary according to the type of solvent employed, the duration of the extraction, the temperature used, as well as the structure and stability of the compounds to be extracted (Brglez Mojzer et al., 2016). On the other hand, Soxhlet extraction of phenolic compounds is time-consuming, and the use of elevated temperatures may cause irreversible damage to some thermolabile polyphenols (Aires, 2017). Thus, it is possible that the low concentration of polyphenols observed for *H. longipes* extract when compared to other species is due to the degradation of these compounds.

8.3.3 CYTOTOXICITY ANALYSES

The ethanolic extract from *H. longipes* exhibited cytotoxic activity against Hep G2 and PC3 on varying degrees. *H. longipes* extract significantly inhibited the growth of Hep G2 cells in a time-dependent manner at a concentration of 180 µM. At 72 h of exposure, cell growth was inhibited by 37.08%. Moreover, from Figure 8.4, it is possible to observe that *H. longipes* extract exhibited less inhibitory activity on PC-3 cells than that shown for Hep G2 cells. The inhibition of cell growth was only evident at 96 h of exposure. At this point, percentage inhibition of PC-3 cell growth corresponding to 66.17% was observed.

To our knowledge, the cytotoxic effect of *H. longipes* extract on hepatocytes has not yet been reported. However, previous studies on the cytotoxicity of affinin have demonstrated that this compound is capable of inhibiting the growth of HEK293 (human embryonic kidney cells). Moreover, it is has been reported that affinin inhibits NKCC2 activation in kidney slices by hindering the phosphorylation of the protein (Gerbino et al., 2016). The

NKCC2 cotransport proteins are expressed specifically on liver cells, where they are responsible for sodium reabsorption. However, isoforms of NKCC2 are present on most animal cells, including hepatocytes. These isoforms are known as NKCC1 (Kim et al., 2001). In general, NKCC cotransporters regulate intracellular Cl (−) concentrations favoring salt transport in epithelial tissue (Russell, 2000). Thus, NKCC1 proteins in Hep G2 cells may be inhibited by affinin through a similar mechanism as that observed on HEK293 cells. This inhibition may lead to an abnormal flux of ionic molecules that result in an osmotic imbalance in the cell which causes its death; however, further studies must be carried out in order to determine the process through which this compound exerts cytotoxicity.

From Figure 8.4, it would appear that affinin may exhibit cytotoxic selectivity against Hep G2 cells. Previous studies by Huang et al. (2019) have reported that affinin stimulates the expression of AMP-activated protein kinase (AMPK) in hepatocytes. APK proteins are capable of inhibiting fatty acid synthesis in liver cells. This property has shown the potential to be used in the treatment of certain diseases, such as diabetes and obesity (Huang et al., 2019).

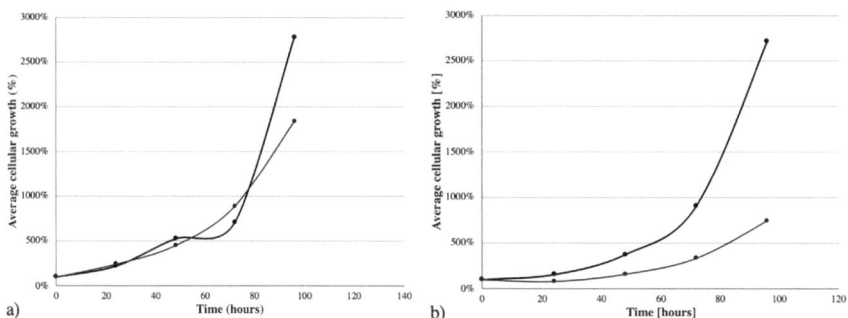

FIGURE 8.4 Effect of *H. longipes* root extract on cellular growth. (a) PC-3 prostate adenocarcinoma. (b) Hep G2 hepatocellular carcinoma.

Moreover, AMPK has been suggested as a pharmacological target for the treatment of cancer. The abnormal behavior of hepatocytes is related to irregularities in metabolism. Hepatocellular carcinoma, the most common type of liver cancer in the United States, is associated with decreased levels of AMPK expression that favor the proliferation of cancerous cells (Li and Chen, 2018). Mechanisms through which AMPK may induce tumor suppression include the regulation of the inflammation process, control of the metabolic pathways, and blocking of the cell cycle. Thus, the growth inhibition effect observed in Hep G2 cells is related to the ability of affinin to induce death on cancerous cells with abnormal metabolic cycles rather than a generalized cytotoxic effect. More studies regarding the effects of affinin on cancerous hepatocytes should be carried out in future investigations. Additionally, the evaluation of cytotoxicity for *H. longipes* extract should be attempted with noncancerous cell cultures.

8.3.4 AFFININ ISOLATION BY PREPARATIVE HPLC

The affinin solid weight collected was 3.5 mg. The calculated yield was 0.7 mg/g dw. The analysis of the collected peak with GC-MS showed a single peak chromatogram (Figure 8.5), suggesting the purity of the sample. The retention time of affinin in the UPLC was 3.38 min. The presence of affinin was confirmed in the MS/MS analysis (Figure 8.6) with the molecular ions 81, 98, 126, and 141; all of them are previously reported (Bae et al., 2010; Barbosa et al., 2016, 2017).

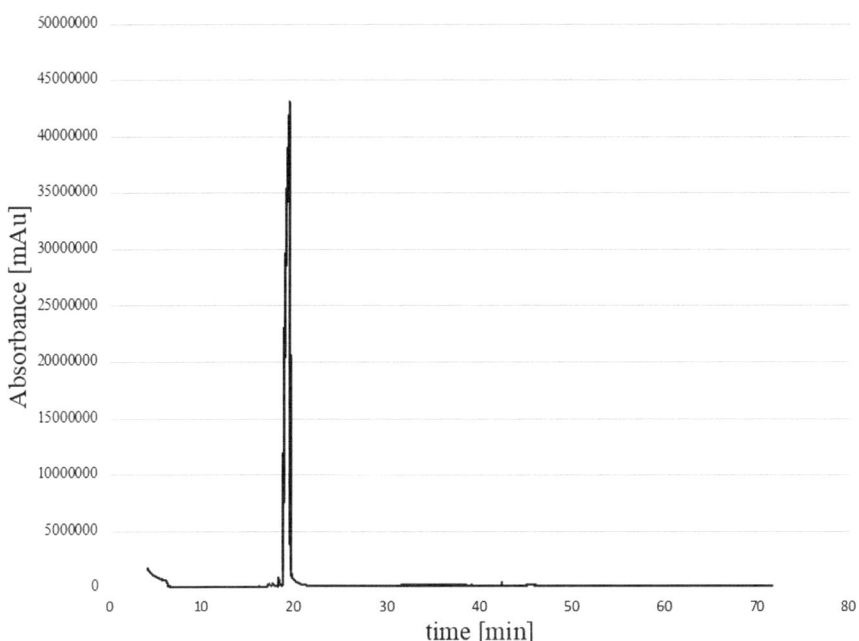

FIGURE 8.5 Chromatogram for GC-MS of purified affinin sample isolated from *H. longipes* extract.

Affinin extraction and isolation have been previously described by Molina-Torres et al. (1996), where the extraction of affinin was carried out from frozen, dried roots through maceration with ethyl acetate. The isolation of the *N*-alkylamide was achieved through TLC using *n*-hexane-ethyl acetate for the development of Si gel plates. Affinin was purified with an overall yield of 7.3 mg/g dw tissue.

Extraction by maceration with acetone as solvent was also used by Déciga-Campos et al. (2012) to purify affinin from *H. longipes* roots. Affinin purification was performed by fractionation of this extract on open column chromatography. The presence of affinin in the collected fractions was evaluated by comparison with a standard sample of affinin by TLC. The total yield of purified from the root sample was 19.617 g of affinin with respect to the dry extract obtained from 4.7 kg of *H. longipes* root tissue. Affinin yield for this method with respect to the initial root tissue was 4.1 mg/g dw tissue.

From our experience, significant factors that affect the yield of affinin in the ethanolic extracts include the extraction method employed including the solvent used, temperature and time of extraction, the storage conditions of root tissue as well as the conditions to which the plant is subjected when grown. External factors, such as abiotic stress, may induce fluctuations in the total affinin production in a specimen.

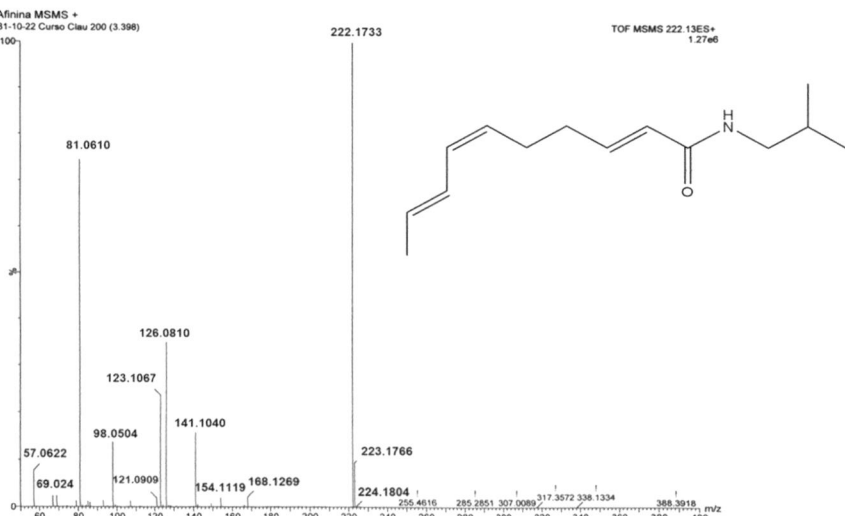

FIGURE 8.6 Ion mass spectra and structure of (2E,6Z,8E)-*N*-Isobutyl-2,6,8-decatrienamide (affinin).

8.3.5 HPLC-DAD QUANTIFICATION

The concentration of affinin in each of the ethanolic extracts is given in Table 8.14.

TABLE 8.14 Concentration of Affinin in *H. longipes* Ethanolic Extracts

Type of Extraction	Affinin Concentration (mg/mL)
Maceration	3.6067 ± 0.0071
Ultrasonication	1.5360 ± 0.0126
Soxhlet	0.6727 ± 0.0054

The statistical analyses showed that the maceration extract yielded a significantly higher concentration of affinin ($p < 0.001$), suggesting that the conditions used during Soxhlet extraction process were inadequate and may contribute to the degradation of this molecule. To our knowledge, no studies have reported the effect of temperature on the stability of affinin; however, it has been previously suggested that alkamides extracted from *Echinacea* species are prone to oxidation when the temperature is increased (Liu and Murphy, 2007). Thus, the use of high temperatures during prolonged time may have resulted in lower affinin yields. Further studies must be carried out regarding the optimum conditions for the extraction of affinin from *H. longipes* roots.

8.3.6 ANTIMICROBIAL ACTIVITY OF H. LONGIPES ETHANOLIC EXTRACTS

Although neither of the ethanolic extracts showed inhibition against Gram-positive bacteria, the maceration extraction sample exhibited inhibition of 0.41× the diameter of chloramphenicol antibiotic discs in *E. coli*. In the buccal microbiota, the Soxhlet ethanolic extract showed no statistical difference ($p < 0.001$) when compared against gentamicin control. Gram-positive bacteria analyzed (*S. thermophilus* and *L. acidophilus*) showed no inhibition with none of the ethanolic extracts. In *E. coli*, the sonication method showed significantly lower inhibition zone, whereas in buccal microbiota, the Soxhlet method produced significantly higher inhibition ($p < 0.001$).

8.3.7 ANTIMICROBIAL ACTIVITY OF A COMBINATION OF H. LONGIPES ETHANOLIC EXTRACTS AND CITRUS FRUIT EXTRACTS

In order to increase the bactericidal potential of a potential formulation, the product can be supplemented with *Citrus* spp. fruit extracts. The maceration ethanolic extract was tested at different concentrations to analyze the synergy between *H. longipes* and *Citrus* spp. fruit extracts. In all of the cases, the inhibition zone diameter of the formulation with 25% of *H. longipes* ethanolic extract was significantly lower. However, in buccal microbiota, the concentration of 17.5% showed to induce significantly greater inhibition, suggesting that the *H. longipes* ethanolic extract may interfere with the bacterial inhibition having an antagonistic interaction when mixed at large concentrations.

8.3.8 HELIOPSIS LONGIPES DNA FINGERPRINTING

8.3.8.1 DNA EXTRACTION

The DNA isolated from *H. longipes* leaves resulted successful in contrast with the extraction from roots when both chemical and membrane-based extractions methods were used. Table 8.15 illustrates the concentrations and purity obtained for leaf extractions. The HLQA and HLQB correspond to the extractions obtained by the chemical method, while HLKA and HLKB are the DNA extractions obtained using the PureLink™ Plant DNA extraction kit.

TABLE 8.15 Concentrations and Purity for Chemical and Kit DNA Extractions From Leaves

Samples ID	Concentration (ng/mL)	A 260/280
HLQA	5.72	0.650
HLQB	41.370	0.851
HLKA	1.688	0.704
HLKB	0.613	0.450

DNA extractions from root tissue are generally more complicated than leaf DNA extraction; this is due because root tissue is lignified and hard to dissolve, translating into a constant rigidity of the cell wall along with the presence of polysaccharides (Boiteux et al., 1999). Polysaccharides coprecipitate with DNA and may suggest higher amounts of DNA than the actual ones (Varma et al., 2007). Thus, a common problem found in genomic DNA isolation is the low yield and purity obtained during the extraction process. DNA yield varies the most according to the type of tissue from which the extraction is performed, and this may be related to "intrinsic tissue-specific variations," including the number and size of cells as well as total extracellular DNA (Boiteux et al., 1999). DNA yield and purity may vary from plant to plant as well, according to the biochemical composition characteristic from each species.

On the other hand, there is evidence that plant roots generally present higher content of contaminant chemicals, including polyphenols, terpenes, and other phenolic compounds (Khan et al., 2007). During DNA extraction, phenolic compounds are released into the cytoplasm and become oxidized. In their oxidized state, phenolic compounds bind irreversibly to DNA. Moreover, the presence of these contaminants may inhibit further reactions, including the amplification and digestion of DNA sequences. Contamination by polyphenols was tackled by the addition of PVP in the extraction buffer; nevertheless, the absence of bands during SAGE suggests the possibility that other plant metabolites may maintain interactions with genomic DNA. The quantity of these secondary metabolites on the plant tissue increases with the age of the plant material and result in viscous extractions that are difficult to visualize when agarose gel electrophoresis is carried out (Varma et al., 2007); thus, it is usually recommended that young plant material is used for DNA extraction. In the present pilot study, root tissue employed for DNA extraction had been stored under unknown conditions for an undetermined amount of time, which may have led to the poor DNA yield and purity observed by spectrophotometry.

8.3.8.2 BARCODING PROFILE

Amplification of barcoding regions *rbcL, matK*, and *trnH-psbA* was successful with the addition of a modified version of the additive reagent described by Samarakoon et al. (2013). As mentioned before, polyphenols and polysaccharides are molecules that bind to DNA and impair the execution of different processes that depend on the DNA template sequence, including the amplification of DNA fragments. However, the molecules identified by GC-MS on *H. longipes* root extract were mainly sesquiterpenes and alkamides. Thus, it would be convenient to study further the interactions that these compounds may have with PCR reagents. BSA is a PCR facilitator that binds to different organic molecules, including fatty acids. Thus, it is expected that both alkamides and sesquiterpenes will interact with BSA enabling the correct amplification of the fragments. Regarding the effects of sesquiterpenes, it is possible that these compounds may form waxes in the DNA preparation, or inclusively, wax by-products. These waxes act as inhibitory agents during PCR (Kodzius et al., 2012). Moreover, it is known that the oxidized form of terpenes binds with the DNA molecule, avoiding further analysis (Katterman and Shattuck, 1983). Oxidized terpenes may interact with the lysine residues of BSA, leaving the DNA template free to react with the PCR components. A proposed mechanism of PCR inhibition for alkamide–DNA interactions is that the polar amine group of the alkamide molecule might be capable of forming hydrogen bonds with certain nucleotides. Markovits et al. (1981) studied the interactions of carboxamides with the DNA molecule; they proved by NMR analysis and space-filling models that the terminal amide group of some carboxamides interacts with the guanine base. Thus, these molecules have specific interactions with GC-rich DNA sequences. It is possible that the alkamides found on *H. longipes* GC-MS analysis form complexes in the DNA molecule that interfere with the replication of DNA during the amplification of the fragments. Regarding the interaction of BSA with alkamides, it is known that the hydrogen atom present in the amide group of such compounds is able to form hydrogen bonds with the hydroxyl groups of tyrosine residues (Raduner et al., 2006). Therefore, the abundance of tyrosine residues in the BSA structure allows the binding of alkamide compounds from *H. longipes* to BSA impeding DNA binding and, consequently, allowing the amplification of DNA fragments.

8.4 CONCLUSIONS

Heliopsis longipes is a promising plant that has the potential for the development of herbal medicines in the treatment of various diseases. Previously reported analgesic and anti-inflammatory properties of *H. longipes* make of it, a plant with the capability to be used as an adjuvant in the symptomatic treatment of pharyngitis. Moreover, the use of medicinal preparations containing its extract might be useful for the symptomatic treatment of acute pharyngitis due to its antimicrobial effect against buccal microbiota. The presence of different compounds with biological activity such as polyphenols and terpenes suggests that *H. longipes* might be useful for the treatment of other diseases. According to our pilot study results, maceration is the most appropriate method for the extraction of affinin. However, further studies need to be carried out regarding the extraction methods in order to determine the optimum conditions at which higher yields of affinin and other compounds of interest are obtained. Additionally, new methods for the isolation of affinin need to be developed that will increase the affinin purification yield with a higher level of purity. Also, the toxicity evaluation of the *H. longipes* extract must be evaluated on non-cancerous cells as well as on animal models, to observe effects closer to those expected if the extract is used to treat human conditions. Finally, the genetic characterization of *H. longipes* is required to discriminate between species that may have similar properties but unknown effects on health. If the ethanolic extract of this plant is considered to be commercialized, the correct identification of subspecies is fundamental to ensure the optimization of its biological activity.

ACKNOWLEDGMENTS

The authors would like to thank Dr M. Anaberta Cardador Martinez, Dr Paola Isabel Angulo Bejarano, Dr Maria Goretti Arvizu, and Anayelli Demeneghi Rivero, for all the help provided during the pilot study and the writing of this chapter. They would also like to thank Andres Zurita for providing the plant material and the photographic material used in this study and chapter. Finally, They want to thank the academic and technological support provided by the Tecnologico de Monterrey CQ.

KEYWORDS

- *Heliopsis longipes*
- pharyngitis
- traditional medicine

REFERENCES

Aalbers, J., O'Brien, K. K., Chan, W. S., Falk, G. A., Teljeur, C., Dimitrov, B. D., Fahey, T. (2011). Predicting streptococcal pharyngitis in adults in primary care: A systematic review of the diagnostic accuracy of symptoms and signs and validation of the Centor score. *BMC Med., 9*, 67. doi:10.1186/1741-7015-9-67.

Acosta-Madrid, I., Castañeda-Hernández, G., Cilia-López, V., Cariño-Cortés, R., Perez-Hernandez, N., Fernández-Martínez, E., Ortiz, M. (2009). Interaction between *Heliopsis longipes* extract and diclofenac on the thermal hyperalgesia test. *Phytomedicine, 16*(4), 336–341.

Acree Jr, F., Jacobson, M., Haller, H. (1945). An amide possessing insecticidal properties from the roots of *Erigeron* affinis DC. *J. Org. Chem., 10*(3), 236–242.

Aires, A. (2017). Phenolics in foods: Extraction, analysis and measurements. *Phenol. Comp.*, 61–88.

Bae, S. S., Ehrmann, B. M., Ettefagh, K. A., Cech, N. B. (2010). A validated liquid chromatography–electrospray ionization–mass spectrometry method for quantification of spilanthol in *Spilanthes acmella* (L.) Murr. *Phytochem. Anal., 21*(5), 438–443. doi:10.1002/pca.1215.

Bakhsh, Z. A., Al-Khatib, T. A., Al-Muhayawi, S. M., ElAssouli, S. M., Elfiky, I. A., Mourad, S. A. (2015). Evaluating the therapeutic efficacy, tolerability, and safety of an aqueous extract of *Costus speciosus* rhizome in acute pharyngitis and acute tonsillitis. A pilot study. *Saudi Med. J., 36*(8), 997–1000. doi:10.15537/smj.2015.8.11377.

Bakkali, F., Averbeck, S., Averbeck, D., Idaomar, M. (2008). Biological effects of essential oils---A review. *Food Chem. Toxicol., 46*(2), 446–475. doi:10.1016/j.fct.2007.09.106.

Bär, K.-J., Natura, G., Telleria-Diaz, A., Teschner, P., Vogel, R., Vasquez, E., Schaible, H.-G., Ebersberger, A. (2004). Changes in the effect of spinal prostaglandin E2 during inflammation: Prostaglandin E (EP1-EP4) receptors in spinal nociceptive processing of input from the normal or inflamed knee joint. *J. Neurosci., 24*(3), 642–651.

Barbosa, A. F., de Carvalho, M. G., Smith, R. E., Sabaa-Srur, A. U. O. (2016). Spilanthol: Occurrence, extraction, chemistry and biological activities. *Revista Brasileira de Farmacognosia, 26*(1), 128–133. doi:https://doi.org/10.1016/j.bjp.2015.07.024.

Barbosa, A. F., Pereira, C. D. S. S., Mendes, M. F., De Carvalho Junior, R. N., De Carvalho, M. G., Maia, J. G. S., Sabaa-Srur, A. U. O. (2017). Spilanthol content in the extract obtained by supercritical CO_2 at different storage times of *Acmella oleracea* L. *J. Food Process Eng., 40*(3), e12441. doi:10.1111/jfpe.12441.

Baser, K. H. C., Buchbauer, G. (2015). *Handbook of Essential Oils: Science, Technology, and Applications*. CRC Press, Boca Raton, FL, USA.

Blainski, A., Lopes, G., de Mello, J. (2013). Application and analysis of the Folin-Ciocalteu method for the determination of the total phenolic content from *Limonium brasiliense* L. *Molecules, 18*(6), 6852–6865.

Boiteux, L., Fonseca, M., Simon, P. (1999). Effects of plant tissue and DNA purification method on randomly amplified polymorphic DNA-based genetic fingerprinting analysis in carrot. *J. Am. Soc. Hortic. Sci., 124*(1), 32–38.

Boonen, J., Baert, B., Burvenich, C., Blondeel, P., De Saeger, S., De Spiegeleer, B. (2010). LC–MS profiling of N-alkylamides in *Spilanthes acmella* extract and the transmucosal behaviour of its main bio-active spilanthol. *J. Pharm. Biomed. Anal., 53*(3), 243–249.

Boonen, J., Bronselaer, A., Nielandt, J., Veryser, L., De Tre, G., De Spiegeleer, B. (2012). Alkamid database: Chemistry, occurrence and functionality of plant N-alkylamides. *J. Ethnopharmacol., 142*(3), 563–590.

Brglez Mojzer, E., Knez Hrnčič, M., Škerget, M., Knez, Ž., Bren, U. (2016). Polyphenols: Extraction methods, antioxidative action, bioavailability and anticarcinogenic effects. *Molecules, 21*(7), 901.

Campos-Cuevas, J. C., Pelagio-Flores, R., Raya-González, J., Méndez-Bravo, A., Ortiz-Castro, R., López-Bucio, J. (2008). Tissue culture of *Arabidopsis thaliana* explants reveals a stimulatory effect of alkamides on adventitious root formation and nitric oxide accumulation. *Plant Sci., 174*(2), 165–173.

Cardador-Martínez, A., Jiménez-Martínez, C., Sandoval, G. (2011). Revalorization of cactus pear (*Opuntia* spp.) wastes as a source of antioxidants. *Food Sci. Technol., 31*(3), 782–788.

Castro-González, V. (2009). *Monografía de la Raíz de Oro, Chilcuague, Heliopsis longipes A. Gray.* TlahuiEdu AC, México.

Castro-Ruiz, J., Rojas-Molina, A., Luna-Vázquez, F., Rivero-Cruz, F., García-Gasca, T., Ibarra-Alvarado, C. (2017). Affinin (spilanthol), isolated from *Heliopsis longipes*, induces vasodilation via activation of gasotransmitters and prostacyclin signaling pathways. *Int. J. Mol. Sci., 18*(1), 218.

Catic, T., Mehic, M., Binakaj, Z., Sahman, B., Cordalija, V., Kerla, A., Martinovic, I., Eskic, H. (2016). Efficacy and safety of oral spray containing lysozyme and cetylpyridinium: Subjective determination of patients with tonsillopharyngitis. *Mater. Sociomed., 28*(6), 459–463. doi:10.5455/msm.2016.28.459-463.

Chen, Q., Liang, C., Sun, X., Chen, J., Yang, Z., Zhao, H., Feng, L., Liu, Z. (2017). H_2O_2-responsive liposomal nanoprobe for photoacoustic inflammation imaging and tumor theranostics via *in vivo* chromogenic assay. *Proc. Nat. Acad. Sci., 114*(21), 5343–5348.

Cheung, L., Cheung, P. C., Ooi, V. E. (2003). Antioxidant activity and total phenolics of edible mushroom extracts. *Food Chem., 81*(2), 249–255.

Cho, K. S., Lim, Y.-R., Lee, K., Lee, J., Lee, J. H., Lee, I.-S. (2017). Terpenes from forests and human health. *Toxicol. Res., 33*(2), 97.

Choby, B. A. (2009). Diagnosis and treatment of streptococcal pharyngitis. *Am. Fam. Physician, 79*(5), 383–390.

Chouhan, S., Sharma, K., Guleria, S. (2017). Antimicrobial activity of some essential oils—Present status and future perspectives. *Medicines (Basel), 4*(3). doi:10.3390/medicines 4030058.

Cilia-López, V., Juárez-Flores, B., Aguirre-Rivera, J., Reyes-Agüero, J. (2010). Analgesic activity of *Heliopsis longipes* and its effect on the nervous system. *Pharm. Biol., 48*(2), 195–200.

Cilia-Lopez, V. G., Aguirre-Rivera, J. R., Reyes-Agüero, J. A., Juárez-Flores, B. I. (2008). Etnobotánica de *Heliopsis longipes* (Asteraceae: Heliantheae). *Boletín de la Sociedad Botánica de México,* (83), 81–87.

Cilia-López, V. G., Reyes-Agüero, J. A., Aguirre-Rivera, J. R., Juárez-Flores, B. I. (2013). Ampliación de la descripción y aspectos taxonómicos de *Heliopsis longipes* (Asteraceae: Heliantheae). *Polibotánica* (36), 1–13.

Cortez-Espinosa, N., Aviña-Verduzco, J. A., Ramírez-Chávez, E., Molina-Torres, J., Ríos-Chávez, P. (2011). Valine and phenylalanine as precursors in the biosynthesis of alkamides in *Acmella radicans*. *Natural Product Commun., 6*(6), 1934578X1100600625.

Cullen, J., Knees, S. G., Cubey, H. S., Shaw, J. (2011). *The European Garden Flora Flowering Plants: A Manual for the Identification of Plants Cultivated in Europe, Both Out-of-Doors and Under Glass* (vol. 1). Cambridge University Press, Cambridge, UK.

Déciga-Campos, M., Rios, M. Y., Aguilar-Guadarrama, A. B. (2010). Antinociceptive effect of *Heliopsis longipes* extract and affinin in mice. *Planta Medica, 76*(7), 665–670.

Déciga-Campos, M., Arriaga-Alba, M., Ventura-Martínez, R., Aguilar-Guadarrama, B., Rios, M. Y. (2012). Pharmacological and toxicological profile of extract from *Heliopsis longipes* and affinin. *Drug Dev. Res., 73*(3), 130–137.

Dunning, L. T., Savolainen, V. (2010). Broad-scale amplification of *mat*K for DNA barcoding plants, a technical note. *Bot. J. Linnean Soc., 164*(1), 1–9.

Dzoyem, J., McGaw, L., Kuete, V., Bakowsky, U. (2017). Anti-inflammatory and antinociceptive activities of African medicinal spices and vegetables. In *Medicinal Spices and Vegetables from Africa* (pp. 239–270). Elsevier, Amsterdam, The Netherlands.

Eddy, N. B., Leimbach, D. (1953). Synthetic analgesics. II. Dithienylbutenyl-and dithienylbutylamines. *J. Pharmacol. Exp. Ther., 107*(3), 385–393.

El Asbahani, A., Miladi, K., Badri, W., Sala, M., Ait Addi, E. H., Casabianca, H., Mousadik, A. El, Hartmann, D., Jilale, A., Renaud, F. N. R., Elaissari, A. (2015). Essential oils: From extraction to encapsulation. *Int. J. Pharm., 483*(1–2), 220–243. doi:10.1016/j.ijpharm.2014.12.069.

Escobedo-Martínez, C., Guzmán-Gutiérrez, S. L., Hernández-Méndez, M. d. l. M., Cassani, J., Trujillo-Valdivia, A., Orozco-Castellanos, L. M., Enríquez, R. G. (2017). *Heliopsis longipes*: Anti-arthritic activity evaluated in a Freund's adjuvant-induced model in rodents. *Revista Brasileira de Farmacognosia, 27*(2), 214–219.

Farah, B., Visintini, S. (2018). *Benzydamine for Acute Sore Throat: A Review of Clinical Effectiveness and Guidelines.* Ottawa. Retrieved from https://www.ncbi.nlm.nih.gov/books/NBK537954/

FDA. (2007). *Guidance for Industry on Complementary and Alternative Medicine Products and Their Regulation by the Food and Drug Administration.* Retrieved from https://www.fda.gov/regulatory-information/search-fda-guidance-documents/complementary-and-alternative-medicine-products-and-their-regulation-food-and-drug-administration

Fidyt, K., Fiedorowicz, A., Strządała, L., Szumny, A. (2016). β-caryophyllene and β-caryophyllene oxide—Natural compounds of anticancer and analgesic properties. *Cancer Med., 5*(10), 3007–3017.

Fischer Walker, C. L., Rimoin, A. W., Hamza, H. S., Steinhoff, M. C. (2006). Comparison of clinical prediction rules for management of pharyngitis in settings with limited resources. *J. Pediatrics, 149*(1), 64–71. doi:10.1016/j.jpeds.2006.03.005.

Fisher, K., Phillips, C. (2008). Potential antimicrobial uses of essential oils in food: Is citrus the answer? *Trends Food Sci. Technol., 19*(3), 156–164.

Ford, C. S., Ayres, K. L., Toomey, N., Haider, N., Van Alphen Stahl, J., Kelly, L. J., Wikström, N., Hollingsworth, P. M., Duff, R. J., Hoot, S. B., Cowan, R. S., Chase, M. W., Wilkinson, M. J. (2009). Selection of candidate coding DNA barcoding regions for use on land plants. *Bot. J. Linnean Soc., 159*(1), 1–11.

Frye, R., Bailey, J., Blevins, A. E. (2011). Clinical inquiries. Which treatments provide the most relief for pharyngitis pain? *J. Fam. Pract., 60*(5), 293–294.

Fujiwara, N., Kobayashi, K. (2005). Macrophages in inflammation. *Curr. Drug Targets-Inflamm. Allergy, 4*(3), 281–286.

Gerbino, A., Schena, G., Milano, S., Milella, L., Barbosa, A. F., Armentano, F., Procino, G., Svelto, M., Carmosino, M. (2016). Spilanthol from *Acmella oleracea* lowers the intracellular levels of cAMP impairing NKCC2 phosphorylation and water channel AQP2 membrane expression in mouse kidney. *PLoS One, 11*(5), e0156021.

Golac-Guzina, N., Novakovic, Z., Sarajlic, Z., Sukalo, A., Dzananovic, J., Glamoclija, U., Kapo, B., Čordalija, V., Mehic, M. (2019). Comparative study of the efficacy of the lysozyme, benzydamine and chlorhexidine oral spray in the treatment of acute tonsillopharyngitis: Results of a pilot study. *Acta Med. Acad., 48*(2), 140–146. doi:10.5644/ama2006-124.252.

González Morales, S., Flores López, M. L., Benavides Mendoza, A., Flores Olivas, A. (2011). Actividad inhibitoria del extracto de *Heliopsis longipes* sobre *Fusarium oxysporum* f. sp *lycopersici*. *Rev. Mexicana Fitopatol., 29*(2), 146–153.

González-Mas, M. C., Rambla, J. L., López-Gresa, M. P., Blázquez, M. A., Granell, A. (2019). Volatile compounds in *Citrus* essential oils: A comprehensive review. *Front Plant Sci, 10*, 12. doi:10.3389/fpls.2019.00012.

Gutierrez-Lugo, M., Barrientos-Benitez, T., Luna, B., Ramirez-Gama, R., Bye, R., Linares, E., Mata, R. (1996). Antimicrobial and cytotoxic activities of some crude drug extracts from Mexican medicinal plants. *Phytomedicine, 2*(4), 341–347.

Hargreaves, K., Dubner, R., Brown, F., Flores, C., Joris, J. (1988). A new and sensitive method for measuring thermal nociception in cutaneous hyperalgesia. *Pain, 32*(1), 77–88.

Hernández, F. (1888). *Cuatro Libros de la Naturaleza y Virtudes Medicinales de Las Plantas y Animales de la Nueva España*: Rosario Bravo.

Hernandez, I., Lemus, Y., Prieto, S., Molina-Torres, J., Garrido, G. (2009a). Anti-inflammatory effect of an ethanolic root extract of *Heliopsis longipes in vitro*. *Boletín Latinoamericano y del Caribe de Plantas Medicinales y Aromáticas, 8*(3), 160–164.

Hernández, I., Márquez, L., Martínez, I., Dieguez, R., Delporte, C., Prieto, S., Molina-Torres, J., Garrido, G. (2009b). Anti-inflammatory effects of ethanolic extract and alkamides-derived from *Heliopsis longipes* roots. *J. Ethnopharmacol., 124*(3), 649–652.

Huang, W.-C., Peng, H.-L., Hu, S., Wu, S.-J. (2019). Spilanthol from traditionally used *Spilanthes acmella* enhances AMPK and ameliorates obesity in mice fed high-fat diet. *Nutrients, 11*(5), 991.

Hubbert, M., Sievers, H., Lehnfeld, R., Kehrl, W. (2006). Efficacy and tolerability of a spray with *Salvia officinalis* in the treatment of acute pharyngitis: A randomised, double-blind, placebo-controlled study with adaptive design and interim analysis. *Eur. J. Med. Res., 11*(1), 20–26.

Hyldgaard, M., Mygind, T., Meyer, R. L. (2012). Essential oils in food preservation: Mode of action, synergies, and interactions with food matrix components. *Front. Microbiol., 3*, 12. doi:10.3389/fmicb.2012.00012.

Jacobson, M. (1954). Constituents of *Heliopsis* species. III. 1 cis–trans isomerism in affinin. *J. Am. Chem. Soc., 76*(18), 4606–4608.

Jasmin, L., Rabkin, S. D., Granato, A., Boudah, A., Ohara, P. T. (2003). Analgesia and hyperalgesia from GABA-mediated modulation of the cerebral cortex. *Nature, 424*(6946), 316.

Jiang, H., Chen, M., Li, T., Liu, H., Gong, Y., Li, M. (2016). Molecular characterization of *Streptococcus agalactiae* causing community- and hospital-acquired infections in Shanghai, China. *Front. Microbiol., 7*, 1308. doi:10.3389/fmicb.2016.01308.

Jiang, Z., Kempinski, C., Chappell, J. (2016). Extraction and analysis of terpenes/terpenoids. *Curr. Protocols Plant Biol., 1*(2), 345–358.

Jiménez-Silva, Á. A. (2017). Medicina tradicional. *BOLETÍN CONAMED*, (13).

Kalra, M. G., Higgins, K. E., Perez, E. D. (2016). Common questions about streptococcal pharyngitis. *Am. Fam. Physician, 94*(1), 24–31.

Katterman, F., Shattuck, V. (1983). An effective method of DNA isolation from the mature leaves of *Gossypium* species that contain large amounts of phenolic terpenoids and tannins. *Preparative Biochem., 13*(4), 347–359.

Kedare, S. B., Singh, R. (2011). Genesis and development of DPPH method of antioxidant assay. *J. Food Sci. Technol., 48*(4), 412–422.

Kelly, G. S. (1999). Squalene and its potential clinical uses. *Alt. Med. Rev., J. Clin. Ther., 4*(1), 29–36.

Khan, S., Qureshi, M. I., Alam, T., Abdin, M. (2007). Protocol for isolation of genomic DNA from dry and fresh roots of medicinal plants suitable for RAPD and restriction digestion. *African J. Biotechnol., 6*(3), 175–178.

Kim, D., Jeong, H., Kwon, J., Kang, S., Han, B., Lee, E. K., Lee, S. M., Choi, J. W. (2019). The effect of benzydamine hydrochloride on preventing postoperative sore throat after total thyroidectomy: A randomized-controlled trial. *Can. J. Anaesth, 66*(8), 934–942. doi:10.1007/s12630-019-01371-2.

Kim, E.-S., Kang, S.-Y., Kim, Y.-H., Lee, Y.-E., Choi, N.-Y., You, Y.-O., Kim, K.-J. (2015). *Chamaecyparis obtusa* essential oil inhibits methicillin-resistant *Staphylococcus aureus* biofilm formation and expression of virulence factors. *J. Med. Food, 18*(7), 810–817.

Kim, J.-A., Kang, Y. S., Lee, Y. S. (2001). Activation of Na+, K+, Cl− -cotransport mediates intracellular Ca_2+ increase and apoptosis induced by Pinacidil in Hep G2 human hepatoblastoma cells. *Biochem. Biophys. Res. Commun., 281*(2), 511–519.

Klostergaard, J. (1985). A rapid extremely sensitive, quantitative microassay for cytotoxic cytokines. *Lymphokine Res., 4*(4), 309–317.

Kodzius, R., Xiao, K., Wu, J., Yi, X., Gong, X., Foulds, I. G., Wen, W. (2012). Inhibitory effect of common microfluidic materials on PCR outcome. *Sens. Actuators B, Chem., 161*(1), 349–358.

Kress, W. J., Erickson, D. L., Jones, F. A., Swenson, N. G., Perez, R., Sanjur, O., Bermingham, E. (2009). Plant DNA barcodes and a community phylogeny of a tropical forest dynamics plot in Panama. *Proc. Nat. Acad. Sci., 106*(44), 18621–18626.

Lalicevic, S., Djordjevic, I. (2004). Comparison of benzydamine hydrochloride and *Salvia officinalis* as an adjuvant local treatment to systemic nonsteroidal anti-inflammatory drug in controlling pain after tonsillectomy, adenoidectomy, or both: An open-label, single-blind, randomized clinical trial. *Curr. Ther. Res. Clin. Exp., 65*(4), 360–372. doi:10.1016/j.curtheres.2004.07.002.

LaLone, C. A., Hammer, K. D., Wu, L., Bae, J., Leyva, N., Liu, Y., Solco, A. K. S., Kraus, G. A., Murphy, P. A., Wurtele, E. S., Kim, O.-K., Seo, K., Widrlechner, M. P., Birt, D. F. (2007). Echinacea species and alkamides inhibit prostaglandin E2 production in RAW264.7 mouse macrophage cells. *J. Agric. Food Chem., 55*(18), 7314–7322.

Levin, R. A., Wagner, W. L., Hoch, P. C., Nepokroeff, M., Pires, J. C., Zimmer, E. A., Sytsma, K. J. (2003). Family-level relationships of Onagraceae based on chloroplast *rbc*L and *ndh*F data. *Am. J. Bot., 90*(1), 107–115.

Li, T., Chen, Z. J. (2018). The cGAS–cGAMP–STING pathway connects DNA damage to inflammation, senescence, and cancer. *J. Exp. Med., 215*(5), 1287–1299.

Little, E. L. (1948). *Heliopsis longipes*, a Mexican insecticidal plant species. *J. Washington Acad. Sci., 38*(8), 269–274.

Liu, Y., Murphy, P. A. (2007). Alkamide stability in *Echinacea purpurea* extracts with and without phenolic acids in dry films and in solution. *J. Agric. Food Chem., 55*(1), 120–126.

Loh, Y. C., Tan, C. S., Ch'ng, Y. S., Yeap, Z. Q., Ng, C. H., Yam, M. F. (2018). Overview of the microenvironment of vasculature in vascular tone regulation. *Int. J. Mol. Sci., 19*(1), 120.

Looze, F. D., Russo, M., Bloch, M., Montgomery, B., Shephard, A., DeVito, R. (2018). Meaningful relief with flurbiprofen 8.75 mg spray in patients with sore throat due to upper respiratory tract infection. *Pain Manag., 8*(2), 79–83. doi:10.2217/pmt-2017-0100.

López-Martínez, S., Aguilar-Guadarrama, A. B., Rios, M. Y. (2011). Minor alkamides from *Heliopsis longipes* SF Blake (Asteraceae) fresh roots. *Phytochem. Lett., 4*(3), 275–279.

Lozano, J. A. (2003). El dolor de garganta y el uso de los bucofaríngeos. *Offarm, 22*(1), 63–68. Retrieved from https://www.elsevier.es/es-revista-offarm-4-articulo-el-dolor-garganta-el-uso-13042366ER

Markovits, J., Gaugain, B., Barbet, J., Roques, B. P., Le Pecq, J. B. (1981). Hydrogen bonding in deoxyribonucleic acid base recognition. 2. Deoxyribonucleic acid binding studies of acridine alkylamides. *Biochemistry, 20*(11), 3042–3048.

McGovran, E., Bottger, G., Gersdorff, W., Fales, J. (1947). Insecticidal action of *Heliopsis longipes* and *Erigeron* spp. *Publ. US Bur. Ent.*

MedlinePlus. (2019). Herbal Medicine. Retrieved from https://medlineplus.gov/herbalmedicine.html

Molina-Torres, J., García-Chávez, A., Ramírez-Chávez, E. (1999). Antimicrobial properties of alkamides present in flavouring plants traditionally used in Mesoamerica: Affinin and capsaicin. *J. Ethnopharmacol., 64*(3), 241–248.

Molina-Torres, J., Salazar-Cabrera, C. J., Armenta-Salinas, C., Ramírez-Chávez, E. (2004). Fungistatic and bacteriostatic activities of alkamides from *Heliopsis longipes* roots: Affinin and reduced amides. *J. Agric. Food Chem., 52*(15), 4700–4704.

Molina-Torres, J., Salgado-Garciglia, R., Ramirez-Chavez, E., Del Rio, R. E. (1996). Purely olefinic alkamnides in *Heliopsis longipes* and *Acmella* (*Spilanthes*) *oppositifolia*. *Biochem. Syst. Ecol., 24*(1), 43–47.

Moreira, D. d. L., Teixeira, S. S., Monteiro, M. H. D., De-Oliveira, A. C. A. X., Paumgartten, F. J. R. (2014). Traditional use and safety of herbal medicines1. *Revista Brasileira de Farmacognosia, 24*(2), 248–257. doi:https://doi.org/10.1016/j.bjp.2014.03.006.

Muller, D., Lindemann, T., Shah-Hosseini, K., Scherner, O., Knop, M., Bilstein, A., Mosges, R. (2016). Efficacy and tolerability of an ectoine mouth and throat spray compared with those of saline lozenges in the treatment of acute pharyngitis and/or laryngitis: A prospective, controlled, observational clinical trial. *Eur. Arch. Otorhinolaryngol., 273*(9), 2591–2597. doi:10.1007/s00405-016-4060-z.

NCCIH. (Ed.) (2019). National Center for Complementary and Integrative Health.

Nikolova, M. (2011). Screening of radical scavenging activity and polyphenol content of Bulgarian plant species. *Pharmacogn. Res., 3*(4), 256.

NIST. (2019a). 4,11-Selinadiene, from U.S. Department of Commerce

NIST. (2019b). B-farnesene, from U.S. Department of Commerce

NIST. (2019c). Carvacrol, from U.S. Department of Commerce

NIST. (2019d). Caryophyllene, from U.S. Department of Commerce

NIST. (2019e). Viridiflorene, from U.S. Department of Commerce

Ogura, M., Cordell, G., Quinn, M., Leon, C., Benoit, P., Soejarto, D., Farnsworth, N. (1982). Ethnopharmacologic studies. I. Rapid solution to a problem—Oral use of *Heliopsis longipes*—By means of a multidisciplinary approach. *J. Ethnopharmacol., 5*(2), 215–219.

Olivas-Quintero, S., López-Angulo, G., Montes-Avila, J., Díaz-Camacho, S. P., Vega-Aviña, R., López-Valenzuela, J. Á., Salazar-Salas, N. Y., Delgado-Vargas, F. (2017). Chemical composition and biological activities of *Helicteres vegae* and *Heliopsis sinaloensis*. *Pharm. Biol., 55*(1), 1473–1482.

Ortiz, M. I., Cariño-Cortés, R., Pérez-Hernández, N., Ponce-Monter, H., Fernández-Martínez, E., Castañeda-Hernández, G., Acosta-Madrid, I. I., Cilia-López, V. G. (2009). *Antihyperalgesia Induced by Heliopsis longipes Extract.* Paper presented at the Proc. West. Pharmacol. Soc.

Osterziel, K., Julius, S. (1982). Vasodilators in the treatment of hypertension. *Compr. Ther., 8*(11), 43–52.

Pandey, K. B., Rizvi, S. I. (2009). Plant polyphenols as dietary antioxidants in human health and disease. *Oxid. Med. Cell. Longev., 2*(5), 270–278.

pubchem.ncbi.nlm.nih.gov.(2019a).4a-methyl-1,3,4,5,8,8a-hexahydronaphthalen-2-one, from PubChem Database

pubchem.ncbi.nlm.nih.gov. (2019b). 4a,8-dimethyl-2-prop-1-en-2-yl-2,3,4,5,6,7-hexahydro-1H-naphthalene, from PubChem Database

pubchem.ncbi.nlm.nih.gov. (2019c). Benzydamine hydrochloride, from PubChem Database
pubchem.ncbi.nlm.nih.gov. (2019d). Chlorhexidine. Retrieved from https://pubchem.ncbi.nlm. nih.gov/compound/Chlorhexidine. PubChem Database. (CID=9552079). Retrieved Dec. 19, 2019, from PubChem Database https://pubchem.ncbi.nlm.nih.gov/compound/Chlorhexidine
pubchem.ncbi.nlm.nih.gov. (2019e). Cyclohexene,6-butyl-1-nitro, from PubChem Database
pubchem.ncbi.nlm.nih.gov. (2019f). Squalene, from PubChem Database
Raduner, S., Majewska, A., Chen, J.-Z., Xie, X.-Q., Hamon, J., Faller, B., Altmann, K.-H., Gertsch, J. (2006). Alkylamides from *Echinacea* are a new class of cannabinomimetics Cannabinoid type 2 receptor-dependent and -independent immunomodulatory effects. *J. Biol. Chem., 281*(20), 14192–14206.
Rios, M. Y., Aguilar-Guadarrama, A. B., del Carmen Gutiérrez, M. (2007). Analgesic activity of affinin, an alkamide from *Heliopsis longipes* (Compositae). *J. Ethnopharmacol., 110*(2), 364–367.
Rubin, S. A., Swan, H. J. (1981). Vasodilator therapy for heart failure: Concepts, applications, and challenges. *JAMA, 245*(7), 761–763.
Russell, J. M. (2000). Sodium-potassium-chloride cotransport. *Physiol. Rev., 80*(1), 211–276.
Samarakoon, T., Wang, S. Y., Alford, M. H. (2013). Enhancing PCR amplification of DNA from recalcitrant plant specimens using a trehalose-based additive. *Appl. Plant Sci., 1*(1), 1200236.
Sang, T., Crawford, D. J., Stuessy, T. F. (1997). Chloroplast DNA phylogeny, reticulate evolution, and biogeography of *Paeonia* (Paeoniaceae). *Am. J. Bot., 84*(8), 1120–1136.
Sharma, A., Flores-Vallejo, R. D. C., Cardoso-Taketa, A., Villarreal, M. L. (2017). Antibacterial activities of medicinal plants used in Mexican traditional medicine. *J. Ethnopharmacol., 208*, 264–329. doi:10.1016/j.jep.2016.04.045.
Slinger, R., Goldfarb, D., Rajakumar, D., Moldovan, I., Barrowman, N., Tam, R., Chan, F. (2011). Rapid PCR detection of group A *Streptococcus* from flocked throat swabs: A retrospective clinical study. *Ann. Clin. Microbiol. Antimicrob., 10*, 33. doi:10.1186/1476-0711-10-33.
Smith, T. J. (2000). Squalene: Potential chemopreventive agent. *Expert Opin. Investigational Drugs, 9*(8), 1841–1848.
Soufli, I., Toumi, R., Rafa, H., Touil-Boukoffa, C. (2016). Overview of cytokines and nitric oxide involvement in immuno-pathogenesis of inflammatory bowel diseases. *World J. Gastrointest. Pharmacol. Ther., 7*(3), 353.
Tanz, R., Gewitz, M., Kaplan, E., Shulman, S. (2019). Stay the course: Targeted evaluation, accurate diagnosis, and treatment of streptococcal pharyngitis prevent acute rheumatic fever. *J. Pediatr., 216*. doi:10.1016/j.jpeds.2019.08.042.
Tate, J. A., Simpson, B. B. (2003). Paraphyly of *Tarasa* (Malvaceae) and diverse origins of the polyploid species. *Syst. Bot., 28*(4), 723–737.
Texeira-Duarte, M. C., Rai, M. (2015). *Therapeutic Medicinal Plants: From Lab to the Market*. CRC Press, Boca Raton, FL, USA.
Thomas, R. (1983). Ventricular fibrillation and initial plasma potassium in acute myocardial infarction. *Postgrad. Med. J., 59*(692), 354–356.
Toepfner, N., Henneke, P., Berner, R., Hufnagel, M. (2013). Impact of technical training on rapid antigen detection tests (RADT) in group A streptococcal tonsillopharyngitis. *Eur. J. Clin. Microbiol. Infect. Dis., 32*(5), 609–611. doi:10.1007/s10096-012-1783-7.
tropicos.org. (2019). *Heliopsis longipes* (A. Gray) S.F. Blake. Retrieved from http://tropicos. org/Name/50180045. Retrieved Dec. 19, 2019, from Missouri Botanical Garden http:// tropicos.org/Name/50180045

Trostchansky, A., Mastrogiovanni, M., Miquel, E., Rodríguez-Bottero, S., Martínez-Palma, L., Cassina, P., Rubbo, H. (2018). Profile of arachidonic acid-derived inflammatory markers and its modulation by nitro-oleic acid in an inherited model of amyotrophic lateral sclerosis. *Front. Mol. Neurosci., 11*, 131.

Tucker, W. D., Mahajan, K. (2018). Anatomy, blood vessels. In *StatPearls [Internet]*: StatPearls Publishing.

Van Brusselen, D., Vlieghe, E., Schelstraete, P., De Meulder, F., Vandeputte, C., Garmyn, K., Laffut, W., Van de Voorde, P. (2014). Streptococcal pharyngitis in children: to treat or not to treat? *Eur J. Pediatr., 173*(10), 1275–1283. doi:10.1007/s00431-014-2395-2.

Van Limbergen, J., Kalima, P., Taheri, S., Beattie, T. F. (2006). *Streptococcus* A in paediatric accident and emergency: Are rapid streptococcal tests and clinical examination of any help? *Emerg. Med. J., 23*(1), 32–34. doi:10.1136/emj.2004.022970.

Varma, A., Padh, H., Shrivastava, N. (2007). Plant genomic DNA isolation: An art or a science. *Biotechnol. J., Health. Nutr. Technol., 2*(3), 386–392.

Velez-Haro, J. M., Buitimea-Cantúa, N. E., Rosas-Burgos, E. C., Molina-Torres, J., Buitimea-Cantúa, G. V. (2018). Effect of the roots extract from *Heliopsis longipes* on *Aspergillus parasiticus* growth. *Biotecnia, 20*(3), 127–134.

Vickers, A., Zollman, C., Lee, R. (2001). Herbal medicine. *West. J. Med., 175*(2), 125–128. doi:10.1136/ewjm.175.2.125.

Wang, R., Wang, Z., Wu, L. (1997). Carbon monoxide-induced vasorelaxation and the underlying mechanisms. *Brit. J. Pharmacol., 121*(5), 927–934.

Ward, E. (1986). Biochemical mechanisms involved in resistance of plants to fungi. In *Biology and Molecular Biology of Plant-Pathogen Interactions* (pp. 107–131). Springer, New York, NY.

WHO. (2001). *Legal Status of Traditional Medicine and Complementary/Alternative Medicine: A Worldwide Review.*

Wijesundara, N. M., Rupasinghe, H. P. V. (2019). Herbal tea for the management of pharyngitis: Inhibition of *Streptococcus pyogenes* growth and biofilm formation by herbal infusions. *Biomedicines, 7*(3). doi:10.3390/biomedicines7030063.

Woodley, M., and Whelan, A. (1993). *Manual de terapéutica médica*. Barcelona: Ediciones científicas técnicas S.A.

CHAPTER 9

Antidiabetic Natural Products: Mechanisms of Action

ERICK P. GUTIÉRREZ-GRIJALVA[1], LAURA A. CONTRERAS-ANGULO[2], MARILYN S. CRIOLLO-MENDOZA[2], SARA AVILÉS-GAXIOLA[2], and J. BASILIO HEREDIA[2*]

[1]*Cátedras CONACYT-Research Center for Food and Development, Culiacán, Sinaloa 80110, México*

[2]*Research Center for Food and Development, Culiacán, Sinaloa 80110, México*

*Corresponding author. E-mail: jbheredia@ciad.mx.

ABSTRACT

There is a worldwide growing interest in the study of natural products to prevent/alleviate several diseases, such as atherosclerosis, cancer, and type 2 diabetes mellitus (T2DM). Natural products are compounds derived from the secondary metabolism of plants. Natural products such as terpenes, phenolic compounds, and alkaloids are aimed to protect plants against biotic (herbivores, insects) and abiotic conditions (UV-light, draught, climate, among others). These compounds have been the interest of many studies due to their reported potential to prevent and treat T2DM. As a result of these studies terpenes, phenolic compounds, and alkaloids are reported as potential compounds to inhibit carbohydrate metabolic enzymes, ameliorate insulin resistance, and prevent nonenzymatic glycation, among others. Thus, in this chapter, the authors have summarized recent information (2010–2019) on the antidiabetic potential of terpenes, phenolic compounds, and alkaloids, and their mechanisms of action.

9.1 INTRODUCTION

According to the International Diabetes Federation, diabetes is a chronic condition that occurs when the body cannot produce any or enough insulin or cannot effectively use the insulin it produces (International Diabetes Federation, 2019). There are two types of diabetes, but the most common is type 2 diabetes, which affects nearly 90% of diabetic patients (International Diabetes Federation, 2019). The UN has estimated that 463 million adults are currently living with diabetes (diagnosed), around 9.3% of the world's population, and is estimated to rise to 578 million by 2030 (International Diabetes Federation, 2019). The primary treatment for diabetes is the prescription of pharmaceutical drugs, such as metformin, acarbose, among others (DeFronzo et al., 2015); however, people from low- and middle-income countries use medicinal plants alone or in combination with their conventional drugs (Cetto, 2015). The ethnobotanical use of medicinal plants to treat metabolic disorders such as diabetes has been documented since ancient times.

Current ethnopharmacological studies have led to the discovery of new plant-based drugs, and some of them are currently used as a treatment for cancer. Most of the medicinal plants that are used by diabetic patients around the world are not well-studied and even their phytochemical constituents are not fully known. This lack of knowledge may lead to plant-induced toxicologic adverse effects in patients. Nonetheless, studies of natural products from medicinal plants are increasing, metabolomic analyses are allowing the elucidation of their constituents, and through a junction of genomic, transcriptomic, and metabolomic studies, their mechanisms of action are increasingly being reported. This chapter focuses on recent reports related to the mechanism of action of some natural products, such as terpenes, phenolic compounds, and alkaloids.

9.2 TERPENES WITH ANTIDIABETIC PROPERTIES

Derives of the secondary metabolism of the plants are generated a diversity of compounds like the terpenes (generic name), also known as isoprenoids. Their basic structure is isoprene units (C5) linked to each other from head-to-tail, or head-to-head, and some by head-to-middle fusions, all they to form a terpenoid molecule (Figure 9.1) (Ludwiczuk et al., 2017).

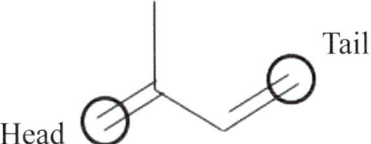

FIGURE 9.1 Basic structure of terpenes: an isoprene unit (C5) showing head and tail.

In the mevalonate pathway (MVA), the terpenes are synthesized in the cytosol, endoplasmic reticulum, and peroxisomes, which start from the mevalonic acid that is converted to isopentenyl diphosphate (IPP) and dimethylallyl diphosphate (DMAPP) by phosphorylation and decarboxylation. IPP and DMAPP are important intermediaries in sesquiterpenes and triterpenes biosynthesis (Singh and Sharma, 2015; Oldfield and Lin, 2012). Another biosynthetic pathway of terpenes is the 2-C-methyl-D-erythritol 4-phosphate (MET) that works in the chloroplasts from pyruvate and glyceraldehyde 3-phosphate as a substrate to form IPP and DMAPP, universal precursors of terpenes (Figure 9.2). By this pathway monoterpenes, diterpenes, and some sesquiterpenes are synthesized (Singh and Sharma, 2015; Tetali, 2019).

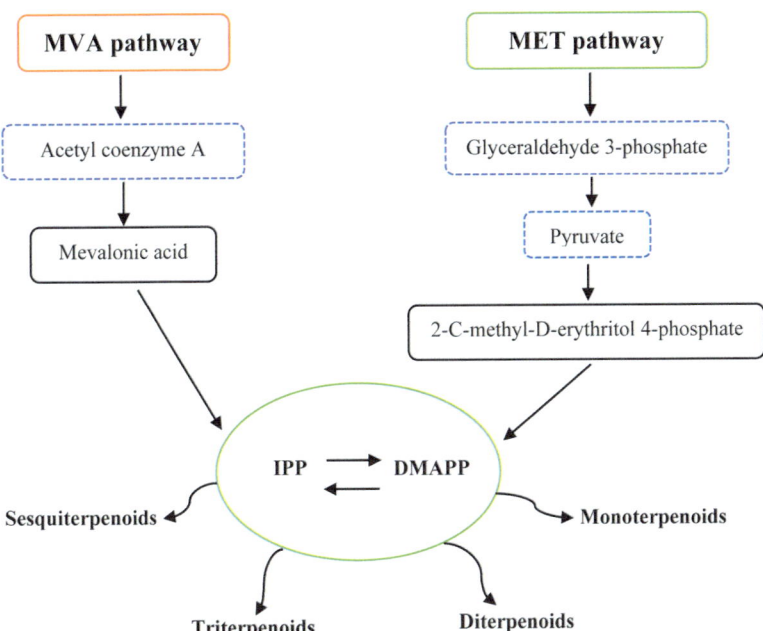

FIGURE 9.2 Biosynthesis of terpenoids. Adapted from Singh and Sharma (2015), Oldfield and Lin (2012) and Tetali (2019).

Terpenes, in base their structure, are classified in monoterpenoids (two isoprene units; $C_{10}H_{16}$) that are found in acyclic, monocyclic, and bicyclic form; sesquiterpenoids (three isoprene unit; $C_{15}H_{24}$) have linear, monocyclic, bicyclic, and tricyclic form; the diterpenoids (four isoprene units; $C_{20}H_{32}$) are linear, bicyclic, tricyclic, tetracyclic, and pentacyclic; the triterpenoids (six isoprene units; $C_{30}H_{48}$) have complex cyclic structure and tetraterpenoids (eight isoprene units; $C_{40}H_{64}$) (Ludwiczuk et al., 2017; Tetali, 2019; Huang et al., 2012).

Terpenes historically have been used since time ago; Egyptians and other cultures used essential oils in religious ceremonies (Brahmkshatriya and Brahmkshatriya, 2013). The main constituents of essential oils are the mono- and sesquiterpenoids, di- and tri-terpenoids are present in gums and resins. Finally, tetraterpenoids constitute a group of compounds named carotenoids, all these compounds are widely distributed in nature (Çitoğlu and Acıkara, 2012).

Each class has specific properties in function of the origin, some of them affect the growth of plants, some others in the metabolism, or the ecology system of the plant (Ringuelet and Viña, 2013). In addition to having various functions in the plant, they have biological activity like antioxidant, antimicrobial, digestive, analgesic, anticancer, anti-inflammatory, anticonvulsive, antidepressant, anxiolytic, neuroprotective, antimutagenic, antiallergic, antibiotic, and antidiabetic (Ludwiczuk et al., 2017; Nuutinen, 2018). The latter represents a significant opportunity to reduce the high incidence of diabetes mellitus (DM) worldwide level. In this sense, several studies have been conducted with terpenes obtained from different sources.

9.2.1 MECHANISMS OF ACTION

Unhealthy lifestyles, such as bad eating habits and sedentarism, are related to the onset of metabolic disorders, such as obesity and DM (Escandon, 2011). DM could be the oldest disease known to man, described by the Egyptians about 3000 years ago (Olokoba et al., 2012). It is defined as a metabolic disorder characterized by an increase in blood glucose and other disturbances of fat and protein metabolism. All these are associated with a deficiency in the production of insulin by the pancreas or its action, or both (Hosseyni et al., 2012; Muruganathan and Srinivasan, 2016; Abdelmeguid et al., 2010). There are three types of DM: type 1 (insulin-dependent, the pancreas lacks cells or contains defective cells), type 2 (noninsulin-dependent or insulin resistance), and gestational (detected during pregnancy) (Abdel-Moneim and Fayez, 2015).

Medicinal plants have been traditionally used as a treatment in DM, which are rich in natural products such as terpene. Their effect might be associated with the regulation of different signaling pathways in the pancreas, liver, and skeletal muscle, all involved in the disease. These might act by several mechanisms such as the promotion of insulin production, reparation of the proliferation of β-cells, and increasing the effects of insulin. The antidiabetic effect of terpenes could be related to their compact molecular structure, low ramification, and less symmetric (Chen et al., 2019; Muruganathan and Srinivasan, 2016).

Some mechanisms studied are related to the inhibition of enzymes involved in glucose metabolism, like α-glucosidase, α-amylase, and aldose reductase. Also with the stimulation of insulin synthesis and release from pancreatic β-cells; the preservation or regeneration of pancreatic β-cells; modulation of glucose uptake; the stimulation of glycogenesis and hepatic glycolysis; and the activation of peroxisome proliferator-activated receptor-gamma (PPAR-γ); the antioxidant effect and diminish β-cells apoptosis (Mabhida et al., 2018; Bahmani et al., 2014) (Figure 9.3).

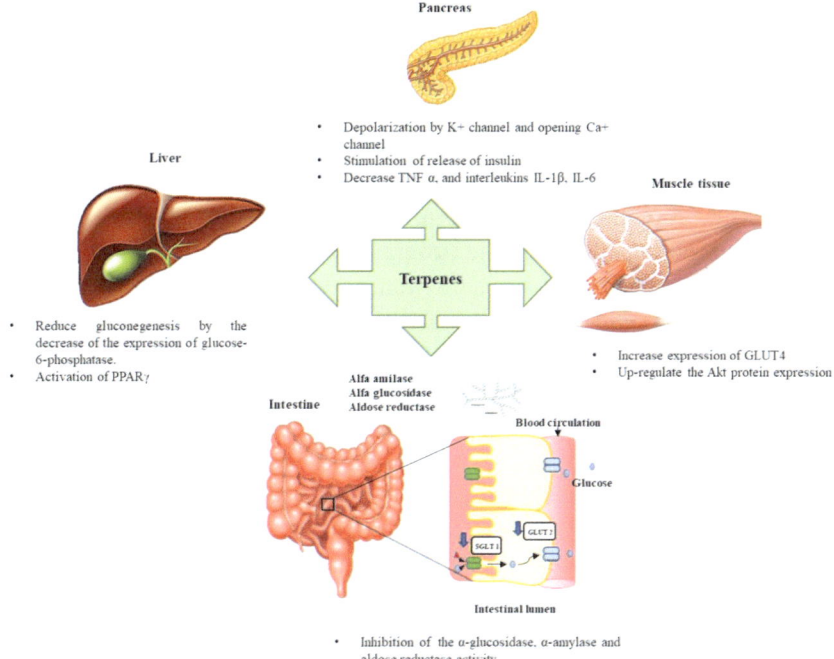

FIGURE 9.3 Overview of the terpenes mechanism of action. Modified from Mabhida et al. (2018) and Bahmani et al. (2014).

Another mechanism is through the inhibition of the sodium-glucose type 1 (SGLT-1) cotransporter, this inhibitory effect was reported for five acyclic terpenes. The SGLTs (located in the luminal membrane of epithelial cells) are proteins that transport Na+ and glucose to the cell using the electrochemical gradient to transport the glucose against the gradient (Valdes et al., 2019).

Terpenoids just like commercial drugs can modulate the expression of the GLUTs (glucose transporters) in the liver (GLUT-2), adipocytes, and skeletal muscles (GLUT-4) and increase the glucose uptake, besides can activate the pathways PI3-K/AKT (phosphatidylinositol 3-kinase/protein kinase B has an important role in the insulin action pathway) and AMP-activated protein kinase (AMPK), which regulates glucose homeostasis (by the activation of proteins GLUTs) and show an effect in the oxidative stress (increasing the expression or activation of Nrf2: nuclear factor [erythroid-derived 2] like 2). Some triterpenoids (ex. lupeol) reduce blood glucose levels and increase insulin due to the activation of the PPAR-γ by increasing the expression of GLUT-4 and translocation in adipocytes (Giacoman-Martínez et al., 2019; El-Abhar and Schaalan, 2014). GLUT-4 translocation is the main step for the glucose uptake and is regulated by two pathways, the AMPK and phosphatidyl inositol 3-kinase (PI3K)/Akt (Tan et al., 2008).

In this sense, the activation of PI3K by some monoterpenes is also involved in the insulin signaling process, due to the activation of the AMPK, which modulates glucose uptake, inhibit gluconeogenesis, and stimulate glycogenolysis (Mabhida et al., 2018; Kaur et al., 2019).

The diminishing of the oxidative and inflammatory stress has a positive effect on the reduction of diabetes and its comorbidities because of its implication in its development and progression. The bicyclic sesquiterpene β-caryophyllene improves the antioxidant enzymatic system, decrease lipid peroxidative markers, and regulates the proinflammatory cytokines (TNF-β: tumor necrosis factor β and IL-6: interleukin 6). The latter is of the utmost importance as they stimulate NF-κB factor (nuclear factor-kappa B), which is involved in the regulation of inflammatory responses (Basha and Sankaranarayanan, 2016; Omara et al., 2010). Another terpene with the antidiabetic effect is ursolic acid showed a reduction of blood glucose levels and preserve the pancreatic cells, regulated the immune imbalance, stimulated T-cell proliferation, and cytokines production (Jang et al., 2009). On the other hand, the monoterpene borneol decreased the HbA1c (glycated hemoglobin) level, which is altered by the effect of the stress oxidative in DM, indicating a reduction in protein glycation (Madhuri and Naik, 2009).

Hepatic gluconeogenesis (pathway of glucose synthesis) is crucial in glucose homeostasis, the enzymes glucose-6-phosphatase and fructose

1-6-phosphatase are keys in the anabolic pathway (Cetto, 2015; Wang et al., 2013). The monoterpenes, D-limonene, and carvone reduce the activity of glucose 6-phospahase and fructose 1,6-biphosphatase, therefore decreasing glucogenesis and stimulating glucose uptake (Muruganathan and Srinivasan, 2016; Murali and Saravanan, 2012; Solayman et al., 2016). Likewise, glycogen, another source of glucose, is stored in the liver, and the muscle is mainly considered a blood glucose buffer. Several studies have shown that some terpenes such as borneol promote an increase in glycogen levels, which is decreased during diabetes (Putta et al., 2016; Madhuri and Naik 2017).

Terpenes can also inhibit the protein tyrosine phosphatase intracellular (PTP1B), which plays an essential role as a negative regulator of the insulin receptor (IR) (Nazaruk and Borzym-Kluczyk, 2015). High expression of PTP1B could induce insulin resistance and leptin resistance, both involved in DM (Sun et al., 2016); thus, its inhibition is very important. Some triterpenes, such as ilekudinol A, ilekudinol B, and lupine, inhibit PTP1B in a noncompetitive form. The inhibitory activity of some triterpenes related to their structure, mainly attributed to the hydroxyl groups at C3 and the carboxyl group C28 or C27 (Nazaruk and Borzym-Kluczyk, 2015; Na et al., 2009, 2010; Hosseyni et al., 2012; Abdel-Moneim and Fayez, 2015; Habtemariam, 2018).

9.3 PHENOLIC COMPOUNDS WITH ANTIDIABETIC PROPERTIES

Phenolic compounds are plant secondary metabolites involved in growth, development, reproduction control, pigmentation, and defense against pathogens (Tanase et al., 2019). The primary precursors for phenolics biosynthesis are phenylalanine (synthesized via the shikimate pathway) and tyrosine, and different enzymes, including erythrose-4-phosphate, phenylalanine ammonia-lyase, cinnamate-4-hydrolxylase, *p*-coumarate-3-hydroxylase, and *o*-methyl transferase, are involved (Shahidi and Yeo, 2018). Phenolic compounds structure is characterized by the presence of a hydroxyl group (–OH) directly bonded to an aromatic hydrocarbon group, which is in turn, associated with more or less complex structures. More than 8000 phenolic compounds have been found in different plant sources, being classified as flavonoids (Isoflavones, chalcones, anthocyanidin, flavanones, flavanols, flavones, and flavonols) or nonflavonoids (Cianciosi et al., 2018).

The profile of phenolic compounds has been associated with plant species, origin, growing condition, harvesting time, and even storage condition (Martins et al., 2016). The consumption of fruits and vegetables containing high amounts of phenolic compounds has increased in recent years since they have been positively correlated to a lower incidence of disorders induced by reactive oxygen species (Olas, 2018). Some studies indicate that dietary polyphenols may be implicated in alleviating and preventing diseases such as cancer and metabolic disorders, such as diabetes. In this chapter, the relationship between polyphenols and their possible effects on diabetes will be discussed.

9.3.1 MECHANISMS OF ACTION

In various human populations, the consumption of particular fruits and vegetables has been associated with a therapeutic effect over DM. Recently, the responsible molecules as well as their mechanisms of action over this disease have been studied. Among these molecules, phenolic compounds stand out. They have been tested as a mixture or individually over in vitro and in vivo model as well as over enzymes involved in diabetes pathology, for example, pancreatic α-amylase, which activity is positively correlated to postprandial glucose levels (Rodríguez et al., 2016). Gallic acid and catechin obtained from black Jamun landraces (*Syzygium cumini* L.) seed have been proposed as efficient antidiabetic compounds since together they restrain starch breakdown by inhibiting porcine pancreatic α-amylase by 98.2% with an IC_{50} value of 12.9 μg mL^{-1}, being lower than that of acarbose (IC_{50} of 24.7 μg mL^{-1}) (Gajera et al., 2017). By a noncompetitive type of inhibition, black carrot cyanidin 3-xylosyl galactoside also inhibits α-amylase with an IC_{50} lower than acarbose (IC_{50} of 16.9 μg mL^{-1}) (Karkute et al. 2018). Other phenolics inhibiting this enzyme are the ones found in millet seed coat extract, they have been considered as safe, and they have been proposed as an ingredient to control food products glycemic index (Shobana et al., 2009). Despite the above, not all tested phenolics effectively inhibit α-amylase activity. For example, the bioactive flavonoid sinestein, isolated from *Orthosiphon stamineus* ethanolic extract, with an IC_{50} of 1.13 mg mL^{-1}, much higher than that of acarbose (Mohamed et al., 2012).

α-glucosidase is another enzyme involved in diabetes and also inhibited with acarbose (IC_{50} of 126 μg mL^{-1}). Although not very effective, this enzyme is inhibited by the mixture of the phenolic compounds maysin, methoxymaysin, and apimaysin extracted from the female inflorescences of different maize races (IC_{50} values from 857 μg mL^{-1} for dark-red silks to 38 460 μg mL^{-1} for green-yellow silks). Molecular docking was used to determine that the tested phenolic acids exhibited lower affinity than acarbose, the above due to a failed interaction with the active site of the human maltase-glucoamylase crystal structure. Instead, it was suggested that phenolics induce a conformational change in the vicinity of the active site, which results in a weak loss of enzymatic activity (Alvarado-Díaz et al., 2019). On the other hand, secoisolariciresinol extracted from nettle (*Urtica Dioica* L.) forms stable complexes with the main enzymes responsible for type 2 diabetes mellitus (T2DM): dipeptidyl peptidase 4, α-amylase and β-glucosidase (Bouchentouf et al., 2020).

The inhibitory effect of phenolics over enzymes involved in diabetes complications have also been tested, for example, aldose reductase, which is involved in diabetes-induced oxidative stress, atherothrombotic cardiovascular disease, and myocardial ischemia/reperfusion injury. Tannic and chlorogenic acid had a strong inhibitory effect over this enzyme, with IC_{50} of 0.5 and 5.47 µM, respectively. On the other hand, gallic acid had a weak inhibitory effect (IC_{50} of 0.176 mM) (Alim et al., 2017). The phenolic compound with the lowest IC_{50} for both α-amylase and α-glucosidase is caffeic acid with IC_{50} of 9.10 and 9.24 µg mL^{-1}, respectively (Oboh et al., 2015).

Phenolics have also been tested in cell lines. Protein tyrosine phosphathase 1B (PTP1B) is known to be a main negative regulator of insulin signaling, so inhibiting or downregulating this enzyme causes enhanced insulin sensitivity. It was found that morin, a phenolic compound isolated mainly from *Maclura pomifera*, *Maclura tinctoria*, and *Psidium guajava,* inhibits PTP1B in HepG2 cells (IC_{50} = 15.0±0.8 µM). This enzymatic inhibition is noncompetitive since morin binds into its noncatalytic phosphotyrosine site, impeding the correct positioning of the hydrolytic water molecule in the second catalytic step. By western blot, it was also found that morin increases the phosphorylation of the IR as well as inhibits gluconeogenesis and enhances glycogen synthesis (Paoli et al., 2013).

Other phenolic compounds, agrimonolide, desmethylagrimonolide, quercetin, luteolin, luteolin-7-*o*-glucoside, kaempferon, and apigenin, were previously isolated from *Agrimonia pilosa* and tested over insulin-resistant induced HepG2 cell model. The effect of these compounds on insulin resistance was comparable to that of metformin, significantly improving glucose uptake by increasing phosphorylation levels of protein kinase B (AKT), which lays downstream of phosphoinositide 3-kinase (PI3K) and facilitates glucose uptake and glycogen synthesis in the liver. Agrimonolide and quercetin had the best results, lowering glucose levels by 51.2 and 55%, respectively (Huang et al., 2015). Besides HepG2, skeletal muscle cells are also very important in vitro model for evaluating the phenolics antidiabetic effect. These cells are responsible for the uptake of more than

75% of glucose in the post-prandial state. Curcumin, a phenolic compound of *Curcuma longa*, has been traditionally used in Indian medicine for its antihyperglycemic effect. In C_2C_{12} mouse myoblast cells, curcumin strongly induced glucose uptake by the activation of PI3K/AKT pathway, which promotes GLUT4 glucose transporter translocation from an intracellular pool to the plasma membrane. Also, an exposition of 40 µM of curcumin for 1 h in combination with insulin produced a synergistic activation of AMPK/acetil-CoA carboxylase (ACC) pathway. This pathway plays a central role in the regulation of glucose and lipid metabolism. These findings indicate that curcumin may be used as a potential antidiabetic therapeutic agent to increase glucose uptake and insulin sensitivity in muscle cells (Kang and Kim, 2010). Another phenolic compound increasing GLUT4-mediated glucose uptake in skeletal muscle cells is octaphlorethol, a compound isolated from the brown alga, *Ishige foliacea*. A dose of 50 µM during 2 h increased GLUT4 translocation to the plasma membrane, increasing up to 87% glucose uptake during 10 h in L6 rat myoblast cells, which was related to the activation of AMPK/ACC and PI3K/AKT pathways (Lee et al. 2012). AMPK/ACC is also activated by caffeic acid by AMPK Thr172 phosphorylation, stimulating skeletal muscle insulin-independent glucose transport in rat epitrochlearis muscle cells (Tsuda et al., 2012). Another in vitro model is pancreatic β-cell since their dysfunction or death is a major determinant for the development of type 2 diabetes. Tyrosol, an olive polyphenol (50 mM), decreased the tunicamycin-induced pancreatic β-cell death after 48 h, and also reduced the expressions of apoptosis-related markers. The effect was mediated by the phosphorylation of c-Jun N-terminal kinases. Therefore, tyrosol could be a potential therapeutic candidate for the amelioration of type 2 diabetes (Lee et al., 2016).

On the other hand, *Phellinus igniarius* phenolic extract containing 7,8-dihydroxycoumarin, 3,4-dihydroxybenzalacetone, 7,3'-dihydroxy-5'-methoxyisoflavone and inoscavin C lowered the glucose level by 50% when given in a dose of 40 mg kg^{-1} day^{-1} during 4 weeks to KK-Ay mice. The results were attributed to a liver enhanced expression of p-AMPKα and an increased expression of GLUT4 in the mice skeletal muscle (Zheng et al., 2018). An extract from *Dimocrpus longan* Lour. pericarp (composed of corilagin, ellagic acid, and gallic acid) also increased GLUT4 expression (by 1.7-fold) when given in a dose of 10 mg kg^{-1} body weight to C57BL/6 mice previously consuming sucrose (3 g kg^{-1} body weight). As a result, serum glucose levels were reduced by 10% more compared to acarbose. The extract significantly inhibited yeast α-glucosidase activity with IC_{50} = 11.68 ± 0.44 ug mL^{-1}, lowered than that of acarbose (IC_{50} = 179.57 ± 7.24 ug mL^{-1})

(Li et al., 2015). The ethanolic extract of *Solanum nigrum* fresh leave also inhibited α-glucosidase, as well as α-amylase with IC_{50} values of 39.725 and 78.8±0.707 µg mL^{-1}. It was proposed as an adjuvant in streamlining diabetes since 375 mg kg^{-1} of the dry extract given to induced diabetic rats during 6 weeks significantly decrease hyperglycemia (by 63%). Furthermore, it increased kidney function, which was reflected in decreased urine output and urine sugar compared to diabetic control rats (Dasgupta et al., 2016). Extracts have also been tested for diabetes prevention. An aqueous extract from the bark of the black wattle tree (*Acacia meansii*), composed mainly by robinetinidol and fisetinidol, prevented KKay obese male mice from developing type 2 diabetes in a 7-week ad libitum diet composed of 5% of this extract. Also, it decreased plasma glucose by 73% and insulin by 83% (Ikarashi et al., 2011).

Phenolics have also been tested alone over in vivo models. Rutin, a flavonoid glycoside found mainly in flowers and fruits of buckwheat, oranges, grapes, lemons, limes, peaches, and berries, reduces blood glucose by 56.2% in Wistar rats (8 mg kg^{-1}, daily for 1 week) (Wang et al., 2015). Eugenol also reduced serum glucose levels (by 43%) when given in a dose of 10 mg kg^{-1} during 45 days to diabetic Sprague–Dawley rats. It also significantly restored the decreased serum levels of insulin and glutathione, as well as increased GLUT4 and AMPK skeletal muscle protein contents (Al-Trad et al., 2019). Green tea polyphenol (−)-epigallocatechin-3-gallate (EGCG) reduces postprandial blood glucose levels by 50% when given to CF-1 mice in a dose of 100 mg kg^{-1} after consuming common corn starch.

Nevertheless, EGCG had no effect over postprandial blood glucose levels following the administration of maltose or glucose, suggesting that EGCG may modulate amylase-mediated starch digestion. It was also found for this phenolic compound to noncompetitively inhibit pancreatic amylase activity by 34% at 20 µM (Forester et al., 2012). EGGG also ameliorated the hepatic morphology and function, alleviated hyperglycemia (by 42.8%), hyperinsulinemia (by 17.6%), and insulin resistance in mice with nonalcoholic liver disease (40 mg kg^{-1} daily). Furthermore, its enhanced insulin clearance and upregulated insulin-degrading enzyme expression in mice liver. The treatment of mice with EGCG improved insulin sensitivity and prevented insulin resistance induced by the high-fat diet (Gan et al., 2015).

A higher blood glucose level reduction (60.76%) was promoted by a 4-week treatment of 100 mg kg^{-1} of protocatechuic acid from *Acaciella angustissima* pods methanolic extract over diabetic rats. Also, serum insulin concentration was restored. One of the possible mechanisms that protocatechuic acid could perform is either the increase of pancreatic insulin secretion from existing β-pancreatic cells or the regeneration of β-pancreatic cells (Rodríguez-Méndez et al., 2018). The lower antiglycemic effect was observed with 7-*O*-galloyl-d-sedoheptulose from corni fructus (*Cornus officinalis*). In a dose of 100 mg kg^{-1} given daily during 6 weeks to diabetic mice, glucose serum level was reduced by 10%, also leptin, C-peptide, resistin, tumor necrosis factor-α (TNF-α), and interleukin (IL)-6 while adiponectin was augmented. 7-*O*-galloyl-d-sedoheptulose also modulated protein expressions of proinflammatory nuclear factor-kappa B (NF-κB), inducible nitric oxide synthase (iNOS), and transforming growth factor-β1 (TGF-β1). The mechanism of this compound was determined to be anti-inflammatory and anti-inflammatory-related anti-oxidative action (Park et al., 2013). Syringaresinol-di-*O*-β-*D*-glucoside, mostly extracted from *Polygonatum sibiricum*, also promote anti-oxidative activity, increasing serum fasting insulin and pancreatic insulin in around 100% and 3.5-fold, respectively.

Moreover, total kidney glucose was reduced by 42.85% when given in a dose of 75 mg kg^{-1} to diabetic mice. Western blot revealed decreased superoxide dismutase (SOD) and catalase (CAT) levels (Zhai and Wang, 2018). Besides rodents, zebrafish have also been used as an in vivo model. In a dose of 1 μg g^{-1} bodyweight, dieckol, a marine algal polyphenol isolated, and purified from *Ecklonia cava*, decreased blood glucose levels of alloxan-induced diabetic zebrafish more than 3.3 times when compared to metformin. Also, a reduction of glucose-6-phosphate and phosphoenolpyruvate carboxykinase (PEPCK) levels were observed in the liver tissues. Dieckol also increased phosphorylation of AKT in the muscle tissue of the zebrafish. AKT activation was involved in mediating the effect of dieckol on glucose transport activation and insulin sensitivity (Kim et al., 2016).

Other phenolic compounds have also been studied in diabetic complications, such as the pathophysiology of cardiometabolic syndrome, which is highly promoted by decreased nitric oxide bioavailability. Morin ameliorates diabetes-induced endothelial dysfunction of diabetic aortas by increasing nitric oxide levels and endothelial-dependent relaxation responses via AKT signaling and upregulated p-AKT (at Ser473 and Thr308) and endothelial NOS (at Ser1177) expression (Taguchi et al., 2016). Another complication of diabetes is oxidative and nitrosative stress, which causes tissue damage being the liver, the most affected organ. An *n*-butanolic fraction from *Annona crassiflora* fruit peel containing chlorogenic acid, epicatechin, procyanidin B2, and caffeoyl-glucoside given to diabetic mice (100 mg kg^{-1} daily for 30 days), reduced lipid peroxidation, protein carbonylation, and nitration and iNOS content as well as in SOD and CAT activities and contents in hepatic tissue. Also, the extract reduced glycemia in 25.6% (Justino, 2016).

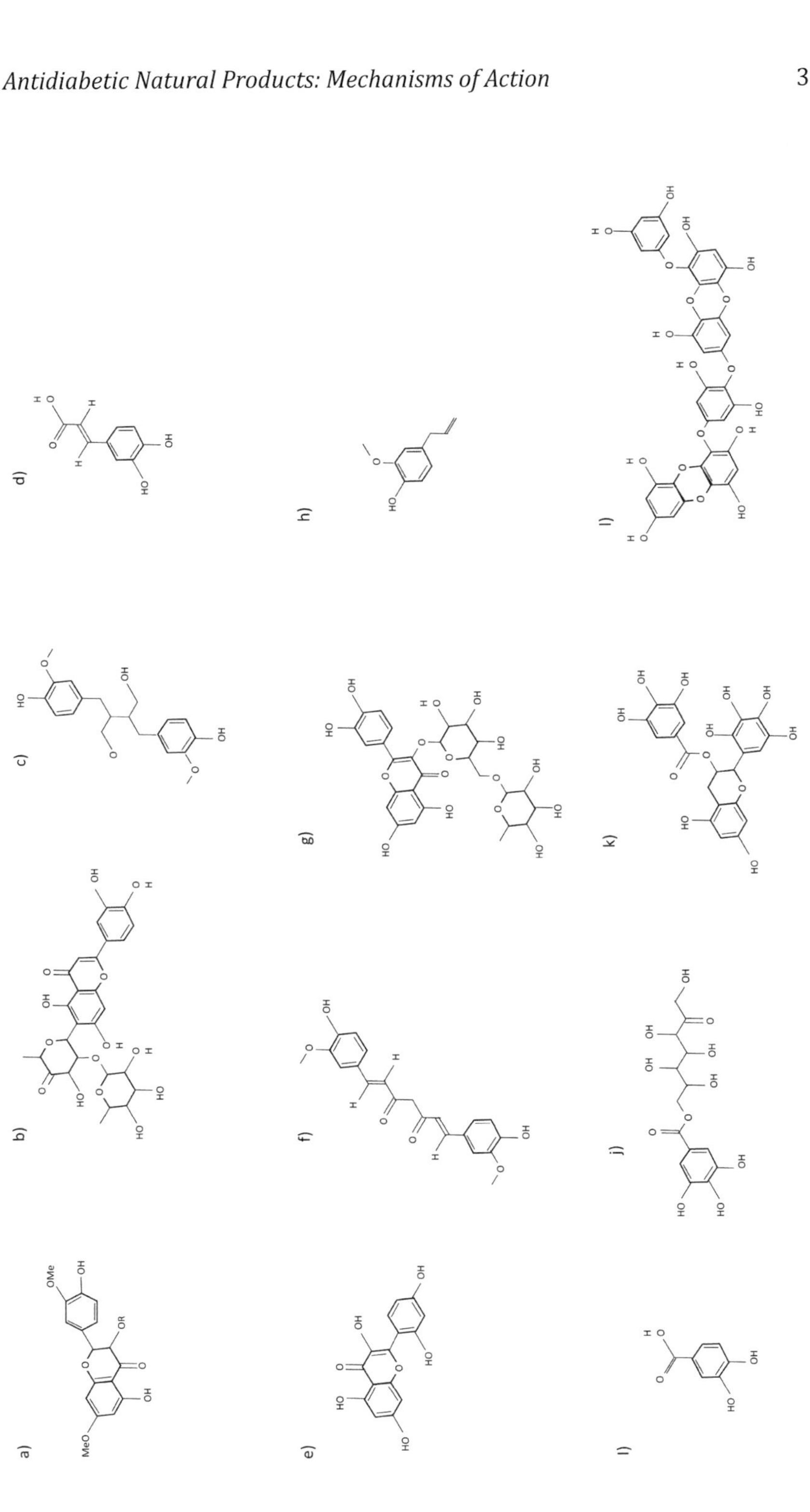

FIGURE 9.4 Chemical structure of phenolic compounds with antidiabetic potential.

Retinopathy is another diabetic complication, and some studies have shown that *Zingiber zerumbet* rhizome ethanol extracts in induced diabetic rats (300 mg kg^{-1} per day for 3 months) effectively preserved the expression of occludin, and claudin-5, leading to less blood-retinal barrier breakdown and less vascular permeability. By molecular analysis, it was found that retinal gene expression of *TNF-α*, IL-1β, IL-6, vascular endothelial growth factor, intercellular adhesion molecule-1, and vascular cell adhesion molecule-1 were all decreased. The protein expression of p38 mitogen-activated protein kinase in the diabetic retina was also downregulated. These results suggest that the retinal protective effects of the extract occur through improved retinal structural change and inhibiting retinal inflammation (Tzeng et al., 2015).

9.4 ALKALOIDS WITH ANTIDIABETIC PROPERTIES

The alkaloids are a group of natural compounds that contain basic heterocyclic nitrogen atoms in their structure. Its name is derived from "alkaline" and was used to describe any base that contains nitrogen. Alkaloids are present in living organisms and have different types of structures, as well as biosynthetic pathways and pharmacological activities. Alkaloids have traditionally been isolated from plants and have been used for hundreds of years as a traditional remedy for some conditions. Alkaloids have also been used as drugs, teas, poultices, and potions, as well as poisons for more than 4000 years. Therefore, this group of compounds has had great importance within the scientific field (Robinson, 1974; Saxena et al., 2013; Snieckus, 1968).

There is a large number of alkaloids, which involve complex molecular structures, so their rational classification is difficult. However, these can be grouped by families, depending on the type of heterocyclic ring system present in the molecule. The names of the individual members of each one are usually derived from the name of the plant in which they have been found, or the characteristic effect they can cause. According to the heterocyclic ring system, they contain pyrrolidine alkaloids (they contain a pyrrolidine ring system), pyridine alkaloids (they have a piperidine ring system), pyrrolidine–pyridine alkaloids (the heterocyclic ring system present in its alkaloids is pyrrolidine–pyridine), pyridine–piperidine alkaloids (containing pyridine linked to a piperidine ring system), quinoline alkaloids (containing a basic quinoline heterocyclic ring), and isoquinoline alkaloids (containing isoquinoline of the heterocyclic platform system) (Pelletier, 1983).

Some alkaloids have stimulating properties, such as caffeine and nicotine, morphine that is used as an analgesic, and quinine as an antimalarial drug.

They have also been linked to antihypertensive, antiarrhythmic, and anticancer activity (Mander and Liu, 2010). The alkaloids have also been linked to the potential to suppress lesions induced by T2DM, including diabetic vascular dysfunction, diabetic heart disease, diabetic hyperlipidemia, diabetic nephropathy, diabetic encephalopathy, diabetic osteopathy, diabetic enteropathy, and diabetic retinopathy (Ran et al., 2019).

9.4.1 MECHANISMS OF ACTION

The described antidiabetic mechanisms of the alkaloids can be grouped mainly into two groups: first, in which there is the ability to repair or stimulate the proliferation of pancreatic β-cells, the stimulation of insulin secretion, increased sensitivity to insulin, decreased resistance, increased glycogenesis, and inhibition of the latter, while in the second group is the ability to decrease the level of glycogenic enzymes (Zhao et al., 2019). It has been suggested that some alkaloids have DPP-4 inhibitory effects, such as the case of berberine (Chakrabarti et al., 2011; Yaribeygi et al., 2019). DPP-4 inhibitors prevent the inactivation of GLP, maintaining the endogenous levels of GLP-1 that causes glucose-induced insulin release from β-cells, as well as glucagon suppression (Dicker, 2011). In this sense, berberine is a bioactive alkaloid used in Chinese medicine that has numerous positive effects on biological systems, has been extensively studied for its diabetes-related effects, such as glucose reduction, as well as inhibition of enzymes related to it. Some works related to this are mentioned.

Berberine showed resistance to insulin-induced by fat and diabetic phenotype in type 2 diabetic hamsters, which may be due to alterations of specific genes related to metabolism and its main regulators: such as liver receptor X (LXR) α, the PPAR-α, and the sterile regulatory element-binding protein (SREBP), which were observed in the liver of diabetic hamsters treated and not treated with this compound, finding a significant decrease in SREBP mRNA levels, as well as an increase in LXRα and PPARα mRNA levels (Liu et al., 2010). Berberubin is the primary metabolite of berberine that has demonstrated a stronger glucose-lowering effect than berberine in vivo. In this sense, (Yang et al., 2017) administered berberubin (50 mg kg^{-1} d^{-1}, ig) for 6 weeks to C57BL/6 mice with high-fat diet (HFD)-induced hyperglycemia, which caused a greater reduction in plasma glucose levels compared to those caused by berberine (120 mg kg^{-1} d^{-1}) or berberubin (25 mg kg^{-1} d^{-1}). In addition, a dose-dependent decrease in the activity of α-glucosidase in the intestine of mice treated with berberubin was found.

In addition, studies conducted on the normal human liver cell line L-O2 in vitro, treatment with berberubin at concentrations of 5, 20, 50 µmol L^{-1} increased glucose consumption, improved glycogenesis, stimulated glucose analog 2 absorption -NBDG, and modulated levels of glucose-6-phosphatase and hexokinase mRNA. Therefore, it is concluded that berberubin has a stronger glucose-lowering effect than berberine in mice with HFD-induced hyperglycemia.

The fraction of the alkaloid extract of *Coptis chinensis* Franch (Huanglian in Chinese) showed activity on α-glucosidase inhibition with an IC50 of 3.528 mg mL^{-1}, which could be used for the treatment of diabetes according to the authors. Within the alkaloids identified in this fraction of the extract were found therein, coptisin, epiberberine, jatrorrizine, and berberin (Ge et al., 2014). On the other hand, β-cell replacement treatment, such as islet or pancreas transplantation, is considered an effective treatment method for type 1 diabetes. However, in most patients and studies in mice, these cells usually die mainly from apoptosis (Christoffersson et al., 2016; Anuradha et al., 2014).

Trigonelline is the main component of some traditional Chinese medicines to treat diabetes, such as pumpkin, fenugreek, and *Mirabilis jalapa* L. (Zhou et al., 2012). This compound decreases the blood glucose concentration in animals and humans, as well as protected streptozotocin-induced type 2 diabetic rats and diabetic peripheral neuropathy (Christodoulou et al., 2019; Folwarczna et al., 2016; Tharaheswari et al., 2014; van Dijk et al., 2009; Hamden et al., 2013). Therefore, a study considered investigating the protection of trigonelline in hyperglycemia, apoptosis of β-cells, and inflammation in type 1 diabetic mice, for which streptozotocin (160 mg kg^{-1}) was injected intraperitoneally in diabetic mice. The murine model was treated with trigonelline and another with insulin for 4 weeks. The levels of blood glucose, serum insulin and inflammatory factors, β-cell apoptosis, insulin content, and oxidative stress parameters in the pancreas were subsequently calculated (Liu et al., 2018). Results showed that trigonellin significantly decreased blood glucose levels, serum TNF-α, IL-6, and IL-1β while increasing serum insulin and adiponectin levels in diabetic mice. In addition, treatment with this alkaloid suppresses apoptosis of β-cells by regulating the decrease in caspase 3 expression, so it is believed that trigonellin protects diabetic mice mediated by decreased blood glucose, increasing of insulin expression in β-cells, as well as regulation of the inflammatory response and suppression of apoptosis of β-cells.

Another alkaloid known for its multiple pharmacological effects is neferin, a compound derived from lotus seeds, which has demonstrated its antidiabetic potential in some studies (Zhao et al., 2014), for example, it has been reported that in insulin-resistant rats, where neferin significantly reduced fasting blood glucose levels, insulin, triglycerides, TNF-α, and caused an increase in insulin sensitivity (Pan et al., 2009). On the other hand, in a different study neferin demonstrated its important role in the translocation of glucose to the cell surface, glucose uptake, and GLUT4 expression; this by inducing the fusion of the plasma membrane of GLUT4 and the increase of intracellular Ca^{2+}, which promoted glucose absorption and eased insulin resistance in an in vitro model of L6 cells. Also, in this work, neferin was found to significantly activate phosphorylation of AMPK and protein kinase C (PKC), so AMPK and PKC inhibitors blocked the expression of induced GLUT4 by nephrine and increased intracellular Ca^{2+}. Nephrine-induced GLUT4 expression and intracellular Ca^{2+} were inhibited by G protein and PLC inhibitors. Only intracellular Ca^{2+} was inhibited by inositol triphosphate receptor (IP3R) inhibitors. Therefore, it is concluded in this work that neferin promotes the expression of GLUT4 through the G-PLC-PKC and AMPK protein pathways, inducing the fusion of the plasma membrane of GLUT4 and the subsequent absorption of glucose and increasing the intracellular Ca^{2+} through the G-PLC-IP3-IP3R protein pathway (Zhao et al., 2019).

High glucose levels can induce endothelial dysfunction, which is related to their apoptosis, which is an important pathological characteristic of diabetic vasculopathy. For this reason, research has been conducted on preventive effects of the neferine alkaloid, on the lesion induced by hyperglycemia of human umbilical vein endothelial cells (HUVECs). Results showed that pretreatment with this compound effectively suppresses HUVECs apoptosis induced by high glucose. Besides, neferin significantly inhibited the high glucose-induced activation of the PI3K/Akt pathway in HUVECs, as well as the high glucose-induced activation of the NF-κB signal, which was suppressed with the pretreatment of the compound. The inhibition of endothelial apoptosis induced by high glucose presented by neferin could be given by blocking the ROS/Akt/NF-κB pathway, so this compound has potential for the treatment of diabetic vasculopathy (Guan et al., 2014).

Oximatrine, a quinolizidine alkaloid, has been widely used for the treatment of hepatitis. It also has been shown to have hypoglycemic and hypolipidemic effects in streptozotocin-induced diabetic rats (Guo et al., 2014). Similar results were observed when diabetic rats were treated with oximatrin to assess the improvement of aortic endothelial dysfunction since they exhibited a markedly reduced body weight and higher plasma glucose levels. In addition, the expression of the NOX4 protein in the aortas of the diabetic rats increased, so that treatment with this compound could improve diabetic endothelial dysfunction through increased bioavailability of NO by positive regulation of eNOS expression and regulation NOX4 expression negative (Wang et al., 2019).

According to the work of Yu et al. (2016), the alkaloids rutaercapine and evodiamine from *Evodia rutacarpae* may have therapeutic potential for the treatment of hyperglycemia and type 2 diabetes, which is based on their capacity to promote CAR-mediated inhibition of the O1 hairpin recruitment (FoxO1) and the hepatocyte nuclear factor 4α (HNF4α) in the PEPCK and glucose-6-phosphatase (G6Pase) gene promoters, in HepG2 hyperlipidemic cells. Also, in vivo demonstrated that the treatment of mice with the ercaercapine route improved tolerance to glucose in a CAR-dependent manner.

The effect of piperine, the main alkaloid present in black pepper, on the blood glucose level of aloxane-induced diabetic mice was studied in acute and subacute study models, finding that subacute administration of piperine has a statistically significant antihyperglycemic activity, while sharply raising blood glucose at high doses (Atal et al., 2012), this could be related to the activity of piperine as an agonist of PPAR-γ, already demonstrated in some of its derivatives in streptozotocin-induced diabetes model (Kharbanda et al., 2016).

According to the literature, there is growing evidence that suggests the potential of alkaloids to intervene in the pathway of insulin signal transduction, reverse molecular defects that result in insulin resistance and glucose intolerance, as well as improve disease complications, even though molecular studies with greater depth are still needed to clarify the specific mechanism of action of these compounds.

9.5 CONCLUSION

Natural products such as terpenes, alkaloids, and phenolic compounds are some of the main natural products found in medicinal plants ethnobotanically used to treat diabetes in folk medicine. These plants have been the object of many studies that have dilucidated some of their potential antidiabetic mechanisms. However, nowadays ethnopharmacological studies have neither yet managed to establish toxicological data or pharmacokinetics for most plants nor the effective doses for any of these molecules, which limits their use as antidiabetic drugs. Thus, it is essential to direct preclinical and clinical studies from natural products obtained from medicinal plants for the discovery of new drugs for diabetic treatment or prevention of its comorbidities.

KEYWORDS

- metabolic syndrome
- antidiabetic potential
- natural products
- phenolics
- terpenes
- alkaloids

REFERENCES

Abdelmeguid, N. E.; Fakhoury, R.; Kamal, S. M.; Al Wafai, R. J., Effects of *Nigella sativa* and thymoquinone on biochemical and subcellular changes in pancreatic β-cells of streptozotocin-induced diabetic rats. *Journal of Diabetes* **2010**, *2* (4), 256–266.

Abdel-Moneim, A.; Fayez, H., A review on medication of diabetes mellitus and antidiabetic medicinal plants. *International Journal of Bioassays* **2015**, *4* (6), 4002–4012.

Alim, Z.; Kilinç, N.; Şengül, B.; Beydemir, Ş., Inhibition behaviours of some phenolic acids on rat kidney aldose reductase enzyme: an in vitro study. *Journal of Enzyme Inhibition and Medicinal Chemistry* **2017**, *32* (1), 277–284.

Al-Trad, B.; Alkhateeb, H.; Alsmadi, W.; Al-Zoubi, M., Eugenol ameliorates insulin resistance, oxidative stress and inflammation in high fat-diet/streptozotocin-induced diabetic rat. *Life Sciences* **2019**, *216*, 183–188.

Alvarado-Díaz, C. S.; Gutiérrez-Méndez, N.; Mendoza-López, M. L.; Rodríguez-Rodríguez, M. Z.; Quintero-Ramos, A.; Landeros-Martínez, L. L.; Rodríguez-Valdez, L. M.; Rodríguez-Figueroa, J. C.; Pérez-Vega, S.; Salmeron-Ochoa, I., Inhibitory effect of saccharides and phenolic compounds from maize silks on intestinal α-glucosidases. *Journal of Food Biochemistry* **2019**, *43* (7), e12896.

Anuradha, R.; Saraswati, M.; Kumar, K. G.; Rani, S. H., Apoptosis of beta cells in diabetes mellitus. *DNA and Cell Biology* **2014**, *33* (11), 743–748.

Atal, S.; Agrawal, R. P.; Vyas, S.; Phadnis, P.; Rai, N., Evaluation of the effect of piperine per se on blood glucose level in alloxan-induced diabetic mice. *Acta Poloniae Pharmaceutica* **2012**, *69* (5), 965–969.

Bahmani, M.; Golshahi, H.; Saki, K.; Rafieian-Kopaei, M.; Delfan, B.; Mohammadi, T., Medicinal plants and secondary metabolites for diabetes mellitus control. *Asian Pacific Journal of Tropical Disease* **2014**, *4*, S687–S692.

Basha, R. H.; Sankaranarayanan, C., β-Caryophyllene, a natural sesquiterpene lactone attenuates hyperglycemia mediated oxidative and inflammatory stress in experimental diabetic rats. *Chemico-Biological Interactions* **2016**, *245*, 50–58.

Bouchentouf, S.; Said, G.; Kambouche, N.; Kress, S., Identification of phenolic compounds from nettle as new candidate inhibitors of main enzymes responsible on type-II diabetes.

Current Drug Discovery Technologies **2020**, *17* (2), doi: 10.2174/1570163815666180829 094831.

Brahmkshatriya, P. P.; Brahmkshatriya, P. S., Terpenes: chemistry, biological role, and therapeutic applications. In *Natural Products: Phytochemistry, Botany and Metabolism of Alkaloids, Phenolics and Terpenes*, Ramawat, K. G.; Mérillon, J.-M., Eds. Springer: Berlin, Heidelberg, 2013; 2665–2691.

Cetto, A. A., Diabetes and Metabolic Disorders: An Ethnopharmacological Perspective. In *Ethnopharmacology*, 1st ed.; Heinrich M, a. J. A. K., Ed. Wiley Blackwell: Hoboken, NJ, 2015; 227–238.

Chakrabarti, R.; Bhavtaran, S.; Narendra, P.; Varghese, N.; Vanchhawng, L.; Mohamed Sham Shihabudeen, H.; Thirumurgan, K., Dipeptidyl peptidase-IV inhibitory activity of *Berberis aristata*. *Journal of Natural Products* **2011**, *4*, 158–163.

Chen, L.; Lu, X.; El-Seedi, H.; Teng, H., Recent advances in the development of sesquiterpenoids in the treatment of type 2 diabetes. *Trends in Food Science & Technology* **2019**, *88*, 46–56.

Christodoulou, M. I.; Tchoumtchoua, J.; Skaltsounis, A. L.; Scorilas, A.; Halabalaki, M., Natural alkaloids intervening the insulin pathway: new hopes for antidiabetic agents? *Current Medicinal Chemistry* **2019**, *26* (32), 5982–6015.

Christoffersson, G.; Rodriguez-Calvo, T.; von Herrath, M., Recent advances in understanding Type 1 Diabetes. *F1000Research* **2016**, *5*, 110.

Cianciosi, D.; Forbes-Hernández, T. Y.; Afrin, S.; Gasparrini, M.; Reboredo-Rodriguez, P.; Manna, P. P.; Zhang, J.; Bravo Lamas, L.; Martinez Florez, S.; Agudo Toyos, P., Phenolic compounds in honey and their associated health benefits: a review. *Molecules* **2018**, *23* (9), 2322.

Çitoğlu, G. S.; Acıkara, Ö. B., Column chromatography for terpenoids and flavonoids. In *Chromatography and its Application*, Dhanarasu, S., Ed. InTech: Rijeka, 2012; 13–49.

Dasgupta, N.; Muthukumar, S.; Murthy, P. S., *Solanum nigrum* leaf: natural food against diabetes and its bioactive compounds. *Research Journal of Medicinal Plants* **2016**, *10*, 181–193.

DeFronzo, R. A.; Ferrannini, E.; Zimmet, P.; Alberti, K. G. M. M., *International Textbook of Diabetes Mellitus*. John Wiley & Sons: Hoboken, NJ, 2015.

Dicker, D., DPP-4 inhibitors: impact on glycemic control and cardiovascular risk factors. *Diabetes Care* **2011**, *34* (Supplement 2), S276–S278.

El-Abhar, H. S.; Schaalan, M. F., Phytotherapy in diabetes: review on potential mechanistic perspectives. *World Journal Diabetes* **2014**, *5* (2), 176.

Escandon, C. J., Prevencion de la diabetes tipo 2: estrategias sugeridas. In *Sindrome cardiometabólico: una vision práctica*, Ortiz V. M. R, T. A., Rubio G., Covarrubias J.M., Ed. Editorial Alfil: México city, 2011; 27–34.

Folwarczna, J.; Janas, A.; Pytlik, M.; Cegiela, U.; Sliwinski, L.; Krivosikova, Z.; Stefikova, K.; Gajdos, M., Effects of trigonelline, an alkaloid present in coffee, on diabetes-induced disorders in the rat skeletal system. *Nutrients* **2016**, *8* (3), 133.

Forester, S. C.; Gu, Y.; Lambert, J. D., Inhibition of starch digestion by the green tea polyphenol, (−)-epigallocatechin-3-gallate. *Molecular Nutrition & Food Research* **2012**, *56* (11), 1647–1654.

Gajera, H.; Gevariya, S. N.; Hirpara, D. G.; Patel, S.; Golakiya, B., Antidiabetic and antioxidant functionality associated with phenolic constituents from fruit parts of indigenous black jamun (*Syzygium cumini* L.) landraces. *Journal of Food Science and Technology* **2017**, *54* (10), 3180–3191.

Gan, L.; Meng, Z.-j.; Xiong, R.-b.; Guo, J.-q.; Lu, X.-c.; Zheng, Z.-w.; Deng, Y.-p.; Zou, F.; Li, H., Green tea polyphenol epigallocatechin-3-gallate ameliorates insulin resistance in non-alcoholic fatty liver disease mice. *Acta Pharmacologica Sinica* **2015**, *36* (5), 597.

Ge, A.-h.; Bai, Y.; Li, J.; Liu, J.; He, J.; Liu, E.-w.; Zhang, P.; Zhang, B.-l.; Gao, X.-m.; Chang, Y.-x., An activity-integrated strategy involving ultra-high-performance liquid chromatography/quadrupole-time-of-flight mass spectrometry and fraction collector for rapid screening and characterization of the α-glucosidase inhibitors in *Coptis chinensis* Franch. (Huanglian). *Journal of Pharmaceutical and Biomedical Analysis* **2014**, *100*, 79–87.

Giacoman-Martínez, A.; Alarcón-Aguilar, F. J.; Zamilpa, A.; Hidalgo-Figueroa, S.; Navarrete-Vázquez, G.; García-Macedo, R.; Román-Ramos, R.; Almanza-Pérez, J. C., Triterpenoids from *Hibiscus sabdariffa* L. with PPARδ/γ Dual Agonist Action: in vivo, in vitro and in silico studies. *Planta Medica* **2019**, *85* (05), 412–423.

Guan, G.; Han, H.; Yang, Y.; Jin, Y.; Wang, X.; Liu, X., Neferine prevented hyperglycemia-induced endothelial cell apoptosis through suppressing ROS/Akt/NF-κB signal. *Endocrine* **2014**, *47* (3), 764–771.

Guo, C.; Zhang, C.; Li, L.; Wang, Z.; Xiao, W.; Yang, Z., Hypoglycemic and hypolipidemic effects of oxymatrine in high-fat diet and streptozotocin-induced diabetic rats. *Phytomedicine: International Journal of Phytotherapy and Phytopharmacology* **2014**, *21* (6), 807–814.

Habtemariam, S., Antidiabetic potential of monoterpenes: a case of small molecules punching above their weight. *International Journal of Molecular Sciences* **2018**, *19* (1), 4.

Hamden, K.; Bengara, A.; Amri, Z.; Elfeki, A., Experimental diabetes treated with trigonelline: effect on key enzymes related to diabetes and hypertension, beta-cell and liver function. *Molecular and Cellular Biochemistry* **2013**, *381* (1-2), 85–94.

Hosseyni, E. S.; Kashani, H. H.; Asadi, M. H., Mode of action of medicinal plants on diabetic disorders. *Life Sciences Journal* **2012**, *9* (4), 2776–2783.

Huang, M.; Lu, J.-J.; Huang, M.-Q.; Bao, J.-L.; Chen, X.-P.; Wang, Y.-T., Terpenoids: natural products for cancer therapy. *Expert Opinion on Investigational Drugs* **2012**, *21* (12), 1801–1818.

Huang, Q.; Chen, L.; Teng, H.; Song, H.; Wu, X.; Xu, M., Phenolic compounds ameliorate the glucose uptake in HepG2 cells' insulin resistance via activating AMPK: antidiabetic effect of phenolic compounds in HepG2 cells. *Journal of Functional Foods* **2015**, *19*, 487–494.

Ikarashi, N.; Toda, T.; Okaniwa, T.; Ito, K.; Ochiai, W.; Sugiyama, K., Anti-obesity and antidiabetic effects of acacia polyphenol in obese diabetic KKAy mice fed high-fat diet. *Evidence-Based Complementary and Alternative Medicine* **2011**, *2011*, 952031.

International Diabetes Federation, *IDF Diabetes Atlas*, 978-2-930229-87-4; 2019.

Jang, S.-M.; Yee, S.-T.; Choi, J.; Choi, M.-S.; Do, G.-M.; Jeon, S.-M.; Yeo, J.; Kim, M.-J.; Seo, K.-I.; Lee, M.-K., Ursolic acid enhances the cellular immune system and pancreatic β-cell function in streptozotocin-induced diabetic mice fed a high-fat diet. *International Immunopharmacology* **2009**, *9* (1), 113–119.

Justino, A. B.; Pereira, M. N.; Vilela, D. D.; Peixoto, L. G.; Martins, M. M.; Teixeira, R. R.; Miranda, N. C.; da Silva, N. M.; de Sousa, R. M.; de Oliveira, A., Peel of araticum fruit (*Annona crassiflora* Mart.) as a source of antioxidant compounds with α-amylase, α-glucosidase and glycation inhibitory activities. *Bioorganic Chemistry* **2016**, *69*, 167–182.

Kang, C.; Kim, E., Synergistic effect of curcumin and insulin on muscle cell glucose metabolism. *Food and Chemical Toxicology* **2010**, *48* (8-9), 2366–2373.

Karkute, S. G.; Koley, T. K.; Yengkhom, B. K.; Tripathi, A.; Srivastava, S.; Maurya, A.; Singh, B., Antidiabetic phenolic compounds of black carrot (*Daucus carota* Subspecies *sativus* var. atrorubens Alef.) inhibit enzymes of glucose metabolism: an in silico and in vitro validation. *Medicinal Chemistry* **2018**, *14* (6), 641–649.

Kaur, K. K.; Allahbadia, G.; Singh, M., Monoterpenes-A class of terpenoid group of natural products as a source of natural antidiabetic agents in the future—A review. *CPQ Nutrition* **2019**, *3* (4), 2–21.

Kharbanda, C.; Alam, M. S.; Hamid, H.; Javed, K.; Bano, S.; Ali, Y.; Dhulap, A.; Alam, P.; Pasha, M. A., Novel piperine derivatives with antidiabetic effect as PPAR-gamma agonists. *Chemical Biology & Drug Design* **2016**, *88* (3), 354–362.

Kim, E.-A.; Lee, S.-H.; Lee, J.-H.; Kang, N.; Oh, J.-Y.; Ahn, G.; Ko, S. C.; Fernando, S. P.; Kim, S.-Y.; Park, S.-J., A marine algal polyphenol, dieckol, attenuates blood glucose levels by Akt pathway in alloxan induced hyperglycemia zebrafish model. *RSC Advances* **2016**, *6* (82), 78570–78575.

Lee, H.; Im, S. W.; Jung, C. H.; Jang, Y. J.; Ha, T. Y.; Ahn, J., Tyrosol, an olive oil polyphenol, inhibits ER stress-induced apoptosis in pancreatic β-cell through JNK signaling. *Biochemical and Biophysical Research Communications* **2016**, *469* (3), 748–752.

Lee, S.-H.; Kang, S.-M.; Ko, S.-C.; Lee, D.-H.; Jeon, Y.-J., Octaphlorethol A, a novel phenolic compound isolated from a brown alga, *Ishige foliacea*, increases glucose transporter 4-mediated glucose uptake in skeletal muscle cells. *Biochemical and Biophysical Research Communications* **2012**, *420* (3), 576–581.

Li, L.; Xu, J.; Mu, Y.; Han, L.; Liu, R.; Cai, Y.; Huang, X., Chemical characterization and anti-hyperglycaemic effects of polyphenol enriched longan (*Dimocarpus longan* Lour.) pericarp extracts. *Journal of Functional Foods* **2015**, *13*, 314–322.

Liu, L.; Du, X.; Zhang, Z.; Zhou, J., Trigonelline inhibits caspase 3 to protect beta cells apoptosis in streptozotocin-induced type 1 diabetic mice. *European Journal of Pharmacology* **2018**, *836*, 115–121.

Liu, X.; Li, G.; Zhu, H.; Huang, L.; Liu, Y.; Ma, C.; Qin, C., Beneficial effect of berberine on hepatic insulin resistance in diabetic hamsters possibly involves in SREBPs, LXRalpha and PPARalpha transcriptional programs. *Endocrine Journal* **2010**, *57* (10), 881–893.

Ludwiczuk, A.; Skalicka-Woźniak, K.; Georgiev, M. I., Chapter 11—Terpenoids. In *Pharmacognosy*, Badal, S.; Delgoda, R., Eds. Academic Press: Boston, MA, 2017; 233–266.

Mabhida, S. E.; Dludla, P. V.; Johnson, R.; Ndlovu, M.; Louw, J.; Opoku, A. R.; Mosa, R. A., Protective effect of triterpenes against diabetes-induced β-cell damage: an overview of in vitro and in vivo studies. *Pharmacological Research* **2018**, *137*, 179–192.

Madhuri, K.; Naik, P. R., Ameliorative effect of borneol, a natural bicyclic monoterpene against hyperglycemia, hyperlipidemia and oxidative stress in streptozotocin-induced diabetic Wistar rats. *Biomedicine & Pharmacotherapy* **2017**, *96*, 336–347.

Mander, L.; Liu, H.-W., *Comprehensive Natural Products II: Chemistry and Biology, vol. 1*. Elsevier: Amsterdam, 2010.

Martins, N.; Barros, L.; Ferreira, I. C., In vivo antioxidant activity of phenolic compounds: facts and gaps. *Trends in Food Science & Technology* **2016**, *48*, 1–12.

Mohamed, E. A. H.; Siddiqui, M. J. A.; Ang, L. F.; Sadikun, A.; Chan, S. H.; Tan, S. C.; Asmawi, M. Z.; Yam, M. F., Potent α-glucosidase and α-amylase inhibitory activities of standardized 50% ethanolic extracts and sinensetin from *Orthosiphon stamineus* Benth as antidiabetic mechanism. *BMC Complementary and Alternative Medicine* **2012**, *12* (1), 176.

Murali, R.; Saravanan, R., Antidiabetic effect of d-limonene, a monoterpene in streptozotocin-induced diabetic rats. *Biomedicine & Preventive Nutrition* **2012**, *2* (4), 269–275.

Muruganathan, U.; Srinivasan, S., Beneficial effect of carvone, a dietary monoterpene ameliorates hyperglycemia by regulating the key enzymes activities of carbohydrate metabolism in streptozotocin-induced diabetic rats. *Biomedicine & Pharmacotherapy* **2016,** *84*, 1558–1567.

Na, M.; Kim, B. Y.; Osada, H.; Ahn, J. S., Inhibition of protein tyrosine phosphatase 1B by lupeol and lupenone isolated from *Sorbus commixta*. *Journal of Enzyme Inhibition and Medicinal Chemistry* **2009,** *24* (4), 1056–1059.

Na, M.; Thuong, P. T.; Hwang, I. H.; Bae, K.; Kim, B. Y.; Osada, H.; Ahn, J. S., Protein tyrosine phosphatase 1B inhibitory activity of 24-norursane triterpenes isolated from *Weigela subsessilis*. *Journal Phytotherapy Research* **2010,** *24* (11), 1716–1719.

Nazaruk, J.; Borzym-Kluczyk, M., The role of triterpenes in the management of diabetes mellitus and its complications. *Phytochemistry Reviews* **2015,** *14* (4), 675–690.

Nuutinen, T., Medicinal properties of terpenes found in *Cannabis sativa* and *Humulus lupulus*. *European Journal of Medicinal Chemistry* **2018,** *157*, 198–228.

Oboh, G.; Agunloye, O. M.; Adefegha, S. A.; Akinyemi, A. J.; Ademiluyi, A. O., Caffeic and chlorogenic acids inhibit key enzymes linked to type 2 diabetes (in vitro): a comparative study. *Journal of Basic and Clinical Physiology and Pharmacology* **2015,** *26* (2), 165–170.

Olas, B., Berry phenolic antioxidants–implications for human health? *Frontiers in Pharmacology* **2018,** *9*, 78.

Oldfield, E.; Lin, F.-Y., Terpene biosynthesis: modularity rules. *Angewandte Reviews* **2012,** *51* (5), 1124–1137.

Olokoba, A. B.; Obateru, O. A.; Olokoba, L. B., Type 2 diabetes mellitus a review of current trends. *Oman Medical Journal* **2012,** *27* (4), 269–273.

Omara, E. A.; Kam, A.; Alqahtania, A.; M Li, K.; Razmovski-Naumovski, V.; Nammi, S.; Chan, K.; D Roufogalis, B.; Q Li, G., Herbal medicines and nutraceuticals for diabetic vascular complications: mechanisms of action and bioactive phytochemicals. *Current Pharmaceutical Design* **2010,** *16* (34), 3776–3807.

Pan, Y.; Cai, B.; Wang, K.; Wang, S.; Zhou, S.; Yu, X.; Xu, B.; Chen, L., Neferine enhances insulin sensitivity in insulin resistant rats. *Journal of Ethnopharmacology* **2009,** *124* (1), 98–102.

Paoli, P.; Cirri, P.; Caselli, A.; Ranaldi, F.; Bruschi, G.; Santi, A.; Camici, G., The insulin-mimetic effect of Morin: a promising molecule in diabetes treatment. *Biochimica et Biophysica Acta (BBA)—General Subjects* **2013,** *1830* (4), 3102–3111.

Park, C. H.; Noh, J. S.; Park, J. C.; Yokozawa, T., Beneficial effect of 7-O-galloyl-D-sedoheptulose, a polyphenol isolated from *Corni fructus*, against diabetes-induced alterations in kidney and adipose tissue of type 2 diabetic db/db mice. *Evidence-based Complementary and Alternative Medicine* **2013,** *2013*, 736856.

Pelletier, S. W., *Alkaloids: Chemical and Biological Perspectives*. Elsevier: Amsterdam, **1983**.

Putta, S.; Yarla, N. S.; Kilari, E. K.; Surekha, C.; Aliev, G.; Divakara, M. B.; Santosh, M. S.; Ramu, R.; Zameer, F.; Mn, N. P.; Chintala, R.; Rao, P. V.; Shiralgi, Y.; Dhananjaya, B. L., Therapeutic potentials of triterpenes in diabetes and its associated complications. *Current Topics in Medicinal Chemistry* **2016,** *16* (23), 2532–2542.

Ran, Q.; Wang, J.; Wang, L.; Zeng, H. R.; Yang, X. B.; Huang, Q. W., *Rhizoma coptidis* as a potential treatment agent for type 2 diabetes mellitus and the underlying mechanisms: a review. *Frontiers in Pharmacology* **2019,** *10*, 805.

Ringuelet, J. A.; Viña, S. Z., Productos Naturales Vegetales. Editorial de la Universidad Nacional de La Plata (EDULP): 2013.

Robinson, T., Metabolism and function of alkaloids in plants. *Science* **1974,** *184* (4135), 430–435.

Rodríguez, J. C.; Gómez, D.; Pacetti, D.; Nunez, O.; Gagliardi, R.; Frega, N. G.; Ojeda, M. L.; Loizzo, M. R.; Tundis, R.; Lucci, P., Effects of the fruit ripening stage on antioxidant capacity, total phenolics, and polyphenolic composition of crude palm oil from interspecific hybrid *Elaeis oleifera* × *Elaeis guineensis*. *Journal of Agricultural and Food Chemistry* **2016**, *64* (4), 852–859.

Rodríguez-Méndez, A.; Carmen-Sandoval, W.; Lomas-Soria, C.; Guevara-González, R.; Reynoso-Camacho, R.; Villagran-Herrera, M.; Salazar-Olivo, L.; Torres-Pacheco, I.; Feregrino-Pérez, A., Timbe (*Acaciella angustissima*) pods extracts reduce the levels of glucose, insulin and improved physiological parameters, hypolipidemic effect, oxidative stress and renal damage in streptozotocin-induced diabetic rats. *Molecules* **2018**, *23* (11), 2812.

Saxena, M.; Saxena, J.; Nema, R.; Singh, D.; Gupta, A., Phytochemistry of medicinal plants. *Journal of Pharmacognosy and Phytochemistry* **2013**, *1* (6), 168–182.

Shahidi, F.; Yeo, J., Bioactivities of phenolics by focusing on suppression of chronic diseases: A review. *International Journal of Molecular Sciences* **2018**, *19* (6), 1573.

Shobana, S.; Sreerama, Y.; Malleshi, N., Composition and enzyme inhibitory properties of finger millet (*Eleusine coracana* L.) seed coat phenolics: mode of inhibition of α-glucosidase and pancreatic amylase. *Food Chemistry* **2009**, *115* (4), 1268–1273.

Singh, B.; Sharma, R. A., Plant terpenes: defense responses, phylogenetic analysis, regulation and clinical applications. *3 Biotech* **2015**, *5* (2), 129–151.

Snieckus, V., The distribution of indole alkaloids in plants. In *The Alkaloids: Chemistry and Physiology, vol. 11*. Elsevier: Amsterdam, 1968; 1–40.

Solayman, M.; Ali, Y.; Alam, F.; Asiful Islam, M.; Alam, N.; Ibrahim Khalil, M.; Hua Gan, S., Polyphenols: potential future arsenals in the treatment of diabetes. *Current Pharmaceutical Design* **2016**, *22* (5), 549–565.

Sun, J.; Qu, C.; Wang, Y.; Huang, H.; Zhang, M.; Li, H.; Zhang, Y.; Wang, Y.; Zou, W., PTP1B, a potential target of type 2 diabetes mellitus. *Journal Molecular Biology* **2016**, *5* (4), 174.

Taguchi, K.; Hida, M.; Hasegawa, M.; Matsumoto, T.; Kobayashi, T., Dietary polyphenol morin rescues endothelial dysfunction in a diabetic mouse model by activating the Akt/eNOS pathway. *Molecular Nutrition & Food Research* **2016**, *60* (3), 580–588.

Tan, M.-J.; Ye, J.-M.; Turner, N.; Hohnen-Behrens, C.; Ke, C.-Q.; Tang, C.-P.; Chen, T.; Weiss, H.-C.; Gesing, E.-R.; Rowland, A.; James, D. E.; Ye, Y., Antidiabetic activities of triterpenoids isolated from bitter melon associated with activation of the AMPK pathway. *Chemistry & Biology* **2008**, *15* (3), 263–273.

Tanase, C.; Coșarcă, S.; Muntean, D.-L., A critical review of phenolic compounds extracted from the bark of woody vascular plants and their potential biological activity. *Molecules* **2019**, *24* (6), 1182.

Tetali, S. D., Terpenes and isoprenoids: a wealth of compounds for global use. *Planta* **2019**, *249* (1), 1–8.

Tharaheswari, M.; Jayachandra Reddy, N.; Kumar, R.; Varshney, K. C.; Kannan, M.; Sudha Rani, S., Trigonelline and diosgenin attenuate ER stress, oxidative stress-mediated damage in pancreas and enhance adipose tissue PPARgamma activity in type 2 diabetic rats. *Molecular and Cellular Biochemistry* **2014**, *396* (1-2), 161–174.

Tsuda, S.; Egawa, T.; Ma, X.; Oshima, R.; Kurogi, E.; Hayashi, T., Coffee polyphenol caffeic acid but not chlorogenic acid increases 5' AMP-activated protein kinase and insulin-independent glucose transport in rat skeletal muscle. *The Journal of Nutritional Biochemistry* **2012**, *23* (11), 1403–1409.

Tzeng, T.-F.; Hong, T.-Y.; Tzeng, Y.-C.; Liou, S.-S.; Liu, I.-M., Consumption of polyphenol-rich Zingiber Zerumbet rhizome extracts protects against the breakdown of the blood-retinal barrier and retinal inflammation induced by diabetes. *Nutrients* **2015**, *7* (9), 7821–7841.

Valdes, M.; Calzada, F.; Mendieta-Wejebe, J., Structure–activity relationship study of acyclic terpenes in blood glucose levels: potential α-glucosidase and sodium glucose cotransporter (SGLT-1) inhibitors. *Molecules* **2019**, *24* (22), 4020.

van Dijk, A. E.; Olthof, M. R.; Meeuse, J. C.; Seebus, E.; Heine, R. J.; van Dam, R. M., Acute effects of decaffeinated coffee and the major coffee components chlorogenic acid and trigonelline on glucose tolerance. *Diabetes Care* **2009**, *32* (6), 1023–1025.

Wang, L.; Li, X.; Zhang, Y.; Huang, Y.; Zhang, Y.; Ma, Q., Oxymatrine ameliorates diabetes-induced aortic endothelial dysfunction via the regulation of eNOS and NOX4. *Journal of Cellular Biochemistry* **2019**, *120* (5), 7323–7332.

Wang, X.; Liu, R.; Zhang, W.; Zhang, X.; Liao, N.; Wang, Z.; Li, W.; Qin, X.; Hai, C., Oleanolic acid improves hepatic insulin resistance via antioxidant, hypolipidemic and anti-inflammatory effects. *Molecular and Cellular Endocrinology* **2013**, *376* (1), 70–80.

Wang, Y. B.; Ge, Z. M.; Kang, W. Q.; Lian, Z. X.; Yao, J.; Zhou, C. Y., Rutin alleviates diabetic cardiomyopathy in a rat model of type 2 diabetes. *Experimental and Therapeutic Medicine* **2015**, *9* (2), 451–455.

Yang, N.; Sun, R. B.; Chen, X. L.; Zhen, L.; Ge, C.; Zhao, Y. Q.; He, J.; Geng, J. L.; Guo, J. H.; Yu, X. Y.; Fei, F.; Feng, S. Q.; Zhu, X. X.; Wang, H. B.; Fu, F. H.; Aa, J. Y.; Wang, G. J., In vitro assessment of the glucose-lowering effects of berberrubine-9-O-beta-D-glucuronide, an active metabolite of berberrubine. *Acta Pharmacologica Sinica* **2017**, *38* (3), 351–361.

Yaribeygi, H.; Atkin, S. L.; Sahebkar, A., Natural compounds with DPP-4 inhibitory effects: implications for the treatment of diabetes. *Journal of Cell Biochemistry* **2019**, *120* (7), 10909–10913.

Yu, L.; Wang, Z.; Huang, M.; Li, Y.; Zeng, K.; Lei, J.; Hu, H.; Chen, B.; Lu, J.; Xie, W.; Zeng, S., Evodia alkaloids suppress gluconeogenesis and lipogenesis by activating the constitutive androstane receptor. *Biochimica et Biophysica Acta* **2016**, *1859* (9), 1100–1111.

Zhai, L.; Wang, X., Syringaresinol-di-O-β-D-glucoside, a phenolic compound from Polygonatum sibiricum, exhibits an antidiabetic and antioxidative effect on a streptozotocin-induced mouse model of diabetes. *Molecular Medicine Reports* **2018**, *18* (6), 5511–5519.

Zhao, C.; Yang, C.; Wai, S. T. C.; Zhang, Y.; Maria, P. P.; Paoli, P.; Wu, Y.; San Cheang, W.; Liu, B.; Carpene, C.; Xiao, J.; Cao, H., Regulation of glucose metabolism by bioactive phytochemicals for the management of type 2 diabetes mellitus. *Critical Reviews in Food Science and Nutrition* **2019**, *59* (6), 830–847.

Zhao, P.; Tian, D.; Song, G.; Ming, Q.; Liu, J.; Shen, J.; Liu, Q. H.; Yang, X., Neferine promotes GLUT4 expression and fusion with the plasma membrane to induce glucose uptake in L6 cells. *Frontiers in Pharmacology* **2019**, *10*, 999.

Zhao, X.; Shen, J.; Chang, K. J.; Kim, S. H., Comparative analysis of antioxidant activity and functional components of the ethanol extract of lotus (*Nelumbo nucifera*) from various growing regions. *Journal of Agricultural and Food Chemistry* **2014**, *62* (26), 6227–6235.

Zheng, S.; Deng, S.; Huang, Y.; Huang, M.; Zhao, P.; Ma, X.; Wen, Y.; Wang, Q.; Yang, X., Antidiabetic activity of a polyphenol-rich extract from *Phellinus igniarius* in KK-Ay mice with spontaneous type 2 diabetes mellitus. *Food & Function* **2018**, *9* (1), 614–623.

Zhou, J. Y.; Zhou, S. W.; Zeng, S. Y.; Zhou, J. Y.; Jiang, M. J.; He, Y., Hypoglycemic and hypolipidemic effects of ethanolic extract of *Mirabilis jalapa* L. root on normal and diabetic mice. *Evidence-based Complementary and Alternative Medicine: eCAM* **2012**, *2012*, 257374.

CHAPTER 10

Ent-Kaurenes: Natural Agents with Potential for the Pharmaceutical Industry

CARLOS CAMACHO-GONZÁLEZ[1],
FRANCISCO FABIÁN RAZURA-CARMONA[1],
MAYRA HERRERA-MARTÍNEZ[2], EFIGENIA MONTALVO-GONZÁLEZ[1],
SONIA SÁYAGO-AYERDI[1], and JORGE ALBERTO SÁNCHEZ-BURGOS[1*]

[1]*Teccnológico Nacional de México/Instituto Tecnológico de Tepic, Av. Tecnológico # 2595, Col. Lagos del Country, 63175 Tepic, México*

[2]*Instituto de Farmacobiología, Universidad de la Cañada, carretera Teotitlán-San Antonio Nanahuatipán Km 1.7, Paraje Titlacuatitla, Teotitlán de Flores Magón, 68540 Oaxaca, México*

*Corresponding author. E-mail: jsanchezb@ittepic.edu.mx.

ABSTRACT

The characterization of natural extracts rich in bioactive compounds with a beneficial effect on human health has taken great importance in the last decades. One of the most promising natural compounds is diterpenoid kaurenes, responsible for beneficial effects on the human organism, highlighting their antitumor, anticancer, and antibacterial properties, among others. In 1999, kaurenes were classified by the IUPAC as the main fundamental structure of this diterpenoid class, where to date they incorporate a fragment of perhydrophenanthrene conjugated with a cyclopentane ring to build the structure of the tetracyclic nucleus. Particularly, tetracyclic diterpenoid belonging to the ent-kaurenes class is found in various natural resources mainly in higher plants, where they can be found in significant concentrations. Diterpenoids studies have focused on their isolation and identification from a plant, on their biosynthetic pathway, as well as on their biological activities to generate pharmaceutical and biomaterial products. To cherimoya (*Annona cherimola* L.), the diterpenoids are in greater proportion (42%–60%), representing between 554 and 1350 mg/kg of dry material; in this way, it constitutes an excellent source of these bioactive compounds. Thus, the production and application of these diterpenes in numerous productive sectors are viable and attractive.

10.1 GENERALITIES OF ENT-KAURENES

10.1.1 DEFINITION AND STRUCTURE OF ENT-KAURENES

The ent-kaurene class in natural matrices belongs to the tetracyclics C_{20}-diterpenoid which also includes ent-beyerenes (Kataev et al., 2011), ent-atisirenos (Giles, 1999), and ent-traquilobanos (Sun et al., 2006). It was in 1999, where kaurene was classified as the fundamental structure of this subclass of diterpenoids by the IUPAC (Giles, 1999). In terms of chemical structure, most ent-kaurene diterpenoids known to date incorporate a fragment of perhydrophenanthrene conjugated with a cyclopentane ring to construct the tetracyclic core structure (see Figure 10.1).

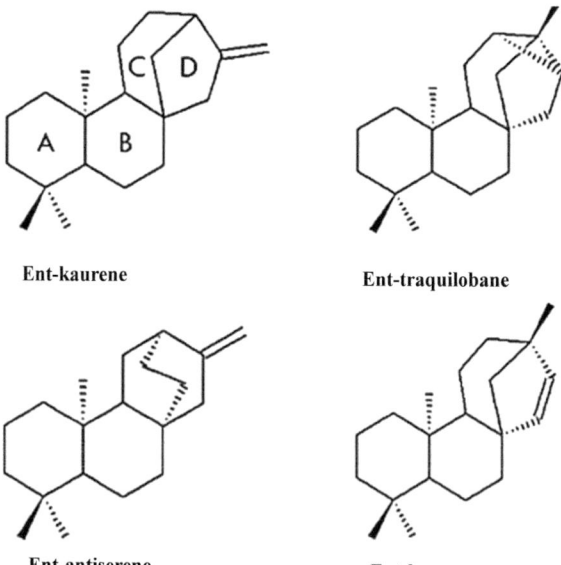

FIGURE 10.1 Tetracyclic diterphenoids: (A) ent-kaurene; (B) ent-traquilobane; (C) ent-atiserene; and (D) ent-beyerene.

10.1.2 STRUCTURE–ACTIVITY RELATIONSHIP OF ENT-KAURENES

The chemical reactivity of these diterpenoids is closely associated with their characteristic structural diversity (cyclic nature, four rings). The biological activity can be related to the electronic density present in its structure, mainly when atoms with considerably high electronegativities (O, N, and F) are present, since they allow affinity and spontaneity against co-reactants, causing a degree of unsaturation substantial, due to the dissociation of the weak bond (π), which implies a site rich in electrons that can interact with compounds in the system other than the co-reactant (formation of reactive oxygen species) and these, in turn, a reactive center of the target compound, as well as the presence of functional groups adhered to highly substituted carbons. This is based on the theory of alkane stability, which mentions that the stability of the formed radical will be in relation to the degree of substitution, that is, when the substituent is bonded to a highly substituted carbon, the bond dissociation energy will be lower and radical formation (H · + RO ·) will result in a less nonspontaneous degree (see Figure 10.2).

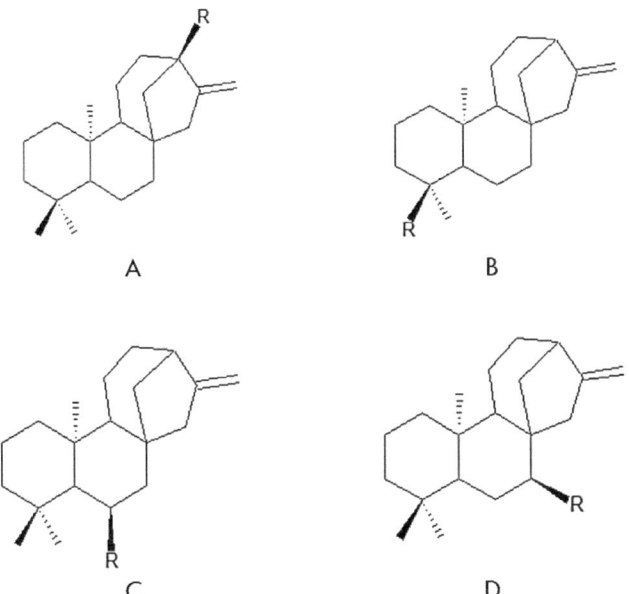

FIGURE 10.2 Structure–activity relationship of ent-kaurenos. Theoretically, the bond dissociation energy for the generation of the radical will be mostly spontaneous when the functional group is linked to a highly substituted carbon (trisubstituted, A and B) than one with a lower degree of substitution (disubstituted, C and D). It will also show significant reactivity when the functional group is more reactive (COOH < CHO < CO < OH < NH_2).

The above was demonstrated in a study carried out in 2011, where an evaluation of three kinds of ent-kaurenes was generated and its relation to the potential damage in malignant cells (carcinomas and leukemia), determining that it is particular functional groups that interact with enzymes involved in cellular exacerbation and that its absence or presence in the structure determines its functionality. Lactonated ent-kaurenes (carboxyl-derived group) represented highly reactive molecules conferred to the specificity in the interaction with nucleophilic compounds (thiol group present in enzymes by cysteine residues) achieving an inhibition of these protein complexes, in addition to the inclusion of a group bulky function (acetyloxy) in diterpenoid promoted low reactivity due to steric hindrance; and on the contrary, when there is the union of two rings by means of a quaternary carbon, it favors the bioactivity of the ent-kaurene attributed to the different flexion angles, exhibiting molecular configurations related to the objective compounds (Ding et al., 2011).

10.2 BIOSYNTHESIS OF ENT-KAURENES

Tetracyclic diterpenes belonging to the ent-kaurene family are derived biosynthetically from geranylgeranyl pyrophosphate (GGPP) through the initial cyclization of GGPP and subsequent modifications of the resulting bicyclic carbon skeleton, such as oxidations, reductions, acetylations, methylations, and glycosylations The mechanism is described below (Toyomasu and Sassa, 2010; Hanson and De Oliveira, 1993).

In the biosynthesis of ent-kaurene, the formation of carbocation begins with the protonation of dimethylated alkene in GGPP. This protonation results in a cascade cyclization sequence to form the bicyclic (−)-copalilPP structure, also known as ent-CPP. A series of transformations from ent-CPP, all catalyzed by (−)-copalyl diphosphate synthase, or ent-kaurene synthase, lead to the formation of the ent-kaurene hydrocarbon core. The initial loss of the leaving group of diphosphate leads to the formation of the ent-primarenyl cation. Subsequent entrapment of the tertiary cation with the terminal alkene in the latter forms a tetracyclic secondary cation, which is called ent-beyeranyl cation. This secondary cation can now undergo a 1,2-alkyl migration. This transformation is also called Wagner–Meerwein rearrangement and results in the formation of the ent-kaurenyl cation after the migration of C_{12} to C_{16} alkyl. The driving force of this transformation is the formation of a tertiary carbocation in C_{13}, which, when unprotected by the enzyme ent-kaurene synthase forms the common hydrocarbon core of ent-kaurenes (see Figure 10.3).

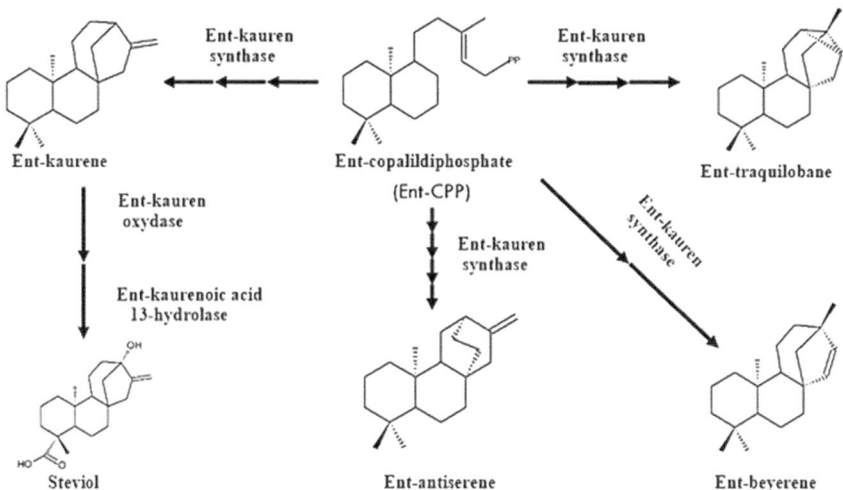

FIGURE 10.3 Biosynthetic route of ent-kaurenes. Dashed arrows represent more than one reaction.

The ent-kaurenes are intermediates in steviol biosynthesis in Stevia rebaudiana. The C_{19} methyl is initially oxidized to a carboxylate to form ent-kaurenoic acid. A final hydroxylation reaction catalyzed by ent-kaurenoic-13-hydroxylase acid introduces a hydroxyl group to form steviol (Riehl et al., 2015).

10.3 BIOLOGICAL ACTIVITY OF ENT-KAURENES

Studies have revealed that some diterpenoid kaurenes are responsible for influencing biological functions in the body presenting antitumor, antiinflammatory, as well as antibacterial properties.

10.3.1 ANTIBACTERIAL ACTIVITY

Ent-kaurenes from species of the genus *Isodon* in traditional folk medicine have had a wide application in recent years in relation to their antibacterial activity, for example, lipophilic extracts of *Isodon eriocalyx* leaves used as antibacterial (Sun et al., 2006) agent conferred to diterpene eriocalixin B kaurene (Table 10.1), with significant inhibitory activity against Gram-positive bacteria (Riehl et al., 2015). This effect has been elucidated and attributed to its numerous substituents that interact with sulfhydryl groups, essential for the enzyme's function, namely unsaturated cyclopentanones α and β (6-member enone in ring A). The hydrogen binding interaction between 6β-OH and finally C_{15} ketone both are considered crucial for the biological activity observed (Leung et al., 2006).

TABLE 10.1 Antibacterial Activity of Eriocalixine B (Sun et al. 2006)

Category	Name	Minimum Inhibitory Concentration (µg/mL)
Gram-positive bacteria	*Staphylococcus aureus*	31
	Staphylococcus epidermidis	62
	Streptococcus sanguis	62
Gram-negative bacteria	*Escherichia coli*	500
	Enterobacter gergoviae	2000
Microzyma	*Candida albicans*	62
Epiphyte	*Zygosaccharomyces bailii*	63

10.3.2 ANTITUMOR ACTIVITY

The antitumor effect of these diterpenoids has been attributed to the easy addition of soft nucleophiles such as alkanothioles in cysteine residues, which consequently achieves a deactivation of sulfhydric enzymes and coenzymes. However, they are specific functional groups that interact with specific residues, where up to date it is known that spirolactone, epoxicone subunits, or a functional group analogous to this because they provide an electrophilic character in diterpene and, therefore, generate high selectivity with cysteine residues (nucleophilic by the thiol group) (Riehl et al., 2015) (see Figure 10.4).

In 2011, a comparative study of DNA damage induced by kaurene diterpenoids present in plants of the *Isodon* genus, it was concluded that the exo-methylene substituent conjugated to a ketone in a five-membered ring in ent-kaurene diterpenoids is essential to that express their potential antitumor effect in DNA, adding that spirolactone-type diterpenoids exhibit greater biological activity attributed primarily to ring A in the molecular structure that exhibits numerous variations of torsion angles, as well as various molecular configurations (Ding et al., 2011).

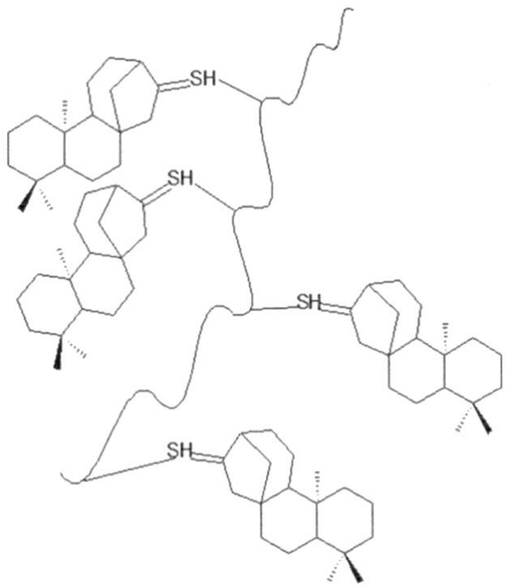

FIGURE 10.4 Ent-kaurene bypass model for specific enzymes in cell tumors. Adapted and modified (Riehl et al., 2015).

10.3.3 ANTI-INFLAMMATORY ACTIVITY

The anti-inflammatory activity of these lipophilic compounds has been investigated under the pathway of inhibition of the nuclear factor kappa-β (NF-kβ). In 2006, it was reported that eriocalixin B was able to inhibit the activation of NF-κβ by blocking the binding of NF-κβ to its response element, possibly through a reversible interaction with the two subunits of NF-κβ, p65, and p50 in an allosteric site in a noncompetitive way (Leung et al., 2006). Experimental data suggested that eriocalixin B may interfere with the binding activity of NF-κβ without blocking the translocation of NF-κβ. It was suggested that the inhibitory process may involve reversible inhibition of the DNA binding activity of the p65 and p50 subunits, although it does not compete with the DNA for the active binding site. In 2009, they investigated the antitumor activity of ponicidin and reported experimental evidence of its ability to bind and cleave DNA (Xu et al., 2009). In the same year, another study revealed that inflexinol inhibited colon cancer cell growth by inducing apoptotic cell death through the interaction of NF-κβ by a direct covalent modification of a p50 cysteine residue in human colon cancer cells (Ban et al., 2009). In the course of the same study, it was found that kamebakaurin interacted with the p50 and p65 subunits of NF-κβ.

However, there are reports that support this effect by other routes of inflammation. The 18-acetoxy-ent-kaur-16-en kaurene from *Annona squamosa* showed significant activity in the cardinal signs of inflammation (edema, hyperalgesia, and erythema) due to the inhibition of proinflammatory agents such as histamine and serotonin (Chavan et al., 2011). A year later, it was found that kaur-16-in-19-oic acid from *Annona reticulata* exhibited an analgesic effect due to the inactivation of the aforementioned proinflammatory targets (Chavan et al., 2012).

10.4 CURRENT APPLICATIONS OF ENT-KAURENES

Diterpene Kaurenes have very varied oxidation patterns that favor fascinating skeletal rearrangements as a result of their various reactive centers. While synthesis via de novo is a possible solution to obtain the required diterpenoid derivatives, the need for multiple steps to build these macrostructures is always limited, so semisynthesis from naturally abundant related molecules is a better alternative. Under this context, stevioside and its hydrolysis products, steviol and isosteviol (see Figure 10.5) are good leaders in the field of medicinal chemistry for the discovery of diterpenoid drugs. The stevioside has a complex diterpenoid glycoside molecule composed of an aglycone, with the ent-kaurene skeleton and three glucose molecules. Hydrolyzed under alkaline or acidic conditions, the stevioside generates a diterpene (steviol) or dyepenoid ent-beyerene (isosteviol) kaurene, respectively (Wang et al., 2018). Except for direct applications, they are widely used to provide central ent-kaurene or ent-beyerene structures for further study of medical chemistry. In addition to a potential industrial character linked to numerous reactive sites (density of functional groups) present in its structure in such a way that its inclusion in various productive sectors is admissible.

10.4.1 ENT-KAURENES IN THE SWEETENER INDUSTRY

Natural sweeteners have received a lot of interest due to the growing health concerns about sugar consumption, as well as the safety-related problems of some nonnutritive artificial sweeteners. In the sense, the sweet taste of Stevia rebaudiana leaves is due to diterpenoid ent-kaurene-type glycosides that commonly contain aglycone and steviol that differ from each other only at the position (C_{13} and/or C_{19}) of the glycosidic constituent. In steviol glycosides, the number of carbohydrate groups at sites C_{13} and C_{19} will determine the degree of sweetness of steviol (Adari et al., 2016; Gerwig et al., 2016). For this characteristic, these chemical substances have entered into a large extent in the sweetener industry, since they present a greater sweetness compared to sucrose (see Table 10.2).

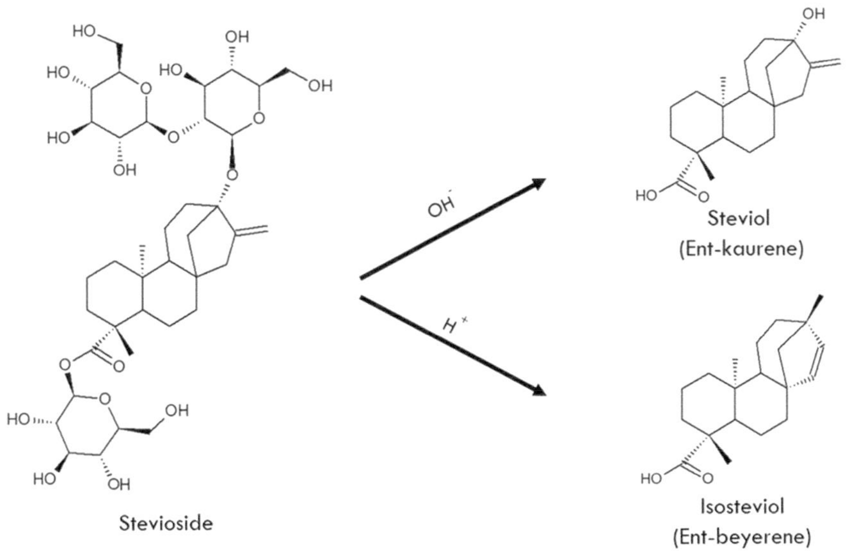

FIGURE 10.5 Chemical structures of stevioside and its hydrolysis products (steviol and isosteviol).

TABLE 10.2 Components of *Stevia Rebaudiana*

Component	Yield (g/100 g Dry Leaf)	Sweetness[a]	Reference
Steviol	Traces (<0.01)	–	(Yadav and Guleria, 2012; Chatsudthipong and Muanprasat, 2009)
Stevioside	5–10	300	
Rebaudioside A	2–5	250–450	(Chatsudthipong and Muanprasat, 2009)
Rebaudioside C	1	50–120	
Sweet A	0.5	50–120	
Rebaudioside D	0.2	250–450	
Rubusoside	Traces (<0.01)	115	(Ko et al., 2012)

[a]Times sweeter than sucrose at a concentration of 0.025%.

10.4.2 ENT-KAURENES AS SOLUBILIZING AGENT

Bioavailability is crucial in the design of oral administration of any medication. There are several factors that affect oral bioavailability, including water solubility, permeability of the drug to cells, presystemic metabolism among others. Today, a low water solubility of the new candidates for pharmaceutically active ingredients has been observed in recent years (Kawabata et al., 2011). Drugs, which are poorly soluble in water, are slowly absorbed into the body. This leads to inadequate bioavailability with toxicity in the stomach and intestinal mucosa, which postpones the clinical development of the drug (Savjani et al., 20012). Studies have revealed that some steviol glycosides such as stevioside and rebaudioside A have solubilizing properties since their incorporation into an active substance maintains and/or increases their biological effects, as shown in Table 10.3.

10.4.3 INHIBITION OF GLUCOSE TRANSPORTERS BY ENT-KAURENES

There are several glucose transporters (GLUTs) in human cells; however, GLUT1 has a greater expression in tissues. Overexpressed GLUT1 is relevant for obesity and diabetes. However, unlike GLUT1, GLUT5 is usually expressed in the small intestine and absorbs fructose (Ellwood et al., 1993). Under this context, excessive consumption of fructose is considered to cause harmful metabolic effects, so GLUT5 becomes an increasingly important goal for human health (Basciano et al., 2005; Rutledge and Adeli, 2007); not to mention that an increase in fructose consumption correlates with activation of lipogenesis and triglycerides at the organism level, and leads to insulin resistance (Basciano et al., 2005; Rutledge and Adeli, 2007). In 2015, the effect of rubusósido inhibition in GLUT1 and GLUT5 of humans expressed in insect cell cultures was studied and found that rubusósido can inhibit both GLUT1 and GLUT5 with IC50 values of 4.6 and 6.7 mM, respectively (Thompson et al., 2015). Finally, for a better understanding of the phenomenon, the in silico coupling method was used, discovering that the rubusósido interacts with the active sites of GLUT1 and GLUT5 in a distinguishable way due to a key tryptophan residue in GLUT1 and alanine in GLUT5 (Thompson et al., 2015).

TABLE 10.3 Solubilization of Insoluble Compounds Using Steviol Glycosides

Steviol Glycoside	Insoluble Compound	Solubility (mg/mL)	Biological Activity of the Complex	References
Rubusoside (10%, p/v)	Curcumin	2.32	Reduction of viability of cancer cells Caco-2, HT-29, MDA-MB-231, and PANC-1	(Zhang et al., 2011)
	Quercetin	7.7	Enhanced inhibition activity against human intestinal maltose—Tyrosinase inhibition activity in fungi maintained	(Nguyen et al., 2015)
Rubusoside (40%, p/v)	Placitaxel	6.26	4-Fold increase in permeability in monocultures of Caco-2 cells	(Liu et al., 2015)
Estevioside (8%, p/v)	Curcuminoids extracted from *Curcuma longa*	11.3	Antioxidant activity maintained and inhibition activity against NS2BNS3pro of dengue virus type 4	(Nguyen et al., 2017)
Rebaudiosid A (8%, p/v)		9.7		

10.5 CONCLUSIONS

The scientific evidence validates the potential use of ent-kaurenes in industry, such as pharmaceuticals. The methods for obtaining these compounds are relatively simple, in addition to the fact that the sources of obtaining can be those fruits that are in the stage of senesence, which could make an integral use of these fruits and it would not be necessary to integrate a new system of culture to only take advantage of these compounds. However, it is important to investigate further regarding the toxicity of these compounds, since their massive or controlled use will depend on it; likewise, the regulations in this regard are still absent, so more scientific information is required.

KEYWORDS

- *Annona cherimola*
- ent-kaurenes
- diterpenes

REFERENCES

Adari B, Alavala S, George SA, Meshram HM, Tiwari AK, Sarma AVS. Synthesis of Rebaudioside-A by Enzymatic Transglycosylation of Stevioside Present in the Leaves of Stevia Rebaudiana Bertoni. *Food Chem.* 2016;200:154–158. doi:10.1016/j.foodchem.2016.01.033.

Ban JO, Oh JH, Hwang BY, et al. Inflexinol Inhibits Colon Cancer Cell Growth Through Inhibition of Nuclear Factor-B Activity via Direct Interaction with p50. *Mol. Cancer Ther.* 2009;8(6):1613–1624. doi:10.1158/1535-7163.mct-08-0694.

Basciano H, Federico L, Adeli K. Nutrition & Metabolism Fructose, Insulin Resistance, and Metabolic Dyslipidemia. *Nutr. Metab.* 2005;2(5):1–14. doi:10.1186/1743-7075-2-5.

Chatsudthipong V, Muanprasat C. Pharmacology & Therapeutics Stevioside and Related Compounds : Therapeutic Benefits Beyond Sweetness. *Pharmacol. Ther.* 2009;121(1):41–54. doi:10.1016/j.pharmthera.2008.09.007.

Chavan MJ, Wakte PS, Shinde DB. Analgesic and Anti-Inflammatory Activities of 18-Acetoxy-ent-kaur-16-ene From Annona Squamosa L. Bark. *Inflammopharmacology.* 2011;19(2):111–115. doi:10.1007/s10787-010-0061-5.

Chavan MJ, Wakte PS, Shinde DB. Analgesic and Anti-Inflammatory Activities of the Sesquiterpene Fraction From Annona Reticulata L. Bark. *Nat. Prod. Res.* 2012;26(16):1515–1518. doi:10.1080/14786419.2011.564583.

Ding L, Zhou Q, Wang L, Wang W, Zhang S, Liu B. Comparison of Cytotoxicity and DNA Damage Potential Induced by Ent-Kaurene Diterpenoids From Isodon Plant. *Nat. Prod. Res*. 2011;25(15):1402–1411. doi:10.1080/14786410802267734.

Ellwood K, Chatzidakis C, Failla M. Fructose Utilization by the Human Intestinal Epithelial Cell Line, Caco-2. *Proc. Soc. Exp. Biol. Med*. 1993;202(4):440-446.

Gerwig GJ, Poele EM, Dijkhuizen L, Kamerling JP. Stevia Glycosides: Chemical and Enzymatic Modifications of Their Carbohydrate Moieties to Improve the Sweet-Tasting Quality. Vol 73. 1st ed. Elsevier Inc: Amsterdam; 2016. doi:10.1016/bs.accb.2016.05.001.

Giles P. Natural Products and Related Compounds. *Pure Appl. Chem*. 1999;71(4):587–643.

Kataev VE, Khaybullin RN, Sharipova RR, Strobykina IY. Ent-Kaurane Diterpenoids and Glycosides: Isolation, Properties, and Chemical Transformations. *Rev. J. Chem*. 2011;1(2): 93–160. doi:10.1134/S2079978011010043.

Kawabata Y, Wada K, Nakatani M, Yamada S, Onoue S. Formulation Design for Poorly Water-Soluble Drugs Based on Biopharmaceutics Classification System: Basic Approaches and Practical Applications. *Int. J. Pharm*. 2011;420(1):1–10. doi:10.1016/j.ijpharm.2011.08.032.

Ko J, Kim Y, Ryu YB, et al. Mass Production of Rubusoside Using a Novel Stevioside-Specific β-Glucosidase From Aspergillus aculeatus. *J. Agric. Food Chem*. 2012;60:6210−6216.

Leung C-H, Grill SP, Lam W, Gao W, Sun H-D, Cheng Y-C. Eriocalyxin B Inhibits Nuclear Factor-B Activation by Interfering with the Binding of Both p65 and p50 to the Response Element in a Noncompetitive Manner. *Mol. Pharmacol*. 2006;70(6):1946–1955. doi:10.1124/mol.106.028480.

Liu Z, Zhang F, Yee G, et al. Cytotoxic and Antiangiogenic Paclitaxel Solubilized and Permeation-Enhanced by Natural Product Nanoparticles. *Anticancer Drugs*. 2015;26:167–179. doi:10.1097/CAD.0000000000000173.

Nguyen T, Si J, Kang C, Chung B, Chung D, Kim D. Facile Preparation of Water Soluble Curcuminoids Extracted From Turmeric (Curcuma Longa L .) Powder by Using Steviol Glucosides. *Food Chem*. 2017;214:366–373. doi:10.1016/j.foodchem.2016.07.102.

Nguyen T, Yu S, Kim J, An E. Enhancement of Quercetin Water Solubility with Steviol Glucosides and the Studies of Biological Properties. *Funct. Foods Health Dis*. 2015;5(12): 437–449.

Riehl PS, DePorre YC, Schindler CS, Groso EJ, Armaly AM. New Avenues for the Synthesis of Ent-Kaurene Diterpenoids. *Tetrahedron* 2015;71(38):6629–6650. doi:10.1016/j.tet.2015.04.116.

Rutledge AC, Adeli K. Fructose and the Metabolic Syndrome: Pathophysiology and Molecular Mechanisms. *Nutr. Rev*. 2007;65(6):13–23. doi:10.1301/nr.2007.jun.S13.

Savjani KT, Gajjar AK, Savjani JK. Drug Solubility: Importance and Enhancement Techniques. *ISRN Pharm*. 2012;2012:1–13. doi:10.5402/2012/195727.

Sun H, Huang S, Han Q. Diterpenoids From Isodon Species and Their Biological Activities. *Nat. Prod. Rep*. 2006; 23(5):673–98. doi:10.1039/b604174d.

Thompson AMG, Iancu CV, Thanh T, Nguyen H, Kim D, Choe J. Inhibition of Human GLUT1 and GLUT5 by Plant Carbohydrate Products; Insights into Transport Specificity. *Sci. Rep*. 2015;5:1–10. doi:10.1038/srep12804.

Toyomasu T, Sassa T. Diterpenes. Comprehensive Natural Products II, Elsevier, 2010, 643-672, ISBN 9780080453828.

Wang M, Li H, Xu F, et al. Diterpenoid Lead Stevioside and its Hydrolysis Products Steviol and Isosteviol: Biological Activity and Structural Modification. *Eur. J. Med. Chem*. 2018;156:1–57. doi:10.1016/j.ejmech.2018.07.052.

Xu X, Ke Y, Zhang Q, Qi X, Liu H. DNA Binding and Cleavage Properties of Certain.pdf. *Tumori* 2009;3:348–351.

Yadav S, Guleria P. Enzymatic Production of Steviol. Elsevier Inc: Amsterdam; 2012. doi:10.1016/B978-0-12-813280-7.00023-2.

Zhang T, Chen J, Zhang Y, Shen Q, Pan W. Characterization and Evaluation of Nanostructured Lipid Carrier as a Vehicle for Oral Delivery of Etoposide. *Eur. J. Pharm. Sci.* 2011; 43(3):174–179. doi:10.1016/j.ejps.2011.04.005.

CHAPTER 11

Anti-Inflammatory, Antioxidant, and Antitumoral Potential of Plants of the Asteraceae Family

MIGUEL ÁNGEL ALFARO-JIMÉNEZ,[1,**] ALEJANDRO ZUGASTI-CRUZ,[1,**] SONIA YESENIA SILVA-BELMARES,[2] JUAN ALBERTO ASCACIO-VALDÉS,[2] and CRYSTEL ALEYVICK SIERRA-RIVERA[1,*]

[1]Laboratory of Immunology and Toxicology, Faculty of Chemistry, Autonomous University of Coahuila, Saltillo, Mexico

[2]Food Research Department, Faculty of Chemistry, Autonomous University of Coahuila, Saltillo, Mexico

**These authors contributed equally to this work

*Corresponding author. E-mail: crystelsierrarivera@uadec.edu.mx

ABSTRACT

Inflammation is a nonspecific protective response of the innate immune system to harmful stimuli, such as microorganisms, burns, trauma, and chemical compounds. In response to tissue damage, the body activates various signaling cascades that allow the removal of the invading agent and tissue repair. The inflammation process is characterized by an acute phase and a chronic phase. In the acute phase, there is an increase in vascular permeability, leukocyte migration, and release of inflammatory mediators, and it is resolved when the foreign agent is removed. However, if the inflammation persists, it is called chronic inflammation. Chronic inflammation has been correlated with cardiovascular, intestinal diseases, diabetes, autoimmunity, and cancer. It is estimated that approximately 15% of human cancers are associated with chronic inflammation because inflammatory mediators, such as cytokines, chemokines, prostaglandins, can stimulate tumor growth and cell proliferation. In particular, various species of the Asteraceae family possess a wide range of secondary metabolites. For this reason, currently, the pharmaceutical industry in alliance with phytochemical research is looking for metabolites obtained from plants of the Asteraceae family that could have beneficial effects on human health, including anti-inflammatory, antioxidant, and antitumor potential.

11.1 INFLAMMATION

Inflammation is a biological defense mechanism that occurs in extravascular connective tissues. Its main function is the destruction and elimination of antigens. The antigens can be classified according to their origin, which can be chemical, infectious, or autoimmune (Valečka et al., 2018), which can activate the immune response (León et al., 2016).

There are five classical signs that are characterized by an inflammatory response: they are heat, redness, swelling, pain, and loss of function of the affected zone (Ashely et al., 2012). These signs are produced due to cellular and biochemical interactions that are regulated by the machinery of the immune system (Vergnolle, 2003).

11.1.1 TYPES OF INFLAMMATION

According to its progression, inflammation can be divided into two types: acute and chronic inflammation (Raposo et al., 2015).

Acute inflammation is the first response produced by the immune system against antigens; it is characterized by a short duration, lasting from hours to days. In this process promotes the exudation of fluids and plasma proteins (edema) and leukocyte extravasation of the bloodstream into the tissues or organs affected by the antigen (León et al., 2016). On the other hand, chronic inflammation occurs as a consequence of the inability of acute inflammation to eliminate pathogens, lasting for prolonged periods of several months to years and is characterized by proliferation of blood vessels, fibrosis, and tissue necrosis (Hasselbalch, 2013).

11.1.2 INFLAMMATION AS A PROCESS OF THE IMMUNE SYSTEM

The inflammatory response is a mechanism of the immune system that is constituted by a complex interaction of cells, organs, and tissues that work together to protect the organism against antigens. Leukocytes, also called white cells, are responsible for activating processes related to antigens elimination. These cells are produced in the bone marrow and move to the bloodstream when they are mature enough. Then, leukocytes are distributed to different parts of the body and remain in lymph nodes and spleen where they are protected until the inflammatory response is activated in response to an antigen (Hasselbalch, 2013).

There are various types of leukocytes, some of them are neutrophils, monocytes, and lymphocytes, which are found regularly in the blood, while other cells such as macrophages, dendritic cells, and mast cells are located in extravascular tissues (León et al., 2016). The main characteristics of each group are given in Table 11.1.

In order to eliminate antigens, this network of cells works together during an inflammatory response. For this, it is necessary that antigens can be detected by phagocyte cells such as macrophages and neutrophils, which ingest and eliminate the antigens by enzymatic digestion. If macrophages are not able to neutralize the invading agent, it is worth mentioning that this process requires the presence of substances called cytokines, which are protein produced by the same leukocytes and its main function is regulating inflammation. According to their effect, cytokines can be classified into two types: proinflammatory and anti-inflammatory cytokines. Proinflammatory cytokines stimulate an inflammatory response, some of them are interleukin 1 (IL-1), interleukin 6 (IL-6), tumor necrosis factor alpha (TNF-α), and interferon gamma, which are released mainly by macrophages and lymphocytes. In contrast, anti-inflammatory cytokines produce the opposite effect by the inhibition of inflammatory mechanisms. Some of the most important are interleukin 4 (IL-4), interleukin 10 (IL-10), and interleukin 13 (IL-13) that are produced generally by lymphocytes (Cassatella et al., 2019). The main characteristics of cytokines are given in Table 11.2.

11.1.3 MECHANISM OF INFLAMMATION

As it has been mentioned, the interaction among antigens, leukocytes, and cytokines induces inflammatory response; however, this process occurs in four consecutive phases (Figure 11.1): silent phase, vascular phase, cellular phase, and climax and are explained below.

TABLE 11.1 Cells involved in inflammatory response

Leukocyte	Characteristics	Reference
Bloodstream		
Neutrophils	It is the most abundant leukocyte in blood. Its size is 9 – 12 µm and has a nucleus divided into 2-5 lobes. It is the first white cell to arrive at the infection site and eliminate pathogens by phagocytosis and enzyme activity.	Silvestre-Roig et al., 2019
Monocytes	It is one of the largest leukocytes. Its diameter is 15 – 17 µm and has a kidney-shaped nucleus. It contains vacuoles, lysosomes and cytoskeleton filaments that support phagocytosis of pathogens.	Lauvaua et al., 2015
Lymphocytes	It is a leukocyte that measures 8 – 10 µm. Its nucleus is prominent and has few cytoplasmic organels. They are divided in three groups according to their function: • T lymphocytes: These cells mature in thymus and participate in the elimination of antigens by the activity of their subgroups: Helper (TCD4$^+$), Cytotoxic (TCD8$^+$) and Natural Killer T cells (NKT). • B lymphocytes: They recognize antigens and produce specific protein molecules called antibodies which contribute to the elimination of microorganisms. • Natural killer cell: They have small protein granules called perforin and granzyme which work together to destroy microorganisms.	Koyasu and Moro, 2012
Extravascular tissue		
Macrophages	It is the largest leukocyte because its average size is 60 – 80 µm. It is the result of the monocyte maturation process and phagocytes antigens that infiltrate into tissues and present them to T lymphocytes. It also produces chemical mediators that activate other white cells.	Ariel et al., 2012
Dendritic cells	It is a phagocyte cell that measures 10 - 15 µm. Its main function is to capture antigens, process them and present epitopes to T lymphocytes and induce their activation.	Lambrecht and Hammad, 2010
Mast cells	It is a white cell located near conjunctive tissues, mainly epidermal tissue. It has a diameter of 8 – 20 µm and it is characterized by eliminating antigens by phagocytosis and producing biochemical molecules during some immune responses such as allergic reactions.	Amin, 2012

TABLE 11.2 Cytokines involved in inflammatory response

Cytokine	Cell origin	Target cells	Biological activity	Reference
Pro-inflammatory cytokines				
TNF-α	Macrophages	Endothelial cells Neutrophils	Activation and expression of transmembrane receptors	Karnes et al., 2015
IL-1	Macrophages Endothelial cells	Endothelial cells	Activation and expression of ligands to integrins	Mantovani et al., 2019
IFN-γ	Th1 lymphocytes TNK lymphocytes	Macrophages T lymphocyte	Phagocytosis activation Differenciation to Th1 lymphocytes	De Pablo-Sánchez et al., 2005
IL-6	Macrophages Endothelial cells T lymphocytes	Lymphocytes	Differenciation to B lymphocytes Activation of CD8 lymphocytes	Rossi et al., 2015
Anti-inflammatory cytokines				
IL-4	T lymphocytes Mast cells	Macrophages T lymphocytes	Inhibition of IFN-γ activation Differentiation to Th2 lymphocytes	Kelly et al., 2003
IL-10	Macrophages T lymphocytes	Macrophages Dendritic cells	Inhibition of IL-12 synthesis	Moore et al., 2001
IL-13	T lymphocytes TNK lymphocytes Mast cells	Macrophages Fibroblasts	Increase of collagen synthesis	Kelly et al., 2003

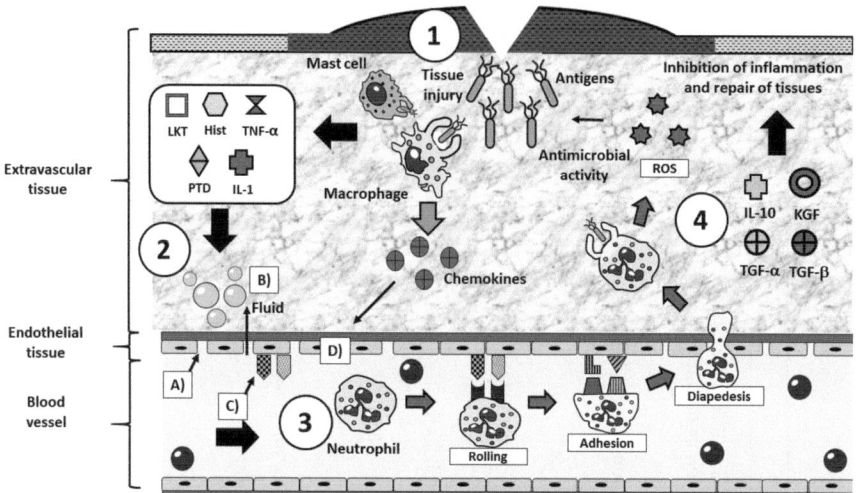

FIGURE 11.1 Mechanism of inflammation. Inflammation is constituted by four consecutive phases: **Phase 1.** Inflammation begins with silent phase, in which occurs a tissue injury that let the entrance of external agents to extravascular tissue. Macrophages and mast cells detect them and release pro-inflammatory mediators such as LKT, TNF-α, PTD, IL-1 and chemokines. **Phase 2.** Vascular phase takes place and occurs the follow phenomenons: (A) Increase of permeability of epithelial tissues. (B) Release of fluids from blood to extravascular medium. (C) Activation and expression of membrane receptors on endothelial cells. (D) Attraction and recruitment of leukocytes in inflammatory zone. **Phase 3.** Cell phase begins and neutrophils bind slightly to endothelial tissues (rolling). Subsequently, neutrophils bind strongly (adhesion) and transmigrate to extravascular tissues (diapedesis). In outside, neutrophils phagocyte inflammatory agents and also produce ROS with antimicrobial activity. **Phase 4:** Antigens are eliminated, and cells release IL-10 to inhibit inflammation and EGF, TGF-α and TGF-β to repair injured tissues. (Alfaro-Jiménez, 2019).

11.1.3.1 SILENT PHASE

Inflammation begins with the silent phase that consists of the release of the first inflammatory mediators by leukocytes located at damaged tissue. Inflammation is induced by the infiltration of antigens through a tissue injury and their detection by macrophages, dendritic, and mast cells. This activates phagocytosis, that is, the process in which leukocytes capture antigens with a plasmatic membrane portion, creating a phagosome inside the cytoplasm followed by a fusion with lysosome granules containing hydrolytic enzymes that are responsible for antigen degradation (Herrera and Wendie, 2014). Furthermore, this process also induces the release of biochemical compounds such as nitric oxide (NO), histamine (Hist), IL-1, leukotrienes, and prostaglandins (PTD), whose effects are related to vasodilation and permeability of blood vessels, two facts that occur in the next steps of inflammation (Ashely et al., 2012).

11.1.3.2 VASCULAR PHASE

The vascular phase is the second part of the inflammatory response in which occurs vasodilation, that is the widening of blood vessels that results from the relaxation of the smooth muscle of the vessels walls and is produced mainly by Hist and NO; these two molecules induce hyperemia and causes an increase of temperature and redness of the affected zone (Aderem, 2003). Subsequently, blood flow decreases at local area due to vascular permeability produced by the activity of chemicals mediators of silent phase and cytokines that also allow the migration of blood plasma into the extravascular tissues generating an increase of blood viscosity and the formation of a swelling called edema (León et al., 2016). Additionally, pain and loss of function of the affected zone can be present by the effects produced by PTD (Herrera and Wendie, 2014).

11.1.3.3 CELL PHASE

The cell phase is the most important step of the inflammatory process due to the arrival of leukocytes from blood to the injured tissue, where they participate in defense mechanisms against antigens. Cell phase is regulated by chemotaxis, that is, the attraction of more leukocytes to the inflammatory zone by the activity of molecules called chemokines, which are produced by other white cells (Chow et al., 2005). Neutrophils are the first cell group to arrive at the zone and their movement from blood to extravascular tissue begin with the activation of endothelial cells of the blood vessels by the action of TNF-α and IL-1, which induce the expression of adhesive membrane molecules such as E and P selectins, that recognize and bind to glycoproteins located on leukocyte membrane, creating a temporary adhesion named rolling (Aznag et al., 2018).

Subsequently, this adhesion is strengthened by the activation of chemokines that stimulate the expression of integrins. These molecules bind to membrane receptors on endothelial cells. When leukocytes have completely adhered to, they start the diapedesis that is defined as the transmigration of the cells from the endothelial wall to the extravascular space (Filippi, 2016). Once located in the tissue, white cells can eliminate antigens by phagocytosis and the activation of other leukocytes by the release of proinflammatory cytokines that increase immune response (Ashely et al., 2012). At the same time, these actions stimulate the production of unstable chemical compounds called free radicals that can react with a lot of biochemical molecules of pathogenic microorganisms, inducing a cytotoxic effect as long as they are present in moderate concentrations (Nelson and Texeira, 2012).

11.1.3.4 CLIMAX

Finally, the inflammatory response finishes when antigens have been eliminated; therefore, the number of leukocytes is restored to normal values. Regarding this process, many regulating molecules participate in the inhibition of inflammation such as IL-10. In addition, the organism starts repairing tissues by a combined action of cytokines and growth factors such as epidermal growth factor (EGF), keratinocytic growth factor, and α and β transforming growth factors (TGF-α and TGF-β), which are produced by fibroblasts, keratinocytes, macrophages, lymphocytes, and endothelial cells. These groups of cells work together to create structural tissue molecules such as fibronectin, keratin, and collagen. Moreover, they also stimulate cell proliferation and blood vessels production (Nelson and Texeira, 2012).

11.1.4 DISEASES RELATED TO INFLAMMATION

Although inflammation is generated as natural protection of the organism, however, persistent inflammation can cause diseases. Some of these alterations are related to overactivation of defense mechanisms against antigens (Viennois et al., 2019) such as the case of autoimmune diseases, a phenomenon in which the immune system identifies the own compounds of the body as strange molecules and starts an inflammatory response with the purpose of eliminating them. This fact causes severe damages to organs and tissues (Ashley et al., 2012) and may also produce the origin of chronic diseases such as arthritis, asthma, and Alzheimer (Figure 11.2). The description of each pathology is shown below.

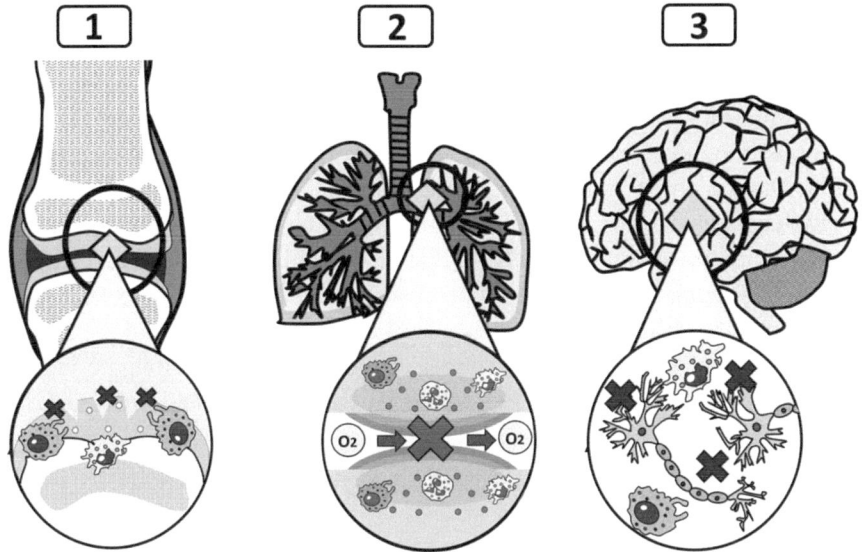

FIGURE 11.2 Effects of inflammation in chronic diseases. The inflammatory response is part of chronic diseases by different mechanisms: (1) In arthritis collagen and cartilage of joints are detected as strange compounds and are attacked by leukocytes, producing their deterioration. (2) In asthma occurs an over activation of inflammatory response due to inhalation of allergenic compounds that produce inflammation of bronchi, avoiding air flow to lungs. (3) In Alzheimer, the formation of Aβ4 stimulate immune response in which neurons are also affected, producing neuronal damage. (Alfaro-Jiménez, 2019).

11.1.4.1 ARTHRITIS

Arthritis is a chronic inflammatory disease that is characterized by inflammation of small, medium, and big peripheral joints (De Armas et al., 2017). Its origin is autoimmune because cartilage and collagen are detected as strange particles by the immune system. This phenomenon induces an inflammatory response in which occurs an increase of TNF-α expression and activation of macrophages and lymphocytes that produce the degradation of structural joints (Coates et al., 2016).

11.1.4.2 ASTHMA

Asthma is a pulmonary pathology that consists of an obstruction of the respiratory tract as a consequence of the development of inflammation and an overproduction of mucus in bronchi obstructing normal airflow and causes difficulty to breathe (Khatami, 2012). It is known that asthma can be produced by the influence of some factors such as inhalation of allergenic compounds such as dust, pollen, and spores. Besides, smoking and genetic predisposition have an important influence as well (Lambrecht and Hammad, 2015). During an asthma attack, there is concentrated a great number of macrophages and mast cells in lung tissues that release bronchoconstrictor mediators and cytokines such as IL-4, IL-5, IL-9, and IL-13. In this case, these interleukins regulate inflammation by eosinophils, a type of leukocytes that are activated specifically in the presence of allergic molecules and parasites, whose function is to induce the release of enzymes that participate in the elimination of antigens but affect epithelial tissue of bronchi and amplify immune response (Rodríguez et al., 2017).

11.1.4.3 ALZHEIMER

Alzheimer is a pathology characterized by the development of neurological damage and a progressive loss of cognitive capacities of patients (Armenteros, 2017). Although it is not considered that inflammation is responsible for its origin, it is known that people with Alzheimer show a chronic inflammatory process on brain tissues due to the presence of β-amyloid peptide, an element that is characterized in Alzheimer disease (Heppner et al., 2015). In the brain are located a certain type of macrophages called microglia. These cells are very sensitive to any change in brain tissues, so when the β-amyloid peptide is detected, microglia starts phagocytosis. However, this produces some negative effects that can be generated as a consequence of the immune response. The principal effect is cytotoxicity produced in neurons due to a high concentration of proinflammatory cytokines such as IL-1, IL-6, and TNF-α that increase inflammatory response (Wang et al., 2015).

11.2 FREE RADICALS DAMAGE

If the inflammation process fails to be regulated by the immune system pathways, it can accumulate free radicals (Zaidun et al., 2018). Free radicals are molecules with a disappeared electron that make them unstable and highly reactive. This fact produces free radicals that can react with other compounds to obtain electrons from them and be stable; however, this causes the formation of new unstable compounds that affect molecules such as lipids, proteins, and deoxyribonucleic acid (DNA), and as a consequence of this, irreversible alterations in cells function occur and cell death is produced. It is known that there are many types of free radicals; some of them are molecules that have oxygen as a basic element and are named reactive oxygen species (ROS). They are compounded by many chemical species but some of the most important are hydroxyl radical (OH), superoxide anion (O_2^-), and hydrogen peroxide (H_2O_2) (Al-Kharashi, 2018).

11.2.1 PRINCIPAL SOURCES OF ROS

The human body has different mechanisms responsible for ROS production, one of them is the mitochondrial respiratory chain complex, which is an important element of the oxidative metabolism of glucose and destinate approximately 10% of the total O_2 consumed by the cell to generate ROS (Maldonado et al., 2010). As it was mentioned before, leukocytes are other important sources of free radicals. In particular, neutrophils, monocytes, and macrophages are the principal cells responsible for ROS production because during phagocytosis, these molecules are produced as secondary compounds, thus their concentration increases significantly on the inflammatory response (Venero, 2002). The presence of ROS in the human body can also be attributed to external factors such as environmental pollution, exposition to ionizing radiation, pesticides, and herbicides. Moreover, smoking and administration of certain types of medical treatments and consumption of chemical additives on food represent other important sources of ROS (Coronado et al., 2015).

11.2.2 PHYSIOLOGICAL FUNCTIONS OF ROS

At normal conditions, ROS produce some physiological functions in the organism. It is known that they participate in synthesis of collagen, PTD, and cholesterol (Guija et al., 2015). Furthermore, ROS contribute to the activation of enzymes of cell membrane involved in signal transduction and regulation of blood vessel tone in which NO has an important role due to its vasodilator properties and capacity of inducing synthesis of vascular smooth muscle (Maldonado et al., 2010). Other functions of ROS are related to inflammatory response where these compounds have cytotoxic effect on pathogenic microorganisms (Sánchez and Méndez, 2013).

11.2.3 ANTIOXIDANT SYSTEMS

Every organism has natural mechanisms that participate in the regulation of ROS production, preventing the development of negative effects on health. Regarding this, antioxidants have an important role. Antioxidants are compounds responsible for neutralizing free radicals avoiding interaction with important cell molecules (Guija et al., 2015). Antioxidants are classified into two types: enzymatic and nonenzymatic antioxidants. Enzymatic antioxidants or also known as antioxidant enzymes, involve enzymes such as catalase (CAT), which converts hydrogen peroxide into water and oxygen, superoxide dismutase, which converts superoxide into oxygen and hydrogen peroxide, and glutathione peroxidase, which catalyzes the reduction of peroxide radicals to alcohols and oxygen (Ighodaro and Akinloye, 2018). These enzymes constitute the first cell defense against oxidative damage. On the other hand, nonenzymatic antioxidants are compounds that neutralize all free radicals that could escape from antioxidant enzymes, some of the most common are transferrin, lactoferrin, bilirubin, and vitamins C and E. It is important to mention that antioxidant compounds are not only produced by cells but also obtained from food (Avello and Suwalsky, 2006; Maldonado et al., 2010). It has been demonstrated that different food products such as wine, vegetables, red fruits, and cereals have a high concentration of antioxidants (Herman et al., 2015).

11.2.4 OXIDATIVE STRESS

The production of high levels of ROS generates many alterations on cells causing a phenomenon called oxidative stress (Herman et al., 2015) in which molecules, especially lipids, proteins, and DNA, are mainly affected (Figure 11.3).

11.2.4.1 LIPIDS

Lipids are molecules of great importance because of the different functions they have in the organism such as storing energy and signaling between cells (Penkauskas and Preta, 2019). Lipids are part of the components of plasma membrane and intracellular membranes of some organelles in which they form a double-layered surface that protects the cell avoiding the free entrance of components from extracellular medium that may produce negative effects inside the cell (Escribá et al., 2015). However, when there is a high concentration of ROS, lipids are one of the first molecules to be affected by them. In this case, ROS causes a lipid peroxidation that is the process in which lipids have an oxidative damage because of the loss of their electrons by the action of free radicals. In this manner, it induces a chain reaction of unstable compounds that produce damage of cell membrane causing alterations of permeability that induces cell death (Gaschler and Stockwell, 2017).

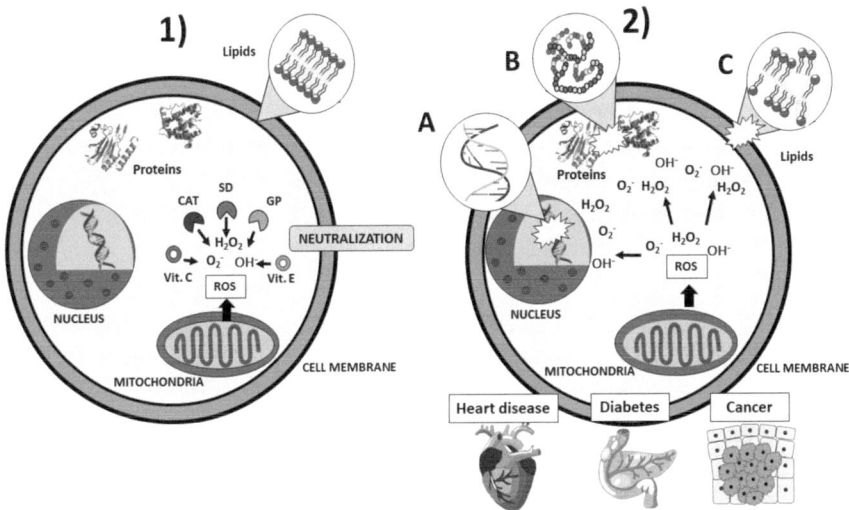

FIGURE 11.3 **Comparison between normal ROS regulation and ROS overproduction in cells**. (1) Cells produce ROS species in mitochondria such as OH-, O_2^- y H_2O_2 which are neutralized by enzymatic antioxidants (CAT, SOD and GP) and non-enzymatic antioxidants (Vit. C and E) that inhibit radical effects. (2) During an overproduction of ROS due to oxidative stress, antioxidants are not enough to inhibit them. Therefore, there are various cell damages such as (A) DNA mutations, (B) oxidation of amino acids and crisscross of peptide chains, C) lipid peroxidation of cell membrane. These alterations along with other factors, induce the development of chronic diseases such as heart disease, diabetes and cancer. (Alfaro-Jiménez, 2019).

11.2.4.2 PROTEINS

Proteins are complex molecules constituted by units called amino acids, which are bound to one another in long chains. The sequence of these amino acids determines the specific structure and functions of proteins (Gupta and Varadarajan, 2018). Also, proteins have an important role in the body because they can work as enzymes, messenger proteins, antibodies, and structural components of cell organelles and tissues. When proteins react with ROS, several alterations on amino acids are produced, especially an oxidation of the amino acids such as phenylalanine, tyrosine, and histidine and the formation of multiple bonds of peptide chains (Coronado et al., 2015). Alterations on proteins cause structural modifications, fragmentation, aggregation, decrease or loss of biological functions, and proteolytic susceptibility (Houée-Levin and Bobrowski, 2013).

11.2.4.3 DNA

DNA is a macromolecule located in the cell nucleus that contains the hereditary material of humans and other organisms. All this information is determined by a specific sequence of four nitrogenous bases, which are adenine, guanine, cytosine, and thymine (Malewicz and Perlmann, 2014). When DNA is exposed to ROS, it may suffer damages on theses bases producing many structural alterations. Normally, the cells have their own protective mechanisms to repair mutations on DNA; however, the presence of a high concentration of ROS may be very dangerous (González-Hunt et al., 2018). Consequently, this fact induces the formation of mutations on the sequence of nucleotides, loss of gene expression, and chromosomal rearrangements that avoid the formal functions of cells and may induce death cell (Cadet and Davies, 2017).

11.2.5 DISEASES RELATED TO OXIDATIVE STRESS

The different alterations produced by the high reactivity of ROS cause the development of several chronic degenerative diseases (Sofi and Dinu, 2016). Some of the most important are heart disease, diabetes, and cancer (Figure 11.3) because these are responsible for the death of millions of people worldwide (Coronado et al., 2015). The description of each disease is shown below.

11.2.5.1 HEART DISEASE

The term heart disease involves a group of sickness related to heart and vascular system whose origin is related to an accumulation of a plaque inside blood vessels that is produced by the interaction between ROS and low-density lipoproteins present on vascular lumen. This plaque produces a disfunction of endothelial tissue and loss of the capacity of modulating vasodilation and vasoconstriction of blood vessels, regulation of platelet aggregation, and adhesion of leukocytes to endothelial tissues, increasing the risk of heart attacks (Jackson et al., 2018).

11.2.5.2 DIABETES

Diabetes is a disease characterized by a high concentration of glucose in blood due to a defect in secretion or activity of insulin, a hormone produced in pancreas that is responsible for keeping normal values of this monosaccharide in blood (Feng et al., 2017). These defects cause an activation of several metabolic pathways on cells that induces the production of a higher amount of ROS which generate cytotoxic effects in different types of cells such as pancreatic cells which are responsible for production of insulin. The destruction of pancreatic cells causes a reduction in insulin production. Therefore, diabetes continues progressing (Maldonado et al., 2010).

11.2.5.3 CANCER

Cancer is a pathology characterized by an abnormal cell proliferation responsible for the formation of tissue mass called tumor. Depending on cell growth, tumors can be classified in benign and malignant tumor. Regarding malignant tumors, the cells that constituted them may appear on a tissue and migrate to different parts of the organisms, creating new tumor focus that can affect the health of patients and even can cause their death. According to World Health Organization (WHO), cancer is considered as the second cause of death in the world. In 2018, there were reported 18 million cases of patients with cancer and an average of 9.6 million of deaths (WHO, 2018). Nowadays, there are reported more than 100 types of cancer of which lung, liver, colon, stomach, and breast cancers are the most common worldwide (WHO, 2018). Inflammation and oxidative stress are two factors correlated with the process of transformation of a normal cell to a tumor cell. This influence is based on the damage that can be produced in DNA, especially in genes responsible for regulating cell cycle (Singh et al., 2019). Cell cycle is a biological process constituted by four consecutive phases (G1, S, G2, and M phases) in which cell prepares itself to divide forming two new cells (Nagano et al., 2017). Normally, this process is regulated by the expression of specific genes such as protooncogenes, tumor suppressor genes, and DNA repair genes. These genes regulate the process of cell division. However, when these genes suffer mutations by free radicals or other cytotoxic compounds, cell cycle becomes abnormal and cells proliferate without control, forming tumors (Singh et al., 2019). Each cell has its own mechanism to prevent the formation of more cancer cells. One of them is the induction of apoptotic cell death. In this process, cell causes its own death through caspases.

11.3 CONVENTIONAL TREATMENTS AND THEIR SIDE EFFECTS

Nowadays, there are different drugs to treat diseases related to inflammatory, oxidative, and tumor processes. Regarding inflammatory diseases, corticosteroids such as cortisone, hydrocortisone, and prednisone are generally used in chronic inflammation because they have the same effect as normal hormones produced by kidney glands with anti-inflammatory activity (García et al., 2010). However, it is reported that the use of these drugs can produce side effects in patients. Some of them are high blood pressure, hyperglycemia, osteoporosis confusion, delirium, hepatic, and kidney damage (Manson et al., 2009).

In the case of cancer, patients are treated with chemotherapy, but it has been reported that chemotherapeutic drugs have a high cytotoxic level that also affects normal cells such as blood cells like erythrocytes and leukocytes. Consequently, this produces side effects such as anemia and immunosuppression (Rodgers et al., 2012; Smallwood et al., 2018).

11.4 ASTERACEAE PLANTS WITH BIOLOGICAL PROPERTIES

Due to the side effects produced by actual treatments, in the last years, the research of new compounds from natural sources has become an important objective of science. Since a long time ago, plants are famous for showing to be useful in the treatment of several health issues (Chemat et al., 2015). In addition, it has been discovered that plants contain compounds that derivate from vegetal metabolism that are not essential in primary functions of plants and only participate in secondary processes. These are called secondary metabolites. These metabolites are classified in families such as saponins, tannins, terpenes, phenolic acids, anthocyanins, and flavonoids and can be extracted with water, methanol, ethanol, and other solvents (Toyanga and Verpoortec, 2013; Piasecka et al., 2015). Previous studies report secondary metabolites that show biological properties in human's health such as anti-inflammatory, antioxidant, and antitumor activities (Chemat et al., 2015).

Asteraceae, also called Compositae, is one of the largest plant families around the world and its members are known to have flower heads composed of many small flowers, called florets, that are surrounded by bracts (García-Risco et al., 2017). Previous reports show that many Asteraceae plants exhibit anti-inflammatory, antioxidant, and antitumor properties in vitro and in vivo. Some of them are *Achillea millefolium, Calendula officinalis, Chromolaena odorata, Cichorium intybus, Conyza canadensis, Eclipta prostrata,* and *Lactuca sativa,* which have demonstrated to have these properties in vitro and in vivo (Table 11.3) and are described below.

11.4.1 ACHILLEA MILLEFOLIUM

Achillea millefolium, commonly known as yarrow, is a plant found in mountain meadows, pathways, crop fields, and home gardens (Benedek and Kopp, 2007). Traditionally, *A. millefolium* is used in a broad range of medical applications such as in treating of inflammatory and spasmodic gastrointestinal disorders (Ahmadi-Dastgerdi et al., 2017), diabetes, hepatobiliary diseases, amenorrhea (El-Kalamouni et al., 2017), wound healing, and skin inflammations (Hatsuko-Baggio, 2015).

In previous reports, different extracts of *A. millefolium* have shown to contain different secondary metabolites with anti-inflammatory activity. Methanolic extracts of aerial parts of yarrow have been found to contain flavonoids (apigenin-7-*O*-glucoside, rutin, and luteolin-7-*O*-glucoside) and dicaffeoylquinic acids involved in the capacity of extracts to inhibit in vitro human neutrophil inflammatory enzymes such as elastase (IC_{50}: 23 µg/mL) and matrix metalloproteinases (IC_{50}: 800 µg/mL) (Benedeck et al., 2007). Inhibition of NO generation in murine macrophage in vitro model has also been found to be produced by guaiane acid derivatives, such as 7βH-guaia-3,9,11(13)-trien-12-oic acid and arteludovicinolide-A isolated from dichloromethane–methanol extracts, exhibiting IC_{50} values of 23.7 and 19.6 µg/mL, respectively (Hegazy et al., 2008). Besides, aqueous extracts have also exhibited an in vitro NO inhibition (IC_{50}: 50 µg/mL) (Burk et al., 2010). On the other hand, in vivo anti-inflammatory activity has been exhibited by hydroalcoholic extracts (10 mg/kg) through the inhibition (68%) of the activity of markers of neutrophil infiltration in chronic gastric ulcer induced rats (Potrich et al., 2010).

TABLE 11.3 Previous studies of biological properties of Asteraceae plants

Plant	Extract	Phytochemicals identified	Biological activity	Assay	Reference
Achillea millefolium	Methanolic	Flavonoids Dicaffeoylquinic acids	Anti-inflammatory	Elastase assay Metalloproteinase assay	Benedek et al., 2007
	Dichloromethane - methanol	Guaiane acid derivatives		Griess assay	Hegazya et al., 2008
	Hydroalcoholic	N/A		Myeloperoxidase assay	Potrich et al., 2010
	Aqueous extract	Phenolic compounds Terpenoids		In vitro inhibitory LPS-induced cytokine secretion in RAW 264.7 cells	Burk et al., 2010
	Essential oils	Eucalyptol Eamphor α-terpineol α-pinene Borneol	Antioxidant	DPPH assay TBARS assay	Candan et al., 2003
	Aqueous extract	Phenolic compounds Terpenoids		DPPH assay DCFH-DA assay	Burk et al., 2010
	Essential oils	Carvacrol Thymol α-pinene		DPPH assay	Kazemi, 2015
	Essential oils	Camphor Terpinen-4-ol Sesquiterpenes Germacrene-D (E)-nerolidol		Oxipress method	El-Kalamouni et al., 2017

TABLE 11.3 *(Continued)*

Plant	Extract	Phytochemicals identified	Biological activity	Assay	Reference
	Aqueous	Apigenin Luteolin, Centaureidin Casticin Sesquiterpenoids	Antitumor	MTT assay	Csupor-Löffler et al., 2009
Achillea millefolium	Aqueous, Methanolic Ethanolic	N/A	Antitumor	MTT assay	Amini Navaie et al. 2015
	Hydroethanolic	Caffeoylquinic acid derivatives Apigenin Luteolin derivatives Quercetin Kaempferol		Sulforhodamine B assay Trypan blue assay Flow cytometry Western blot analysis	Pereira et al., 2018
Ambrosia artemisiifolia	Aqueous	Flavonoids Isorhamnetin kaempferol	Antioxidant	DPPH assay ABTS assay	Mihajlovic et al., 2014
Ambrosia spp. A. Tenuifolia A. scabra A. elatior	N/A	Sesquiterpene lactones Psilostachyin Psilostachyin C, Peruvin Cumanin	Antioxidant Antitumor	Fluorescent probe H2DCF-DA MTT assay	Martino et al., 2015
Ambrosia psilostachya	Methanolic	Sesquiterpene lactones Ambrosanolide	Anti-inflammatory	Griess assay	Lastra et al., 2004
Artemisia biennis	Hydroethanolic	N/A	Antioxidant	Flow cytometry	Mojarrab et al. 2016

TABLE 11.3 (Continued)

Plant	Extract	Phytochemicals identified	Biological activity	Assay	Reference
Artemisia herba-alba	Essential oils	Monoterpenes β-thujone α-thujone camphor 1,8 cineole	Antioxidant	DPPH assay FRAP assay	Younsi et al., 2016
	Methanolic	Phenolic compounds Apigenin-6-C-glycosyl Chlorogenic acid 1,4 dicaffeoylquinic acid			
Baccharis trimera	Hydroethanolic	Flavonoids Quercetin Caffeoylquinic acid Hyperoside Luteolin Hispidulin	Anti-inflammatory	COX-1 screening kit	Chagas et al., 2015
Baccharis uncinella	Dichloromethane	Flavonoids Triterpenes Pectolinaringenin Oleanolic acid Ursolic acids	Anti-inflammatory	In vivo phospholipase A2 induced hind paw edema in rats	Zalewski et al., 2011
Brickellia veronicaefolia	Chloroform	Flavonols Centaureidin	Antioxidant	DPPH assay	Pérez et al., 2004
Calendula officinalis	CO_2	α-amyrin Lupeol Taraxasterol Calenduladiol	Anti-inflammatory	In vivo croton oil ear test in mice	Loggia et al., 1994

TABLE 11.3 (Continued)

Plant	Extract	Phytochemicals identified	Biological activity	Assay	Reference
	Methanolic	Helianol Taraxerol, Dammaradienol, β-Amyrin, Cycloartenol Taraxerol tirucalla -7,24-dieno dammaradienol.		TPA-induced inflammation in mice	Akihisa et al., 1996
		Triterpene glycosides Oleanane Calendulaglycoside Isorhamnetin Quercetin			Ukiya et al., 2006
	Butanolic	Flavonoids terpenoids.	Antioxidant	TRAP assay TBARS assay	Cordova et al., 2002
Calendula officinalis	Ethyl-alcohol.	Flavonoids Triterpenoids Saponins Carotenoids Coumarins	Antioxidant	NBT reduction method TBARS assay DPPH assay ABTS assay Griess assay *In vivo* enzymatic antioxidant activity in mice (SOD, CAT, GP)	Preethia et al., 2006
	Aqueous	N/A	Antitumor	BrdU kit	Jiménez et al., 2006.
	Methanolic	Triterpene glycosides Oleanane Calendulaglycoside Isorhamnetin Quercetin		Sulforhodamine B assay	Ukiya et al., 2006

TABLE 11.3 (Continued)

Plant	Extract	Phytochemicals identified	Biological activity	Assay	Reference
Centaurea americana	Methanolic	Phenolic compounds	Antioxidant	DPPH assay	Salazar et al., 2008
	Hydroalcoholic	N/A	Antitumor	MTT assay	Torres et al., 2011
Centaurea depressa	Methanolic	Flavonoids	Antioxidant	DPPH assay	Hosseinimehr et al., 2007
Centaurea pamphylica		Sesquiterpenes			Shoeb et al., 2007
Centaurea melitensis		Sesquiterpene lactones Isobutyroylsaloniteloide Actiopicrin Onopordicrin Melitensine			Ayad et al., 2012
Chromolaena odorata	Aqueous	Flavonols Quercetin kaempferol	Anti-inflammatory	*In vivo* inhibitory activity of carrageenan induced paw edema and cotton pellet granulome in rats	Owoyele et al., 2005
	Dichloromethane n-Butanol Ethyl acetate	Flavonoids			Owoyele et al., 2007
	Ethyl acetate	Chalcones Flavanones		Griess assay	Dhar et al., 2018
	Aqueous Methanolic	Tannins Steroids Terpenoids Flavonoids Cardiac glycosides	Antioxidant	DPPH assay	Akinmoladun et al., 2007

TABLE 11.3 (Continued)

Plant	Extract	Phytochemicals identified	Biological activity	Assay	Reference
	Chloroform	Phenolic compounds		DPPH assay ABTS assay TBARS assay Griess assay	Rao et al., 2010
	n-hexane	Flavanones Chalcones	Antitumor	CellTiter-Glo luminescent cell viability assay	Kouamé et al., 2012
	Methanolic	Phenolic compounds		MTT assay	Adedapo et al., 2016
Chrysactinia mexicana	Ethanolic	N/A	Anti-inflammatory	In vivo inhibitory activity of LPS induced inflammation in chicks	García-López et al., 2017
Cichorium intybus		Sesquiterpene lactones		In vivo carrageenan induced paw edema in rats Quantitative real-time expression of proinflammatory cytokines by RT-PCR	Ripoll et al., 2007
Cichorium intybus	Aqueous Ethanolic	Phenolic compounds Flavonoids Sterols Glycosides Tannins Terpenoids	Anti-inflammatory	In vivo inhibitory activity of carrageenan induced paw edema and cotton pellet granulome in rats Cytokines production TBARS assay In vitro antioxidant enzyme activity by spectrophotometry	Rizvi et al., 2014

TABLE 11.3 (Continued)

Plant	Extract	Phytochemicals identified	Biological activity	Assay	Reference
	Ethanolic	Hydroxycinnamic acids Quercetin Kaempferol Luteolin Apigenin glycosides	Antioxidant	DPPH assay	Heimler et al., 2009
	Hydroalcoholic	Tannins Saponins Flavonoids			Abbas et al., 2015
	Hydroalcoholic	Phenolic compounds	Antitumor	Sulforhodamine B assay	Conforti et al., 2008
	n-hexane	Triterpenoids		Trypan blue exclusion MTS assay Annexin V-FITC	Saleem et al., 2014
	Methanolic	N/A		MTT assay	Mehrandish et al., 2017
Conyza canadensis	Methanolic	Flavonoids Terpenoids Tannins	Antioxidant	DPPH assay	Hayet et al., 2009
Conyza canadensis	Methanolic	Saponins Tannins Flavonoids Steroids Glycosides Diterpenoids Triterpenoids	Antioxidant	DPPH assay	Shah et al., 2012

TABLE 11.3 (Continued)

Plant	Extract	Phytochemicals identified	Biological activity	Assay	Reference
	Essential oils	Limonene (Z)-lachnophyllum ester (E)-β-ocimene β-pinene (E)-β-farnesene		DPPH assay TBARS assay.	Lateef et al., 2018
	Ethyl acetate	Flavonoids Terpenoids Tannins	Antitumor	Methylene blue assay	Hayet et al., 2009
	Methanolic	Conyzapyranone derivatives Lactone derivatives Epifriedelanol Apigenin β-sitosterol		MTT assay	Csupor-Löffer et al., 2011
	n-hexane fraction of methanolic extract	Conyzapyranone A Conyzapyranone B 2 γ-lactone acetylene derivatives Triterpenes Sterols Flavonoids		Sulforhodamine B assay	Ayaz et al., 2016
Conyza bonariensis	Essential oils	Monoterpenes and sesquiterpenes	Anti-inflammatory	In vivo inhibitory activity of LPS induced inflammation in mice	Souza et al., 2003
Conyza bonariensis	Essential oils	Monoterpenes and sesquiterpenes	Anti-inflammatory	Griess assay ELISA assay	Souza et al., 2003

TABLE 11.3 *(Continued)*

Plant	Extract	Phytochemicals identified	Biological activity	Assay	Reference
Coreopsis tinctoria	Ethanolic	Phenolic compounds Chlorogenic acid Caffeic acid Marein Luteolin Apigenin	Antioxidant	DPPH assay ABTS assay	Zălaru et al., 2014
	Essential oils	Terpenes Limonene Carvone α-pinene Linalool Linalyl acetate α-bergamotene Carveol α-curcumene		DPPH assay ABTS assay FRAP assay	An et al., 2018
Cosmos bipinnatus	Methanolic	Sesquiterpene lactones Dihydrocallitrisin Isohelenin	Anti-inflammatory	Griess assay iNOS and COX-2 expression ELISA assay	Sohn et al., 2013.
		N/A	Antioxidant	DPPH assay ABTS assay	Jang et al., 2008
Cynara cardunculus	Methanolic	Flavonoids Tannins	Antioxidant	DPPH assay	Falleh et al., 2008

TABLE 11.3 (Continued)

Plant	Extract	Phytochemicals identified	Biological activity	Assay	Reference
	Chloroform Ethyl acetate, Ethanol Buthanol	Phenolic compounds		DPPH assay FRAP assay	Kukic et al., 2008
Eclipta prostrata	Methanolic	Steroids Triterpenoids Flavonoids	Anti-inflammatory	In vivo carrageenan and egg white induced hind paw edema in rats	Arunachalam et al., 2009
Eclipta prostrata	Dichloromethane-methanol	Terthiophene derivatives Orobol Wedelolactone Ecliptal	Anti-inflammatory	In vitro evaluation of decrease of proinflammatory cytokines levels produced by RAW264.7 cells	Tewtrakul et al., 2011
	Ethyl acetate	Triterpenoids			Ryu et al., 2012
	Ethanolic Ethyl acetate	Flavanones Flavanoids	Antioxidant	Ferric thiocynate assay	Karthikumar et al., 2007
	Aqueous	Phenolic compounds		TBARS assay DPPH assay TBARS assay	Rao et al., 2009
	Aqueous Ethanolic	Flavonoids Hydrolysable tannins		DPPH assay Protective oxidative protein damage assay	Pukumpuang et al., 2014
	Methanolic	Saponins Dasyscyphin C	Antitumor	MTT assay	Khanna and Kannabiran, 2009
	Aqueous Ethanolic	Lactones Flavones Saponins			Liu et al., 2012

TABLE 11.3 (Continued)

Plant	Extract	Phytochemicals identified	Biological activity	Assay	Reference
	Ethanolic	Terthiophenes Saponins Triterpenoids Flavonoids			Kim et al., 2015
Flaveria trinervia	Chloroform	N/A	Anti inflammatory	In vivo carrageenan induced hind paw edema in rats	Malathi et al., 2012
Fluorensia mycrophylla	Acetone Ethanolic	Phenolic compounds Flavonoids Salicilic acid Gallic acid Ascorbic acid Caffeic acid	Anti-inflammatory Antioxidant Antitumor	Western blot assay DPPH assay ORAC assay MTT assay	Jasso et al., 2017
Gymnosperma glutinosum	Hexane Methylene Chloride	N/A	Antitumor	MTT assay In vivo tumor induced in mice	Gómez-Flores et al., 2009
	Hexane, chloroform Methanol	Diterpenes		MTT assay	Gómez et al., 2012
Heteroteca subaxillaris	Petroleum ether, Dichloromethane Methanol	Flavonoids Santin Pectolinaringenin Hispidulin	Anti-inflammatory	In vivo carrageenan-induced paw edema and TPA induced ear edema in mice	Gorzalczany et al., 2009
Lactuca sativa	Methanolic	Triterpene lactones		In vivo inhibitory activity of carrageenan induced paw edema in rats	Araruna and Carlos, 2010

TABLE 11.3 (Continued)

Plant	Extract	Phytochemicals identified	Biological activity	Assay	Reference
	Aqueous	Flavonoids		Griess assay	Ismail and Miza, 2015
	Methanol-Hydrochloric acid	Polyphenolic compounds		In vivo evaluation of antioxidant enzymes	Adesso et al., 2016
	Ethanolic	Phenolic compounds	Antioxidant	In vivo evaluation of antioxidant enzymes	Garg et al., 2004
	Aqueous	Anthocyanins Cyanidin derivates Quercetin glucoside Ferulic acid Caffeic acid		Fluorescent assay	Mulabagal et al., 2010
Lactuca sativa	Aqueous Methanolic	Phenolic compounds	Antioxidant	DPPH assay	Edziri et al., 2011
	Aqueous	Phenolic compounds	Antitumor	In vitro proliferative inhibition assay	Gridling et al., 2010
		Anthocyanins Flavones Phenolic acids		MTT assay Trypan blue staining	Qin et al., 2018
Parthenium argentatum	Resin	Triterpenes Argentatin A Argentatin B	Anti-inflammatory	Griess assay	Romero et al., 2014
Pluchea odorata	Hydroalcoholic	Phenolic compounds Flavonoids	Antioxidant	DPPH assay ABTS assay ORAC assay	Perera et al., 2009

TABLE 11.3 (Continued)

Plant	Extract	Phytochemicals identified	Biological activity	Assay	Reference
Senesio saligmus	Chloroform	N/A	Anti-inflammatory	In vivo inhibitory activity of carrageenan induced paw edema in rats	Pérez et al., 2013
Sonchus asper	Hexane Ethyl acetate, Chloroform Methanolic	Phenolic compounds Flavonoids	Antioxidant	DPPH assay ABTS assay Hydroxyl scavenging assay Superoxide scavenging assay Hydrogen peroxide scavenging assay	Khan et al., 2012

N/A: Not available

Aqueous extracts and essential oils have demonstrated antioxidant activity in different in vitro assays. This biological property is correlated with their phytochemical composition. The aqueous extract of the flowers (300 µg/mL) inhibits LPS-induced intracellular ROS production in vitro (45%). In addition, previous reports determined the presence of phenolic compounds and terpenoids as antioxidant metabolites in aqueous extract (Burk et al., 2010). Likewise, essential oils from aerial parts of the plant have shown antioxidant activity by inhibition of 2,2-diphenyl-1-picryl-hydrazyl-hydrate (DPPH) (IC_{50}: 1.56 µg/mL) and hydroxyl (IC_{50}: 2.70 µg/mL) radicals and decrease of in vitro oxidative damage (IC_{50}: 1.51 µg/mL) Regarding phytochemical constituents, terpenes (eucalyptol, borneol, α-pinene, thymol, camphor, α-terpineol, carvacrol) are identified as principal secondary metabolites found in essential oils of *A. millefolium* (Candan et al., 2003, Kazemi, 2015; El-Kalamouni et al., 2017).

Furthermore, extracts from *A. millefolium* have demonstrated cytotoxic effect on tumor cells. It is reported the inhibition of cell proliferation of human breast cancer cells (MCF-7) in vitro by aqueous (IC_{50}: 92.04 µg/mL), methanolic (IC_{50}: 14.60 µg/mL), and ethanolic (IC_{50}: 14.60 µg/mL) extracts (Amini Navaie et al., 2015). In addition, sesquiterpenoids such as paulitin and isopaulitin, isolated from aqueous extract, have demonstrated to decrease cell proliferation of cervix adenocarcinoma (HeLa) and lung cancer (A-431). Paulitin exhibits IC_{50} values of 4.76 and 11.82 µM, respectively, while isopaulatin exhibits 1.48 and 6.95 µM (Csupor-Löffler et al., 2009). On the other hand, hydroethanolic extracts decrease cell viability on nonsmall cell lung cancer (NCI-H460) and human colorectal adenocarcinoma (HCT-15) with IC_{50} values of 187.3 and 70.83 µg/mL. Besides, hydroethalic extract (100 µg/mL) has shown to cause an arrest in G2/M phase of cell cycle in both cell lines and an increase of the expression of p53 (150%) and p21(200%) apoptotic proteins, but only in NCI-H460 cells. This biological activity of hydroethanolic extract is attributed to the phytochemical composition constituted mainly by phenolic acids (3,5-*O*-dicaffeoylquinic acid and 5-*O*-caffeoylquinic acid) and flavonoids (luteolin-*O*-acetylhexoside and apigenin-O-acetylhexoside) (Pereira et al., 2018).

11.4.2 CALENDULA OFFICINALIS

Calendula officinalis, also known as marigold, is an annual herbaceous plant native of the Mediterranean climate areas (Martin et al., 2016), which has been employed for traditional medicinal due to its anti-inflammatory, antioxidant antibacterial, antifungal, and antiviral properties (Chandran and Kuttan, 2008; Babaee et al., 2013).

Various studies have reported biological properties of phytochemical compounds isolated from methanolic extracts of *C. officinalis*. In vivo assays have shown the anti-inflammatory activity of triterpene alcohols from *C. officinalis* flowers (helianol, taraxasterol, α-amyrin, lupeol, taraxerol, and cycloartenol) by reducing 12-*O*-tetradecanoylphorbol-13-acetate ear-induced inflammation in mice. Of these metabolites, helianol (ID_{50}: 0.1 mg per ear) and α-amyrin (ID_{50}: 0.2 mg per ear) have shown similar anti-inflammatory activity than indomethacin (ID_{50}: 0.3 mg per ear) as reference compound (Akihisa et al., 1996). Moreover, triterpene glycosides, such as calendulagycoside A (ID_{50}: 0.06 mg per ear) and calendulagycoside B (ID_{50}: 0.05 mg per ear), have also shown anti-inflammatory effects (Ukiya et al., 2006). Likewise, a fraction isolated from the CO_2 extract of Calendula flowers (900 µg/cm^2), enriched in triterpenoid compounds (faradiol, arnidiol, calenduladiol, lupeol, taraxasterol, α-amyrin, β-amyrirn), has exhibited an anti-inflammatory activity by its capacity of reducing (54.8 %) ear inflammation induced by croton oil in mice (Loggia et al., 1994).

On the other hand, antioxidant activity has also been evaluated in different extracts obtained from Calendula flowers. The presence of glycoside flavonoids and terpenoids has been correlated with an antioxidant effect of butanolic extracts, which have shown the capacity of inhibiting in vitro lipid peroxidation in liver microsomes induced by Fe^{2+}/ascorbate (IC_{50}:0.15 mg/mL). Besides, extracts also decrease superoxide (IC_{50}:1.0 mg/mL) and hydroxyl radical levels (IC_{50}:0.50 mg/mL) in vitro (Cordova et al., 2002). Likewise, ethyl-alcoholic extracts show an in vitro scavenging activity of DPPH (IC_{50}: 100 mg/mL), 2,2′-azino-bis (3-ethylbenzothiazoline-6-sulfonic acid) (ABTS) (IC_{50}: 6.5 mg/mL) and NO (575 mg/mL) radicals. Moreover, inhibitory effects of lipid peroxidation (IC_{50}: 480 mg/mL) and superoxide production (IC_{50}: 500 mg/mL) have also been found in these extracts. Ethyl-alcoholic extracts (250 mg/kg) have shown to increase CAT activity (42.84%) and glutathione levels (66.49%) in blood and liver of mice (Preethi et al., 2006). Concerning that, previous studies have reported carotenoids (flavoxanthin, luteoxanthin, lycopene, and β-carotene), flavonoids (lutein, quercetin, and protocatechuic acid), triterpenoids, coumarins, and saponins as secondary metabolites commonly found in *C. officinalis* flowers and are responsible for antioxidant activity (Matysik et al., 2005; Kishimoto et al., 2005).

Triterpene glycosides (calendulaglycoside and calenduloside derivatives) and flavonol glucosides (isorhamnetin and quercetin derivatives) isolated from the flowers *of C. officinalis* by methanolic extraction have also exhibited cytotoxic activity in human tumor cell lines derived from cancer colon (HCC-2998), melanoma (LOX IMVI, SK-MEL-5, and UACC-62), renal (CAKI-1 and UO-31), ovarian (IGR-OV1), breast (MCF-7), and leukemia (MOLT-4 and RPMI-8226). Of all compounds presented in methanolic extracts, calenduloside F 6'-O-n butyl ester and calenduloside G 6'-O-methyl ester exhibited IC$_{50}$ values less than 10 and 20 µM, respectively, in all tumor cell lines (Ukiya et al., 2006). Similarly, aqueous flower extracts (250 µg/mL) also showed in vitro growth inhibition (expressed in percentages) in various tumor cell lines derived from lymphoma (U-937, 21%), leukemia (Jurkat T, 100%), melanoma (ANDO-2, 100%), fibrosarcoma (B9, 72%), and cancers of breast (MB-231, 100%), prostate (DU-145, 72%), cervix (HeLa, 100%), lung (A-549, 100%), pancreas (IMIN-PC1, 100%), and colorectal (DLD1, 83%). The inhibitory mechanism is based on cell cycle arrest in G0/G1 phase and release of Caspase-3-induced apoptosis (Jiménez et al., 2006). Studies report that polysaccharides, proteins, fatty acids, carotenoids, flavonoids, triterpenoids, and saponins are main chemical components found in aqueous extracts of *C. officinalis* and have been strongly related to antitumor activity (Akihisa et al., 1998; Neukirch et al., 2004).

11.4.3 CHROMOLAENA ODORATA

Chromolaena odorata (Siam weed) is a perennial shrub native to South and Central America and introduced into regions of Asia, Africa, and other parts around the world (Dhar et al., 2017). *C. odorata* is known to be used for practitioners of traditional medicine due to anti-inflammatory, astringent, hepatotropic, and diuretic activities (Akinmoladun et al., 2007). Moreover, other uses include the treatment of wounds by its antiseptic properties (Kouamé et al., 2012).

It is reported that aqueous leaf extracts from *C. odorata* show a phytochemical composition constituted by flavonoids, saponins, flavonols (tamarixetin and kaemferide), flavonones, tannins, chalcones, among others (Nguyen and Le, 1993; Phan et al., 2001, Owoyele et al., 2005). These metabolites are responsible for the biological properties such as inflammatory activity, and based on this fact, aqueous leaf extracts (200 mg/kg) have demonstrated the capacity of reducing carrageenan-induced paw edema (80.5%) and cotton pellet granuloma (35.71%) in rats after 5 h of carrageenan injection and cotton implantation, respectively (Owoyele et al., 2005). Dichloromethane and ethyl acetate fractions (50 mg/kg) from an ethanolic extract of aerial parts of *C. odorata* have also exhibited an anti-inflammatory effect by reducing induced edema in rats (72% and 82%, respectively) after 5 h of carrageenan injection; being flavonoids, the compounds considerate as the anti-inflammatory agents (Owoyele et al., 2008). It is also reported the anti-inflammatory activity of a chalcone compound isolated from the ethyl acetate leaf extract named 2′,4-dihydroxy-3′,4′,6′-trimethoxychalcone, which has shown the capacity of decreasing the production of NO, TNF-α, IL-1, and IL-6, in LPS-activated RAW 264.7 macrophages in vitro, with IC_{50} values of 7.8, 6.46, 6.06, and 5.23 µM, respectively (Dhar et al., 2017).

On the other hand, extracts from *C. odorata* also exhibited antioxidant properties. Various secondary metabolites have been found in methanolic leaf extracts, such as alkaloids, tannins, steroids, terpenoids, and flavonoids, that have been associated with radical scavenging activity exhibited by these extracts (250 µg/mL), showing a significant (*$P<0.001$) percentage of antioxidant activity (30%) but lower than those obtained by gallic acid and ascorbic acid (80%) as references, in DPPH assay (Akinmoladu et al., 2007). Moreover, chloroform extracts with a phytochemical composition compounded mainly by phenolic compounds demonstrate to have the same radical scavenging capacity in different antioxidant in vitro assays, exhibiting in thiobarbituric acid reactive substances (TBARS), DPPH, ABTS, and Greiss assays, IC_{50} values of 0.43, 0.31, 0.28, and 1.32 mg/mL, respectively (Rao et al., 2010).

Antitumor activity is another biological property that has been reported in n-hexane fraction from ethanolic leaf extracts of *C. odorata* in which a chalcone compound (20-hydroxy-4,40,50,60-tetramethoxychalcone) has been isolated. This secondary metabolite (20 µM) shows to decrease cell viability in breast cancer cell lines, such as Cal51 (19%) and MCF-7 (18%), and cervical cancer HeLa cell line (30%). This antiproliferative effect is correlated with the induction of cell death by apoptosis (Kouamé et al., 2012). Methanolic leaf extracts (200 µg/mL) also exhibited in vitro cell proliferative inhibition (> 50%) in human colorectal adenocarcinoma cell line HT-29. Polyphenolic content of these extracts has been associated with their antitumor activity (Adepapo et al., 2016).

11.4.4 CICHORIUM INTYBUS

Cichorium intybus, commonly known as chicory, is an erect perennial herb native to Europe, North Africa, and Western Asia (Ripoll et al., 2007) and is considerate as an important medicinal herb for its uses in treatment of diseases of hepatobiliary and renal system (Zaman and Basar, 2013). In addition, chicory plant has also been reported to be used as digestive aid, diuretic, laxative, and mild sedative (Rizvi et al., 2014).

From all the parts of *C. intybus*, root is known to have biological properties such as anti-inflammatory activity. Ethanolic extracts of roots have shown to decrease in vitro pro-inflammatory gene expression of cytokines such as TNF-α, IL-1β, NO, and inducible nitric oxide synthase (iNOS), in macrophage cultures elicited with LPS, showing IC_{50} values of 48, 22, 21, and 39 µg/mL, respectively. Besides, ethanolic extracts (100 and 200 mg/kg) have demonstrated in vivo anti-inflammatory activity by decreasing carrageenan-induced edemas (76%) and collagen-induced arthritis (71%) in rat models. The presence of sesquiterpene lactones is associated with this anti-inflammatory activity (Ripoll et al., 2007). Furthermore, a previous study reported that both aqueous and ethanolic root extracts (500 mg/kg) also decrease IL-6 levels (35.13% and 23.25%, respectively) in serum of blood collected from carrageenan-induced paw edema rats (Rizvi et al., 2014). This fact can be attributed to polyphenols, flavonoids, sterols, glycosides, tannins, and terpenoids, which have been previously reported in chicory roots (Street et al., 2013; Mehmood et al., 2012).

Likewise, it has been found that *C. intybus* root extracts have antioxidant activity in radical scavenging in vitro assays. Ethanolic root extracts have shown DPPH radical scavenging activity (IC_{50}: 19.30 mg extract/mg DPPH), that is attributed to polyphenolic compounds such as caffeoyl tartaric acid, chlorogenic acid, caffeoyl and quercetin derivatives, kaempferol, luteolin, and apigenin (Heimler et al., 2009). On the other hand, hydroalcoholic leaf extract also shows a DPPH scavenging activity (IC_{50}: 67.2 µg/mL) influenced by the presence of tannins, saponins, and flavonoids identified in phytochemical screenings (Abbas et al., 2014).

Regarding antitumor activity, hydroalcoholic root extracts of *C. intybus* have demonstrated to produce an in vitro antiproliferative activity in various cancer cell lines, such as amelanotic melanoma (C32), renal adenocarcinoma (ACHN), and breast cancer (MCF-7), showing IC_{50} values of 30.78, 14.93, and 12.65 µg/mL, respectively. These effects have been correlated with the presence of phenolic compounds (Conforti et al., 2008). Moreover, n-hexane extracts from aerial parts of *C. intybus* have been reported to inhibit in vitro cell proliferation of lymphoblastic leukemia cells (Jurkat cells) with an IC_{50} value of 100 µg/mL (Saleem et al., 2014). The presence of volatile oils, fatty acids, and triterpenoids are associated with the antitumor activity of the extracts (Nandagopal and Kumari, 2007). Likewise, methanolic root extracts have been found to decrease cell viability (IC_{50}: 300 µg/mL) of human breast cancer cell line (SKBR3) in vitro (Mehrabdish et al., 2017).

11.4.5 CONYZA CANADENSIS

Conyza canadensis, also known as Canadian horseweed, is a plant species native to America but distributed around the world (Hayet et al., 2009). *C. canadensis* has been traditionally used for its pharmacological applications, including treatment of skin diseases such as sore smallpox, wounds, swellings, and pain produced by arthritis disease, and for diarrhea and dysentery, acting as a diuretic agent (Csupor-Loffler et al., 2011; Shah et al., 2012).

In vitro assays have demonstrated the antioxidant activity of methanolic extracts of *C. canadensis* due to its DPPH radical scavenging activity (IC_{50}: 120 μg/mL) correlated with the presence of flavonoids, terpenoids, tannins, saponins, steroids, diterpenoids, and triterpenoids (Hayet et al., 2009). In addition, hexane, ethyl acetate, and aqueous fractions obtained from methanolic extracts have been found to reduce DPPH radicals, with IC_{50} values of 50.35, 46.34, and 44.55 μg/mL, respectively (Shah et al., 2012). On the other hand, essential oils from aerial parts also exhibited antioxidant activity in DPPH and hydroxyl radicals, with IC_{50} values of 34.5 and 53.4 μg/mL, respectively. This is attributed to the presence of monoterpene and sesquiterpene constituents such as limonene, (Z)-lachnophyllum ester, (E)-β-ocimene, β-pinene, (E)-β-farnesene, arcurcumene, and myrcene (Lateef et al., 2018).

Petroleum ether and ethyl acetate extracts from aerial parts of *C. canadensis* cause cytotoxicity in laryngeal carcinoma cell line (Hep-2) in vitro models, with IC_{50} values of 45 and 50 μg/mL, respectively, and according to phytochemical screening, flavonoids, terpenoids, and tannins are some of the compounds present in both extracts and which may be correlated with induction of tumor cell death (Hayet et al., 2009). Furthermore, secondary metabolites isolated from methanolic extracts from roots of *C. canadensis* such as epifriedelanol and spinasterol have exhibited in vitro antiproliferative activity in cervix adenocarcinoma (HeLa), breast cancer (MCF-7), and epidermoid carcinoma (A-431) cell lines. Epifriedelanol showed IC_{50} values of 16.39, 61.43, and 5.40 μg/mL, respectively, while in spinasterol were 13.93, 26.50, and 13.66 μg/mL (Csupor-Loffler et al., 2011). *n*-Hexane fraction obtained from methanolic root extracts also shows in vitro cytotoxic effects in human lung adenocarcinoma cell lines such as A-549 and H1299, with IC_{50} values of 94.73 and 84.85 μg/mL, respectively. In addition, phytochemical screening of n-hexane fraction exhibits the presence of γ-dihydropyranone derivatives conyzapyranone A and conyzapyranone B, 2 γ-lactone acetylene derivatives (4E,8Z-matricaria-γ-lactone), triterpenes, sterols, and flavonoids, considered as responsible for antitumor activity (Ayaz et al., 2016).

11.4.6 ECLIPTA PROSTRATA

Eclipta prostrata, commonly known as Trailing Eclipta or Bhangra, is a prostrate annual herb that grows widely in tropical areas (Tewtrakul et al., 2011) and has a traditional reputation due to its use as a medicinal agent for treating of various human diseases such as kidney and liver weakness (Arunachalam et al., 2009), inflammatory conditions, and ophthalmic and digestive disorders (Ryu et al., 2013).

In previous in vivo studies, methanolic leaf extracts (200 mg/mL) have exhibited anti-inflammatory activity by decreasing carrageenan and egg white induced hind paw edemas in rats, with inhibition values of 38.80% and 38.23%, respectively. In addition, the phytochemical composition of methanolic leaf extracts of *E. prostrata* has been evaluated, showing the presence of steroids, alkaloids, triterpenoids, flavonoids, tannins, reducing sugars, and saponins, which are identified as responsible for their anti-inflammatory effects (Arunachalam et al., 2009). Furthermore, dichloromethane–methanol extract from aerial parts of the plant has been found to contain terthiophene derivatives and flavones such as orobol. In vitro studies have shown that orobol inhibits NO and prostaglandin E_2 release in LPS stimulated RAW264.7 cells, with IC_{50} values of 4.6 and 49.6 µM, respectively. Besides, inhibition of iNOS expression has been associated with a decrease of NO (Tewtrakul et al., 2012). Ethyl acetate fraction from an ethanolic extract from aerial parts of *E. prostrata* (30 µg/mL) also exhibited the capacity of inhibiting LPS-induced production of proinflammatory cytokines such as TNF-α and IL-6 (more than 50%) in macrophages cells in vitro (Ryu et al., 2012).

Antioxidant activity has been observed in ethanolic and ethyl acetate extracts from aerial parts of *E. prostrata* (500 µg/mL) and decreased peroxide radicals in vitro (77.62% and 54.86%, respectively) (Karthikumar et al., 2007). Aqueous leaf extracts of *E. prostrata* (100 mg/mL) have also shown an in vitro inhibitory effect on DPPH radical (5%) hydroxyl radical (18.10%) and lipid peroxidation (18.40%) (Rao et al., 2009). Moreover, ethanolic extracts from aerial parts of *E. prostrata* also show an antioxidant activity by decreasing DPPH radical levels (IC_{50}: 0.213 mg/mL) (Pukumpuang et al., 2014). In all cases, phenolic compounds, such as polyphenols, flavanones, and flavonoids, are the main components identified as responsible for the antioxidant capacity.

Methanolic leaf extracts have been found to contain a saponin compound, named Dasyscyphin C, that has exhibited a cytotoxic activity in HeLa cells in vitro (IC_{50}: 50 µg/mL) (Khanna et al., 2009). Besides, aqueous and ethanolic fractions obtained from an extract of the whole plant have been reported to contain secondary metabolites such as wedelolactone, flavones (luteolin-7-*O*-glucoside), and saponins (eclalbasaponin), which produce in vitro cytotoxicity in hepatoma cells (smmc-7721), showing IC_{50} values of 1,100, 350, and 100 µg/mL, respectively, being eclalbasaponin the compound with the highest cytotoxic activity (Liu et al., 2012). Moreover, secondary metabolites isolated from aerial parts of *E. prostate* have shown antitumor activity. One of them is 3'-methoxy-2,2':5',2''-terthiophene, characterized by inhibiting cell proliferation of human ovarian cancer cells (SKOV3) in vitro (IC_{50}: 7.73 µM) (Kim et al., 2015).

11.4.7 *LACTUCA SATIVA*

Lactuca sativa (Lettuce) is a famous vegetable consumed around the world not only for nutrimental purposes but also for its medicinal properties. Traditionally, *L. sativa* has been used as a remedy for pain, inflammation, and digestive disorders (Ismail and Mirza, 2015) but recently has also been associated as a possible treatment for neuronal disorders and inflammatory ailments (Araruna and Carlos, 2010).

Anti-inflammatory activity has been observed in lactones, such as 3,14-Dihydroxy11,13-dihydrocostunolide and 8-tigloyl-15-deoxyl-actucin isolated from methanolic extracts of aerial parts of *L. sativa*, showing significant ($p < 0.05$) decrease of carrageenan-induced paw edema mice model, at doses of 5, 10, and 20 mg/kg, compared to negative controls (Araruna et al., 2010) In addition, aqueous leaf extracts (1 g/kg) also demonstrate a decrease (77.39%) of carrageenan-induced paw edema, determining that flavonoids are responsible for this effect (Ismail and Miza, 2015). On the other hand, methanol–hydrochloric acid leaf extracts (250 µM) decrease the LPS-induced NO production (58.33%) and the release of cytokines TNF-α (73.38 %) and IL-6 (48.27 %) in a macrophages cell line (J774Q.1). Besides, phytochemical screening shows the presence of polyphenolic compounds such as chlorogenic acid, feruloyl tartaric acid, quercetin, esculin, and caffeoylferuloylquinic acid, which are correlated with the inhibition of NO production (Adesso et al., 2016).

Previous studies have reported in vitro antioxidant activity of aqueous and methanolic extracts of *L. sativa* due to their DPPH radical scavenging activity, exhibiting IC_{50} values of 4.1 and 3.5 µg/mL, respectively (Edziri et al., 2001). On the other hand, ethanolic leaf extracts (400 mg/kg) have also shown in vivo antioxidant activity by increasing CAT (40.71%) and SOD (62.34%) levels on serum collected from rats treated with the extracts (Garg et al., 2004). Moreover, aqueous leaf extracts from red lettuce have two anthocyanins compounds named cyanidin-3-O-(6″-malonyl-β-glucopyranoside) and cyanidin-3-O-(6″-malonyl-β-glucopyranoside methyl ester) that are correlated with inhibition of in vitro lipid peroxidation by 88% and 91.5%, respectively, at 0.25 µM concentration. Besides, both metabolites inhibit cyclooxygenase enzymes COX-1 (64% and 65.8%) and COX-2 (78.9% and 84.3%) at 5 µM concentration (Mulabagal et al., 2010).

Regarding antitumor activity, aqueous leaf extracts from *L. sativa* have shown an in vitro cytotoxic effect in human tumor cell lines such as breast cancer (MCF-7), leukemia (HL-60), hepatoma (Bel7402), colon cancer (HCT-9) with IC_{50} values of 1094.4 µg/mL, 451.4 µg/mL, 100 mg/mL, and 100 mg/mL, respectively. Also, aqueous leaf extract (200 µg/mL) exhibits cytotoxicity on lung adenocarcinoma (A549) and colorectal adenocarcinoma (HCT-8) cell lines, with inhibition values of 70% and 60%, respectively. Anthocyanins, flavones, and phenolic acids have been identified in aqueous extracts from *L. sativa* and have demonstrated to produce in vitro antitumor effect by inducing expression of tumor suppressor p21 protein and downregulating proto-oncogene cyclin D1 (Griding et al., 2010; Qin et al., 2018).

11.5 CONCLUSIONS

Currently, the number of cases of patients with diseases such as cancer and inflammatory and autoimmune diseases has increased, which have in common a lack of regulation in the cell cycle, an increase in the production of free radicals, and a failure in the modulation of the inflammatory mechanisms. There are treatments to combat these diseases; however, they have side effects.

Therefore, secondary metabolites obtained from a natural source, such as the plants of the Asteraceae family, have demonstrated an anti-inflammatory, antioxidant, and antitumor potential in in vitro and in vivo models; this has been correlated with the presence of saponins, tannins, terpenes, phenolic acids, anthocyanins, and flavonoids.

11.6 PERSPECTIVES

The Asteraceae family comprises more than species worldwide, of which only a small fraction has been studied with respect to its present metabolites and its biological effects. In addition to obtain evidence on the functionality of phytochemicals, it is also very important to determine their toxicity in different biological models in vitro and in vivo.

The outlook for alternate treatment of diseases involve with inflammation and cancer has remarkable brightened with the advancement of knowledge about the potent antioxidants, such as polyphenols, presented in some plants and fruits, against oxidative stress, and has become the focus of interest for therapy for degenerative disorders, inflammation, and cancer. However, a major limitation of the therapeutic relevance of certain polyphenols is their poor bioavailability (for example, resveratrol). However, plant polyphenols and other secondary metabolites are bioactive compounds with potential in preventing and treating several chronic disorders, mainly due to their ability to modulate key proinflammatory and pro-oxidant signaling pathways.

However, it is important that plant extracts are standardized in the contents of chemical constituents, as the species have many varieties and cultivars and can vary, particularly the active principles.

Due to the relatively low number of publications, it is concluded that further development of optimum methods for extraction, analysis, and isolation of secondary metabolites of the Asteraceae family should be carried over to assess their antioxidant capacity and their food technological and pharmaceutical industry.

KEYWORDS

- medicinal plants
- Alzheimer's disease
- acetylcholinesterase
- inhibitors.

REFERENCES

Abbas, Z. K., Saggu, S., Sakeran M. I., Zidan, N., Rehman, H., Ansari, A. A. Phytochemical, antioxidant and mineral composition of hydroalcoholic extract of chicory (*Cichorium intybus* L) leaves. *Saudi J. Biol. Sci.* 2015, 22, (3), 322–326.

Adedapo, A. A., Oyagbemi, A. A., Fagbohun, O. A., Omobowale, T. O., Yakubu, M. A. Evaluation of the anticancer properties of the methanol leaf extract of *Chromolaena odorata* on HT-29 cell line. *J. Pharmacogn. Phytochem.* 2016, 5, (2), 52–57.

Aderem, A. Phagocytosis and the inflammatory response. *J. Infect. Dis.* 2003, 187, (2), 340–345.

Adesso, S., Pepe, G., Sommella, E., Manfra, M., Scopa, A., Sofo, A., Tenore, G. C., Russo, M., Di Gaudio, F., Autore, G., Campiglia, P., Marzocco, S. Anti-inflammatory and antioxidant activity of polyphenolic extracts from *Lactuca sativa* (var. Maravilla de Verano) under different farmin methods. *J. Sci. Food Agric.* 2016, 96, (12), 4194–4206.

Ahmadi-Dastgerdi, A., Ezzatpanah, H., Asgary, S., Dokhani, S., Rahimi, E. Phytochemical, antioxidant and antimicrobial activity of the essential oil from flowers and leaves of *Achillea millefolium* subsp. *Millefolium. J. Essent. Oil Bear Pl.* 2017, 20, (2), 395–409.

Akihisa, T., Yasukawa, K., Oinuma, H., Kasahara, Y., Yamanouchi, S., Takido, M., Kumaki, K., Tamura, T. Triterpene alcohols from the flowers of Compositae and their anti-inflammatory effects. *Phytochemistry.* 1996, 43, (6), 1255–1260.

Akinmoladun, A.C., Ibukun, E.O., Dan-Ologe, I.A. Phytochemical constituents and antioxidant properties of extracts from the leaves of *Chromolaena odorata. Sci. Res. Essays.* 2007, 2, (6), 191–194.

Al-Kharashi, A.S. Role of oxidative stress, inflammation, hypoxia and angiogenesis in the development of diabetic retinopathy. *Saudi Med. J.* 2018, 32, (4), 318–323.

Amin, K. The role of mast cells in allergic inflammation. *Respir. Med.* 2012, 106, (1), 9–14.

Amini-Navaie, B., Kavoosian, S., Fattahi, S., Hajian-Tilaki, K., Asouri, M., Bishekolaie, R., Akhavan-Niaki, N. Antioxidant and cytotoxic effect of aqueous and hydroalcoholic extracts of the *Achillea millefolium* L. on MCF-7 breast cancer cell line. *IBBJ.* 2015, 1, (3), 119–125.

An, L., Sun, Y., Huang, J., Liu, Y., Yuan, H., Zhang, R., Sun, Y. Chemical compositions and in vitro antioxidant activity of the essential oil from *Coreopsis tinctorial* Nutt. flower. *J. Essent. Oil-Bear. Plants.* 2018, 21, (4), 876–885.

Araruna, K., Carlos, B. Anti-inflammatory activities of triterpene lactones from *Lactuca sativa. Phytopharmacology.* 2010, 1, (1), 1–6.

Ariel, A., Maridonneau-Parini, I., Rovere-Querini, P., Levine, J. S., Mühl, H. Macrophages in inflammation and its resolution. *Front. Immunol.* 2012, 3, (324), 1–3.

Armenteros-Borrell, F. M. Alzheimer's disease and environmental risk factors. *Rev. Cuba. Enferm.* 2017, 33, (1), 1–17.

Arunachalam, G., Subramanian, N., Pazhani, G. P., Ravichandran, V. Anti-inflamamtory activity of methanolic extract of *Eclipta prostrata* L. (Asteraceae). *Afr. J. Pharm. Pharmacol.* 2009, 3, (3), 97–100.

Ashely, N.T., Weil, Z. M.., Nelson, R. J. Inflammation: Mechanisms, costs, and natural variation. *Annu. Rev. Ecol. Evol. Syst.* 2012, 43, (1), 385–406.

Avello, M., Suwalsky, M. Radicales libres, antioxidantes naturales y mecanismos de protección. *Atenea.* 2006, 494, (1), 161–172.

Ayad, R., Ababsa, Z., Belfadel, F., Akkal, S., León, F., Brouard, I., Medjoubi, K. Phytochemical and biological activity of Algerian *Centaurea melitensis. Int. J. Med. Arom. Plants.* 2012, 2, (1), 151–154.

Ayaz, F., Sarimahmut, M., Küçükboyaci, N., Ulukaya, E. Cytotoxic effect of *Conyza canadensis* (L.) cronquist on human lung cancer cell lines. *Turk. J. Pharm. Sci.* 2016, 13, (3), 342–346.

Aznag, F. Z., Kadmiri, N. E., Izaabel, E. H. Tumor necrosis factor-alpha and tumor necrosis factor beta polymorphisms and risk of breast cancer: Review. *Gene Rep.* 2018, 12, (1), 317–323.

Babaee, N., Moslemi, D., Khalilpour, M., Vejdani, F., Moghadamnia, Y., Bijano, A., Baradaran, M., Kazemi, M., Khalilpour, A., Pouramir, M., Moghadamnia, A. Antioxidant capacity of *Calendula officinalis* flowers extract and prevention of radiation induced oropharyngeal mucositis in patients with head and neck cancers: A randomized controlled clinical study. *Daru*. 2013, 21, (1), 1–8.

Benedek, B., Kopp, B., Melzig, M. *Achillea millefolium* L. s.l.—Is the anti-inflammatory activity mediated by protease inhibition? *J. Ethnopharmacol*. 2007, 113, (2), 312–317.

Benedek, B., Kopp, B. *Achillea millefolium* L. s.l. revisited: Recent findings confirm the traditional use. *WMW*. 2007, 157, (13), 312–314.

Burk, D. P., Cichacz, Z. A., Daskalova, S. M. Aqueous extract of *Achillea millefolium* L. (Asteraceae) inflorescences suppresses lipopolysaccharide-induced inflammatory responses in RAW 264.7 murine macrophages. *J. Med. Plants Res*. 2010, 4, (3), 225–234.

Cadet, J., Davies, K. J. A. Oxidative DNA damage & repair: An introduction. *Free Radical Bio. Med*. 2017, 107, (2), 1–25.

Candan F, Unlu M, Tepe B, Daferera D, Polissiou M, Sökmen A, Akpulat A. Antioxidant and antimicrobial activity of the essential oil and methanol extracts of *Achillea millefolium* subsp. *millefolium* Afan. (Asteraceae). *J. Ethnopharmacol*. 2003, 87, (2), 215–220.

Chagas-Paula, D. A., Branquinho-Oliveira, T., Vasconcelos-Faleiro, D. P., Barbosa-Oliveira, R., Batista-Da Costa, F. Outstanding anti-inflammatory potential of selected Asteraceae species through the potent dual Inhibition of cyclooxygenase-1 and 5-lipoxygenase. *Planta Med*. 2015, 81, (1), 1296–1307.

Chandran, P. K., Kuttan, R. Effect of *Calendula officinalis* flower extract on acute phase proteins, antioxidant defense mechanism and granuloma formation during thermal burns. *J. Clin. Biochem. Nutr*. 2008, 43, (2), 58–64.

Chemat, F., Fabiano-Tixier, S., Vian, M. A., Allaf, T., Vorobiey, E. Solvent-free extraction of food and natural products. *TrAC*. 2015, 71, (1), 157–168.

Cassatella, M. A., Östberg, N. K., Tamassia, N., Soehnlein, N. T. Biological roles of neutrophil-derived granule proteins and cytokines. *Trends Immunol*. 2019. 40, (7), 648–664.

Chow, C., Clermont, G., Kumar, R., Lagoa, C., Tawadrous, Z., Gallo, D., Betten, B., Bartels, J., Constantine, G., Fink, M., Billiar, T., Vodovotz, Y. The acute inflammatory response in diverse shock states. *Shock*. 2005, 24, (1), 74–84.

Coates, L.C., FitzGerald, O., Helliwella, P. S., Paul, C. Psoriasis, psoriatic arthritis, and rheumatoid arthritis: Is all inflammation the same? *Semin. Arthritis Rheu*. 2016, 46, (3), 291–304.

Conforti, F., Loele, G., Statti, G. A., Marrelli, M., Ragno, G., Menichini, F. Antiproliferative activity against human tumor cell lines and toxicity test on mediterranean dietary plants. *Food Chem. Toxicol*. 2008, 46, (10), 3325–3332.

Cordova, C. A. S., Siqueira, J. R., Netto, C. A., Yunes, R. A., Volpato, A. M., Filho, V. C., Curi-Pedrosa, R., Creczynski-Pasa, T. B. Protective properties of butanolic extract of the *Calendula officinalis* L.(marigold) against lipid peroxidation of rat liver microsomes and action as free radical scavenger. *Redox Rep*. 2002, 7, (2), 95–102.

Coronado, M., Vega-León, S., Gutiérrez, R., Vázquez, M., Radilla, C. Antioxidants: Present perspective for the human health. *Rev. Chil. Nutr*. 2015, 42, (2), 206–212.

Csupor-Loffer, B., Hajdú, Z., Zupkó, I., Molnár, J., Forgo, P., Vasas, A., Kele-Hohmann, J. Antiproliferative constituents of the roots of *Conyza canadensis*. *J. Med. Plant Res*. 2011, 77, (11), 1183–1188.

Dhar, R., Kimseng, R., Chokchaisiri, R., Hiransai, P., Utaipan, T., Suksamrarn, A., Chunglok, W. 2′,4-Dihydroxy-3′,4′,6′-trimethoxychalcone from *Chromolaena odorata* possesses anti-inflammatory effects via inhibition of NF-κB and p38 MAPK in lipopolysaccharide-activated RAW 264.7 macrophages. *Immunopharmacol. Immunotoxicol*. 2018, 40, (1), 43–51.

De Armas-Hernández, A., Solís-Cartas, U., Prada-Hernández, D. M., Benítez-Falero, Y., Vázquez-Abreu, R. L. Atherosclerotic risk factors in patients with rheumatoid arthritis. *Rev. Cub. Med. Mil*. 2017, 46, (1), 51–63.

De Pablo-Sánchez, R., Montserrat-Sanz, J., Prieto-Martín, A., Reyes-Martín, E., Álvarez de Mon Soto, M., Sánchez-García, M. The balance between pro-inflammatory and anti-inflammatory citokines in septic states. *Med. Intensiva*. 2005, 29, (3), 151–158.

Edziri, H. L., Smach, M. A., Ammar, S., Mahjoub, M. A., Mighri, Z., Aouni, M., Mastouri, M. Antioxidant, antibacterial, and antiviral effects of *Lactuca sativa* extracts. *Ind. Crops Prod*. 2011, 34, (1), 1182–1185.

El-Kalamouni, C., Venskutonis, P. R., Zebib, B., Merah, O., Raynaud, C., Talou, T. Antioxidant and Antimicrobial Activities of the Essential Oil of *Achillea millefolium* L. Grown in France. *Medicines*. 2017, 4, (30), 1–9.

Escribá, P. V.; Busquets, X.; Inokuchi, J.; Balogh, G.; Torok, Z.; Horváth, I.; Harwood, J.; Vigh, L. Membrane lipid therapy: Modulation of the cell membrane composition and structure as a molecular base for drug discovery and new disease treatment. *Prog. Lipid Res*. 2015, 59, (1), 38–53.

Falleh, H., Ksouri, R., Chaieb, K., Karray-Bouraoui, N., Trabelsi, N., Boulaaba, M., Abdelly, C. Phenolic composition of *Cynara cardunculus* L. organs, and their biological activities. *C. R. Biol*. 2008, 331, (5), 372–379.

Feng, A., Peña, Y., Li, W. Ischemic heart disease in diabetic and non-diabetic patients. *Rev. Haban Cienc. Méd*. 2017, 16, (2), 217–228.

Filippi, M. Mechanism of diapedesis, importance of the transcellular route. *Adv. Immunol*. 2016, 129, (1), 25–53.

García-Arieta, A., Hernández-García, C., Avendaño-Solá, C. Regulación de los medicamentos genéricos: Evidencias y mitos. *Inf. Ter. Sist. Nac. Salud*. 2010, 34, (3), 71–82.

García-López, J. C., Álvarez-Fuentes, G., Pinos-Rodríguez, J. M., Jasso-Pineda, Y., Contreras-Treviño, H. I., Camacho-Escobar, M. A., López-Aguirre, S., Lee-Rangel, H. A., Rendón-Huerta, J. A. Anti-inflammatory effects of *Chrysactinia mexicana* gray extract in growing chicks (*Gallus gallusdomesticus*) challenged with LPS and PHA. *Int. J. Curr. Microbiol. Appl. Sci*. 2017, 6, (1), 550–562.

García-Risco, M. R., Mouhid, L., Salas-Pérez, L., López-Padilla, A., Santoyo, S., Jaime, L., Ramírez de Molina, A., Reglero, G., Fornari, T. Biological activities of Asteraceae (*Achillea millefolium* and *Calendula officinalis*) and Lamiaceae (*Melissa officinalis* and *Origanum majorana*) plant extracts. *Plant Foods Hum. Nutr*. 2017, 72, (1), 96–102.

Garg, M., Garg, C., Mukherjee, P. K., Suresh, B. Antioxidant potential of *Lactuca sativa*. *Anc. Sci. Life*. 2004, 24, (1), 6–10.

Gaschlera, M. M., Stockwell, B. R. Lipid peroxidation in cell death. *Biochem. Bioph. Res. Co*. 2017, 482, (3), 419–425.

Gomez-Flores, R., Quintanilla-Licea, R., Verde-Star, M. J., Morado-Castillo, R., Vázquez-Díaz, D., Tamez-Guerra, R., Tamez-Guerra, P., Rodríguez-Padilla C. Long-chain alkanes and ent-labdane-type diterpenes from *Gymnosperma glutinosum* with cytotoxic activity against the murine lymphoma L5178Y-R. *Phytother. Res*. 2012, 26, (11), 1632–1636.

Gómez-Flores, R., Verástegui-Rodríguez, L., Quintanilla-Licea, R., Tamez-Guerra, P., Monreal-Cuevas, E., Tamez-Guerra, R., Rodríguez-Padilla, C. Antitumor properties of *Gymnosperma glutinosum* leaf extracts. *Cancer Investigation*. 2009, 27, (1), 149–155.

González-Hun, C. P., Wadhwa, M., Sanders, W. L. DNA damage by oxidative stress: Measurement strategies for two genomes. *Current Opinion Toxicol.* 2018, 7, (1), 87–94.

Gorzalczany, S., Rosella, M. A., Spegazzini, E. D., Acevedo, C., Debenedetti, S. L. Anti-inflammatory activity of *Heterotheca subaxillaris* var. *latifolia* (Buckley) Gandhi & R.D. Thomas, Asteraceae. *Braz. J. Pharmacogn.* 2009, 19, (4), 876–879.

Gridling, M., Popescu, R., Kopp, B., Wagner, K-H., Krenn, L., Kruptiza, G. Anti-leukaemic effects of two extract types of *Lactuca sativa* correlate with the activation of Chk2, induction of p21, downregulation of cyclin D1 and acetylation of α-tubulin. *Oncol. Rep.* 2010, 23, (4), 1145–1151.

Guija-Poma, E., Inocente-Camones, M. A., Ponce-Pardo, J., Zarzosa-Norabuena, E. Evaluación de la técnica 2,2-difenil-1-picrilhidrazilo (DPPH) para determinar capacidad antioxidante. *Horiz. Med.* 2015, 15, (1), 57–60.

Gupta, K., Varadarajan, R. Insights into protein structure, stability and function from saturation mutagenesis. *Curr. Opin. Struc. Biol.* 2018, 50, (1), 117–125.

Hasselbalch, H. C. Chronic inflammation as a promotor of mutagenesis in essential thrombocythemia, polycythemia vera and myelofibrosis. A human inflammation model for cancer development? *Leukemia Res.* 2013, 37, (2), 214–220.

Hatsuko-Baggio, C., De Martini-Otofuji, G., Setim-Freitas, C., Mayer, B., Andrade-Marques, M. C., Mesia-Vela, S. Modulation of antioxidant systems by subchronic exposure to the aqueous extract of leaves from *Achillea millefolium* L. in rats. *Nat. Prod. Res.* 2015, 30, (5), 613–615.

Hayet, E., Maha, M., Samia, A., Ali, M. M., Souhir, B., Abderaouf, K., Mighri, Z., Mahjoub, A. Antibacterial, antioxidant and cytotoxic activities of extracts of *Conyza canadensis* (L.) Cronquist growing in Tunisia. *Med. Chem. Res.* 2009, 18, (6), 447–454.

Heimler, D., Isolani, L., Vignolini, P., Romani, A. Polyphenol content and antiradical activity of *Cichorium intybus* L. from biodynamic and conventional farming. *Food Chem.* 2009, 114, (3), 765–770.

Hegazya, E. F., Abdel-Lateffb, A., Gammal-Eldeeenc, A. M., Turkyb, F., Hiratad, T., Parée, P. W., Karchesyf, J., Kamelb, M. S., Ahmed, A. A. Anti-inflammatory activity of new guaiane acid derivatives from *Achillea Coarctata*. *Nat. Prod. Commun.* 2008, 3, (6), 851–856.

Heppner, F.L., Ransohoff, R.M., Becher, B. Immune attack: the role of inflammation in Alzheimer disease. *Nat. Rev. Neurosci.* 2015, 16, (6), 358–372.

Herman-Lara, H., Alanís-Garza, E. J., Estrada Puente, M. F., Mureyko, L. L., Alarcón-Torres, D. A., Ixtepan-Turrent, L. Nutrición que previene el estrés oxidativo causante del Alzheimer. Prevención del Alzheimer. *Gac. Med. Mex.* 2015, 151, (1), 245–251.

Herrera, V., Wendie, E. Inflamación I. *Rev. Act. Clin.* 2014, 43, (1), 2261–2265.

Hosseinimehr, S., Pourmorad, F., Shahabimajd, N., Shahrbandy, K., Hosseinzadeh, R. *In vitro* antioxidant activity of *Polygonium hyrcanicum, Centaureae depressa, Sambucus ebulus, Mentha spicata* and *Phytolacca americana*. *PJBS.* 2007, 10, (4), 637–640.

Houée-Levina, C., Bobrowski, K. The use of the methods of radiolysis to explore the mechanisms of free radical modifications in proteins. *J. Proteomics.* 2013, 92, (1), 51–62.

Ighodaro, O. M., Akinloye, O. A. First line defense antioxidants-superoxide dismutase (SOD), catalase (CAT) and glutathione peroxidase (GPX): Their fundamental role in the entire antioxidant defense grid. *Alexandria Med. J.* 2018, 54, (4), 287–293.

Ismail, H., Mirza, B. Evaluation of analgesic, anti-inflammatory, anti-depressant and anti-coagulant properties of *Lactuca sativa* (CV. Grand Rapids) plant tissues and cell suspension in rats. *BMC Complement Altern. Med.* 2015, 15, (199), 1–7.

Jackson, J. L., Leslie, C. E., Hondorp, S. N. Depressive and anxiety symptoms in adult congenital heart disease: Prevalence, health impact and treatment. *Prog. Cardiovasc. Dis.* 2018, 61, (3), 294–299.

Jang, I., Park, J., Park, E., Park, H., Lee, S. Antioxidative and antigenotoxic activity of extracts from Cosmos (*Cosmos bipinnatus*) flowers. *Plant Food Hum. Nutr.* 2008, 63, (4), 205–210.

Jasso de Rodríguez, D., Carrillo-Lomelí, D. A., Rocha-Guzmán, N. E., Moreno-Jiménez, M. R., Rodríguez-García, R., Díaz-Jiménez, M. L. V., Flores-López, M. L., Villarreal-Quintanilla, J. A. Antioxidant, anti-inflammatory and apoptotic effects of *Flourensia microphylla* on HT-29 colon cancer cells. *Ind. Crops Prod.* 2017, 107, (1), 472–481.

Jiménez, E., García, A., Paco, L., Algarra, I., Collado, A., Garrido, F. A new extract of the plant *Calendula officinalis* produces a dual *in vitro* effect: cytotoxic anti-tumor activity and lymphocyte activation. *BMC Cancer*. 2006, 6, (119), 1–14.

Khanna, V, G., Kannabiran, K. Anticancer-cytotoxic activity of saponins isolated from the leaves of *Gymnema sylvestre and Eclipta prostrata* on HeLa cells. *Int. J. Green Pharm.* 2009, 3, (3), 227–229.

Karnes, J. M., Daffner, S. D., Watkins, C. M. Multiple roles of tumor necrosis factor-alpha in fracture healing. *Bone*. 2015, 78, (1), 87–93.

Karthikumar, S., Vigneswari, K., Jegatheesan, K. Screening of antibacterial and antioxidant activities of leaves of *Eclipta prostrata* (L.). *Sci. Res. Essays*. 2007, 2, (4), 101–104.

Kazemi, M. Phytochemical and antioxidant properties of *Achillea millefolium* from the eastern region of Iran. *Int. J. Food Prop.* 2015, 18, (10), 2187–2192.

Kelly, A., Hanson, E., Boothby, M., Keegan, A. Interleukin-4 and interleukin-13 signaling connections maps. *Science*. 2003, 300, (1), 1527–1528.

Khan, R. A., Khan, M. R., Sahreen, S., Ahmed, M. Evaluation of phenolic contents and antioxidant activity of various solvent extracts of *Sonchus asper* (L.) Hill. *Chem. Cent. J.* 2012, 6, (12), 1–7.

Kishimoto, S., Maoka, T., Sumitomo, K., Ohmiya, A. Analysis of carotenoid composition in petals of Calendula (*Calendula officinalis* L). *Biosci. Biotechnol. Biochem*. 2005, 69, (11), 2122–2128.

Kim, H. Y., Kim, H. M., Ryu, B., Lee, J. S., Choi, J. H., Jang, Constituents of the aerial parts of *Eclipta prostrata* and their cytotoxicity on human ovarian cancer cells *in vitro*. *Arch. Pharm. Res*. 2015, 38, (11), 1963–1969.

Kouamé, P. B. F., Jacques, C., Bedi, G., Silvestre, V., Loquet, D., Barillé-Nion, S., Robins, R. J., Teal, I. Phytochemicals isolated from leaves of *Chromolaena odorata*: Impact on viability and clonogenicity of cancer cell lines. *Phytother. Res*. 2012, 27, (6), 835–840.

Koyasu, S., Moro, K. Role of innate lymphocytes in infection and inflammation. *Front. Immunol*. 2012, 3, (101), 1–14.

Kukić, J., Popović, V., Petrović, S., Mucaji, P., Ćirić, A., Stojković, D., Soković, M. Antioxidant and antimicrobial activity of *Cynara cardunculus* extracts. *Food Chem*. 2008, 107, (2), 861–868.

Lambrecht, B., Hammad, H. The role of dendritic and epithelial cells as master regulators of allergic airway inflammation. *Lancet*. 2010, 376, (9743), 835–843.

Lambrecht, B., Hammad, H. The immunology of asthma. *Nature Immunol*. 2015, 16, (1), 45–56.

Lastra, A. L., Ramírez, T. O., Salazar, L., Martínez, M., Trujillo-Ferrera, J. The ambrosanolide cumanin inhibits macrophage nitric oxide synthesis: Some structural considerations. *J. Ethnopharmacol*. 2004, 95, (1), 221–227.

Lateef, R., Bhat, K, A., Chandra, S., Banday, J. A. Chemical composition, antimicrobial and antioxidant activities of the essential oil of *Conyza canadensis* growing wild in Kashmir valley. *Am. J. Essent. Oil*. 2018, 6, (1), 35–41.

Lauvaua, G., Loke, P., Hohl, T. M. Monocyte-mediated defense against bacteria, fungi, and parasites. *Semin. Immunol.* 2015, 27, (6), 397–409.

León-Regal, M. L., Alvarado-Borges, A., De Armas-García, J. O., Miranda-Alvarado, L., Varens-Cedeño, J. A., Cuesta-del Sol, J. A. Respuesta inflamatoria aguda. Consideraciones bioquímicas y celulares. *Rev. Finlay*. 2016, 5, (1), 41–62.

Liu, Q. M., Zhao, H. Y., Zhong, X. K., Jiang, J. G. *Eclipta prostrata* L. phytochemicals: Isolation, structure elucidation, and their antitumor activity. *Food Chem. Toxicol*. 2012, 50, (11), 4016–4022.

Loggia, R., Tubaro, A., Sosa, S., Becker, H., Isaac, O. The role of triterpenoids in the topical anti-inflammatory activity of *Calendula officinalis* flowers. *Planta Med*. 1994, 60, (6), 516–520.

Malathi, M., Sudarshana, M., Niranjan, M. Anti-inflammatory effect of chloroform extract of *Flaveria trinervia* (Sprengel) C. Mohr: A medicinal herb. *IJPBS*. 2012, 3, (4), 218–221.

Maldonado-Saavedra, O., Jiménez-Vázquez, E. N., Guapillo-Vargas, M. R. B., Ceballos-Reyes, G. M., Méndez-Bolainal, E. Free radicals and their role in chronic-degenerative diseases. *Rev. Med. UV*. 2010, 10, (2), 32–39.

Malewicza, M., Perlmann, T. Function of transcription factors at DNA lesions in DNA repair. *Exp. Cell Res*. 2014, 15, (1), 94–100.

Manson, S. C., Brown, R., Cerulli, A., Fernández-Vidaurre, C. The cumulative burden of oral corticosteroid side effects and the economic implications of steroid use. *Respir. Med*. 2009, 103, (7), 975–994.

Mantovani, A., Dinarello, C.A., Molgora, M., Garlanda, C. Interleukin-1 and related cytokines in the regulation of inflammation and immunity. *Immunity*. 2019, 50, (4), 778–795.

Martin, D., Navarro del Hierro, J., Villanueva Bermejo, D., Fernández-Ruiz, R., Fornari, T., Reglero, G. Bioaccessibility and antioxidant activity of *Calendula officinalis* supercritical extract as affected by *in vitro* codigestion with olive oil. *J. Agric. Food Chem*. 2016, 64, (46), 8828–8837.

Martino, R., Florencia-Beer, M., Elso, O., Donadel, O., Sülsen, V., Anesini, C. Sesquiterpene lactones from *Ambrosia* spp. are active against a murine lymphoma cell line by inducing apoptosis and cell cycle arrest. *Toxicol. Vitro*. 2015, 29, (1), 1529–1536.

Matysik, G., Wojciak-Kosior, M., Paduch, R. The influence of Calendula officinalis flowers extract on cell cultures and the chromatographic analysis of extracts. *J. Pharm. Biomed. Anal*. 2005, 38, (2), 285–292.

Mehmood, N., Zubair, M., Rizwan, K., Rasool, N., Shahid, M., Uddin Ahmad, V. Antioxidant, antimicrobial and phytochemical analysis of *Cichorium intybus* seeds extract and various organic fractions. *Iran J. Pharm. Res*. 2012, 11, (4), 1145–1151.

Mehrandish, R., Mellati, A. A., Rahimipour, A., Nayeri, N. D. Anti-cancer activity of methanol extracts of *Cichorium intybus* on human breast cancer SKBR3 cell line. *Razavi Int. J. Med*. 2017, 5, (1), 1–4.

Mihajlovic, L., Radosavljevic, J., Burazer, L., Smiljanic, K., Cirkovic-Velickovic T. Composition of polyphenol and polyamide compounds in common ragweed (*Ambrosia artemisiifolia* L.) pollen and sub-pollen particles. *Phytochemistry*. 2015, 109, (1), 125–132.

Mojarrab, M., Mehrabi, M., Ahmadi, F., Hosseinzadeh, L. Protective effects of fractions from *Artemisia biennis* hydroethanolic extract against doxorubicin-induced oxidative stress and apoptosis in PC12 cells. *IJBMS*. 2016, 19, (5), 503–510.

Moore K. W., de Waal Malefyt, R., Coffman, R. L., O'Garra, A. Interleukin-10 and the interleukin-10 receptor. *Annu. Rev. Immunol.* 2001, 19, (1), 683–765.

Mulabagal, V., Ngouajio, M., Nair, A., Zhang, Y., Gottumukkala, A. L., Nair, M. G. *In vitro* evaluation of red and green lettuce (*Lactuca sativa*) for functional food properties. *Food Chem*. 2010, 118, (2), 300–306.

Nagano, T., Lubling, Y., Varnai, C., Dudley, C., Leung, W., Baran, Y., Cohen, N. M., Wingett, S., Fraser, P., Tanay, A. Cell-cycle dynamics of chromosomal organization at single-cell resolution. *Nature*. 2017, 547, (7661), 61–67.

Nandagopal, S., Ranjitha-Kumari, B. D. Phytochemical and antibacterial studies of Chicory (*Cichorium intybus* L.): A multipurpose medicinal plant. *Adv. Biol. Res.* 2007, 1, (2), 17–21.

Nelson, P. J., Teireirab, M. M. Dissection of inflammatory processes using chemokine biology: Lessons from clinical models. *Immunol. Lett.* 2012, 145, (1), 55–61.

Neukirch, H., D'Ambrosio, M., Dall-Via, J., Guerriero, A: Simultaneous quantitative determination of eight triterpenoid monoesters from flowers of 10 varieties of *Calendula officinalis* L. and characterization of a new triterpenoid monoester. *Phytochem. Anal*. 2004, 15, (1), 30–35.

Nguyen, T., Le, V. Chalcones of *Eupatorium odoratum* L. from Vietnam. *J. Chem. Vietnam*. 1993, 2, (1), 79–81.

Owoyele, V., Adediji, J., Soladoye, A. Anti-inflammatory activity of aqueous leaf extract of *Chromolaena odorata*. *Inflammopharmacology*. 2005, 13, (5), 479–484.

Owoyele, V., Oguntoye, S., Dare, K., Ogunbiyi, B., Aruboula, E., Soladoye, A. Analgesic, anti-inflammatory and antipyretic activities from flavonoid fractions of *Chromolaena odorata*. *J. Med. Plants Res*. 2008, 2, (9), 219–225.

Penkauskas, T., Preta, G. Biological applications of tethered bilayer lipid membranes. *Biochimie*. 2019, 157, (1), 131–141.

Perera, W. H., Tabart, J., Gómez-Quesada, A., Sipel, A. Antioxidant capacity of three cuban species of the genus pluchea cass. (Asteraceae). *J. Food Biochem*. 2009, 34, (1), 249–261.

Pereira, J. M., Peixoto, V., Teixeira, A., Sousa, D., Barros, L., Ferreira, I. C., Vasconcelos, M. H. *Achillea millefolium* L. hydroethanolic extract inhibits growth of human tumor cell lines by interfering with cell cycle and inducing apoptosis. *Food Chem. Toxicol*. 2018, 118, (1), 635–644.

Pérez-González, C., Serrano-Vega. R., González-Chávez, M., Zavala-Sánchez, M. A., Pérez-Gutiérrez, S. Anti-inflammatory activity and composition of *Senecio salignus* Kunth. *Biommed Res Int*. 2013, 2013, 814693. Doi: 10.1155/2013/814693.

Pérez G, R. M., Vargas S, R., Martínez M., F. J., Córdova R, I. Antioxidant and free radical scavenging activities of 5,7,3′-trihydroxy-3,6,4′-trimethoxyflavone from *Brickellia veronicaefolia*. *Phytother. Res*. 2004, 18, (1), 428–430.

Peters, U., Dixon, A. E., Forno, E. Obesity and asthma. *J. Allergy Clin. Immunol*. 2018, 141, (4), 1169–1179.

Phan, T. T., Wang, L., See, P., Grayer, R. J., Chan, S. Y., Lee, S. T. Phenolic compounds of *Chromolaena odorata* protect cultured skin cells from oxidative damage: Implication for cutaneous wound healing. *Biol. Pharm. Bull*. 2002, 24, (12), 1373–1379.

Piasecka, A., Jedrzejczak-Rey, N., Bednarek, P. Secondary metabolites in plant innate immunity: Conserved function of divergent chemicals. *Tansley Rev*. 2015, 206, (3), 1–17.

Potrich, F. B., Allemand, A., Motada-Silva, L., Cristinados-Santos, A., Hatsuko-Baggio, C., Setim-Freitas, C., Bueno-Méndes, D. A. G., Andre, E., Paula-Werner, M. F., Andrade-Marques, M. C. Antiulcerogenic activity of hydroalcoholic extract of *Achillea millefolium* L.: Involvement of the antioxidant system. *J. Ethnopharmacol*. 2010, 130, (1), 85–92.

Preethia, K. C., Kuttan, G., Kuttan, R. Antioxidant potential of an extract of *Calendula officinalis*. flowers *in vitro* and *in vivo*. *Pharm. Biol.* 2006, 44, (9), 691–697.

Pukumpuang, W., Chansakaow, S., Tragoolpua, Y. Antioxidant activity, phenolic compound content and phytochemical constituents of *Eclipta prostrata* (Linn) Linn. *Chiang Mai J. Sci.* 2014, 41, (413), 568–576.

Qin, X. X., Zhang, M. Y., Han, Y. Y., Hao, J. H., Liu, C. J., Fan, S. X. Beneficial phytochemicals with anti-tumor potential revealed through metabolic profiling of new red pigmented lettuces (*Lactuca sativa* L). *Int. J. Mol. Sci.* 2018, 19, (4), 1–15.

Rao, D. B., Kiran, C. R., Madhavi, Y., Rao, R. K., Rao, T. R. Evaluation of antioxidant potential of *Clitoria ternate* L. and *Eclipta prostrata* L. *Ind. J. Biochem. Biophys.* 2009, 46, (3), 247–252.

Rao, K., Chaudhury, P., Pradhan, A. Evaluation of antioxidant activities and total phenolic content of *Chromolaena odorata*. *Food Chem. Toxicol.* 2010, 48, (2), 729–732.

Raposo, T. P., Beirão, B. C. B., Pang, L. Y., Queiroga, F. L., Argyle, D. J. Inflammation and cancer: Till death tears them apart. *Veterinary J.* 2015, 205, (2), 161–174.

Ripoll, C., Schmidt, B. M., Ilic, N., Poulev, A., Dey, M., Kurmukov, A. G., Raskin, I. Anti-inflammatory effects of a sesquiterpene lactone extract from chicory (*Cichorium intybus* L.) roots. *Nat. Prod. Commun.* 2007, 2, (7), 717–722.

Rizvi, W., Favazuddin, M., Shariq, S., Singh, O., Moin, S., Akhtar, K., Kumar, A. Anti-inflammatory activity of roots of *Cichorium intybus* due to its inhibitory effect on various cytokines and antioxidant activity. *Anc. Sci. Life*. 2014, 34, (1), 44–49.

Rodgers, G., Becker, P., Blinder, M., Cella, D., Khan, A., Cleeland, C., Coccia, P. Cancer- and chemotherapy induced anemia. *J. Nat. Compr. Canc. Netw*. 2012, 10, (5), 629–653.

Rodríguez-Rodríguez, M., Barbarroja-Escudero, A. J. B., Sánchez-González, M. J. An update on asthma. *Medicine*. 2017, 12, (1), 1745–1756.

Romero, J. C., Martínez-Vázquez, A., Herrera, M. P., Martínez-Mayorga, K., Parra-Delgado, H., Pérez-Flores, F. J., Martínez-Vázquez, M. Synthesis, anti-inflammatory activity and modeling studies of cycloartane-type terpenes derivatives isolated from *Parthenium argentatum*. *Bioorg. Med. Chem.* 2014, 22, (24), 6893–6898.

Rossi, J. F., Lu, Z. Y., Jourdan, M., Klein, M. Interleukin-6 as a therapeutic target. *Clin. Cancer Res.* 2015, 21, (6), 1248–1257.

Ryu, S, Shin, J. S., Jung, J. Y., Cho, Y. W., Kim, S. J., Jang, S., Lee, K. T. Echinocystic acid isolated from *Eclipta prostrata* suppresses kipopolysaccharide-induced Inos, TNF-α, and IL-6 expressions via NF-kB inactivation in RAW 264.7 macrophages. *Planta Med*. 2013, 79, (2), 1031–1037.

Salazar, R., Pozos, M., Cordero, P., Pérez, J., Salinas, M., Waksman, N. Determination of the antioxidant activity of plants from Northeast Mexico. *Pharm. Biol.* 2008, 46, (3), 166–170.

Saleem, M., Abbas, K., Naseer, F., Ahmad, M., Syed, N.H., Javed, F., Hussain, K., Asima, S. Anticancer activity of n-hexane extract of *Cichorium intybus* on lymphoblastic leukemia cells (Jurkat cells). *S. Afr. J. Plant Soil*. 2014, 8, (6), 315–319.

Sánchez-Valle, V., Méndez-Sánchez, N. Estrés oxidativo, antioxidantes y enfermedad. *Rev. Invest. Med. Sur. Mex.* 2013, 20, (3), 161–168.

Shah, N. Z., Muhammad, N., Azeeem, S., Raud, A. Studies on the chemical constituents and antioxidant profile of *Conyza canadensis*. *MEJMPR*. 2012, 1, (2), 32–35.

Shoeb, M., Macmanus, S., Kong, P., Celik, S., Jaspars, M., Nahar, L., Sarker, S. Bioactivity of the extracts and isolation of lignans and a sesquiterpene from the aerial parts of *Centaurea pamphylica* (Asteraceae). *DARU*. 2007, 15, (3), 118–122.

Sohn, S. H., Yun, B. S., Kim, S. Y., Choi, W. S., Jeon, H. S., Yoo, J. S., Kim, S. K. Anti-inflammatory activity of the active components from the roots of *Cosmos bipinnatus* in lipopolysaccharide stimulated RAW 264.7 macrophages. *Nat. Prod. Res.* 2013, 27, (11), 1037–1040.

Silvestre-Roig, C., Fridlender, Z. G., Glogauger, M., Scapini, P. Neutrophil diversity in health and disease. *Trends Immunol.* 2019, 40, (7), 565–583.

Singh, N., Baby, D., Rajguru, J. P., Patil, P. B., Thakkannavar, S. S., Purari, V. B. Inflammation and Cancer. *Ann. Afr. Med.* 2019, 18, (3), 121–126.

Smallwood, M. J., Nissim, A., Knight, A. R., Whiteman, M., Haigh, R., Wynyard, P. Oxidative stress in autoimmune rheumatic diseases. *Free Radical Bio. Med.* 2018, 125, (1), 3–14.

Sofi, F., Dinu, M. R. Nutrition and prevention of chronic-degenerative diseases. *Agric. Agric. Sci. Procedia.* 2016, 8, (1), 713–717.

Souza, M. C., Siani, A. C., Ramos, M. F. S., Menezes de Lima, O., Henriques, M. G. Evaluation of anti-inflammatory activity of essential oils from two Asteraceae species. *Pharmazie.* 2003, 58, (1), 582–586.

Street, R. A., Sidana, J., Prinsloo, G. *Cichorium intybus*: Traditional uses, phytochemistry, pharmacology, and toxicology. *Evid Based Complement Alternat. Med.* 2013. Doi: 10.1155/2013/579319. 1–13.

Tewtrakul, S., Subhadhirasakul, S., Tansakul, P., Cheenpracha, S., Karalai, C. Antiinflammatory constituents from *Eclipta prostrata* using RAW 264.7 macrophage cells. *Phytother. Res.* 2001, 25, (9), 1313–1316.

Torres-González, L., Muñoz-Espinosa, L. E., Rivas-Estilla, A. M., Trujillo-Murillo, K., Salazar-Aranda, R., Waksman De Torres, N., Cordero-Pérez, P. Protective effect of four Mexican plants against CCl_4-induced damage on the Huh7 human hepatoma cell line. *Ann. Hepatol.* 2011, 10, (1), 73–79.

Toyanga, N. T., Verpoortec, R. V. A review of the medicinal potentials of plants of the genus *Vernonia* (Asteraceae). *J. Ethnopharmacol.* 2013, 146, (3), 681–723.

Ukiya, M., Akihisa, T., Yasukawa, K., Tokuda, H., Suzuki. T., Kimura, Y. Anti-Inflammatory, anti-tumor-promoting, and cytotoxic activities of constituents of marigold (*Calendula officinalis*) flowers. *J. Nat. Prod.* 2006, 69, (12), 1692–1696.

Valečka, J., Almeida, C. R., Su, B., Pierre, P., Gatti, E. Autophagy and MHC-restricted antigen presentation. *Mol. Immunol.* 2018, 99, (1), 163–170.

Venero, J. Daño oxidativo, radicales libres y antioxidantes. *Rev. Cubana Med. Milit.* 2002, 31, (2), 126–133.

Vergnolle, N. The inflammatory response. *Drug Dev. Res.* 2003, 59, (4), 75–381.

Viennois, E., Gewirtz, A. T., Chassaing, B. Chronic inflammatory diseases: are we ready for microbiota-based dietary intervention? *CMGH.* 2019, 8, (1), 61–71.

Wang, W., Tan, M., Yu, J., Tan, L. Role of pro-inflammatory cytokines released from microglia in Alzheimer's disease. *Ann. Transl. Med.* 2015, 3, (10), 1–15.

World Health Organization. Cancer. http://www.who.int/mediacentre/factsheets/fs297/es. (accessed Dic. 22, 2019).

Younsi, F., Trimech, R., Boulila, A., Ezzine, O., Dhahri, S., Boussaid, M., Messaoud, C. Essential Oil and phenolic compounds of *Artemisia herba-alba* (Asso.): Composition, antioxidant, antiacetylcholinesterase, and antibacterial activities. *Int. J. Food Prop.* 2016. 19, (1), 1425–1438.

Zaidun, N. H., Thent, Z. C., Latif, A. A. Combating oxidative stress disorders with citrus flavonoid: Naringenin. *Life Sci.* 2018, 208, (1), 111–122.

Zălaru, C., Crisan, C. C., Călinescu, I., Moldovan, Z., Țârcomnicu, I., Litescu, S. C., Tatia, R., Moldovan, L., Boda, D., Iovu, M. Polyphenols in *Coreopsis tinctoria* Nutt. fruits and the plant extracts antioxidant capacity evaluation. *Cent. Eur. J. Chem.* 2014, 12, (8), 858–867.

Zalewski, C., Passero, L., Melo, A., Corbett, C., Laurent, I. M., Toyama, M., Toyama, D., Romoff, P., Fávero, O., Lago, J. Evaluation of anti-inflammatory activity of derivatives from aerial parts of *Baccharis uncinella*. *Pharm. Biol.* 2011, 49, (6), 602–607.

Zaman, R., Basar, S. N. A review article of Beekhe Kasni (*Cichorium intybus*) its traditional uses and pharmacological actions. *Respir. J. Pharmaceutical Sci.* 2013, 2, (8), 1–4.

Index

A

Acai (*Euterpe oleracea*), 160
Acarbose, 71
   Adverse effects of, 86
   compounds derived from, 87–91
Acerola juice, 165
Acetylcholine (ACh), 230, 234
Acetylcholinesterase (AChE), 230, 231–233, 234–235
Acetylcholinesterase inhibitors (AChEIs), 230, 234–238, 239
Achillea millefolium, 347, 361
Acidic beverages, 172
Actinoplanes uthanesis, 85
Affinin, 250–253
   antimicrobial properties, 253–254, 265–266
   bactericidal activity of, 254
   chemical characteristics, 252
   fungicidal activity of, 254
   identification using UPLC-MS analysis, 264–265
   isolation using preparative HPLC, 264, 275–276
   quantification using HPLC-DAD analysis, 265, 277
Agouti-related neuropeptide expression (AgRP) neurons, 212–214
Agrimonia pilosa, 299
alanine aminotransferase enzyme (ALT), 21
alkaline phosphatase (ALP), 21
Alkaloids, 304–305
   mechanisms of action, 305–308
Aloe vera (*Aloe barbadensis/Aloe vera* L.), 95, 156–157
Alzheimer's disease (AD), 6–7, 156, 230, 233–236, 340
   alternative medicine for, 236
   common causes, 233
   costs of care for patients with, 233
Amaranth (*Amaranthus caudatus* L.), 95

Amaryllidaceae family, 237–238
Amino acids, 160–161
α-amylase, 70–73
   from *Aspergillus oryzae*, 82
   bacterial, 81–82
   characteristics of, 75–77
   effect of betulin, 97
   effect of temperature, pH, and salt on, 77–79
   fungal, 82
   inhibitory activity of, 87, 98
   isoenzymes, 85
   from *Monascus sanguineus*, 82
   from plants, 93–98
   polyphenolic compounds, 92–93
   production, 79–81, 82
   from *Pseudoalteromonas* sp., 81–82
   from *Rhizopus oryzae*, 82
   salivary, 83–84
α-amylase inhibitors (IAA), 71
Amylolytic enzymes, 71
Analysis of Variance (ANOVA), 181
Annona crassiflora fruit peel, 302
Annona reticulata, 324
Annona squamosa, 324
Antioxidants, 341
Anxiety, 5
   pathophysiology, 6
   -related disorders, 5
Anxiolytic substances, 5
Apigenin, 20
Arthritis, 339
Asian medicinal herbs, 110–111
Asteraceae plants, 346–360
Asthma, 339–340
Azadirachta (*Azadirachta indica* A. Juss), 96
Azapirones, 5

B

Bacterial amylase, 81–82
Basil (*Ocimum tenuiflorum* L.), 96

Benzodiazepines (BZDs), 5, 14
Betula pendula Roth. bark, 97
Black pepper (*Piper nigrum* L.), 95
Black wattle tree (*Acacia meansii*), 301
β-blockers, 5
Botanicals, 155–156
Box and Wilson methodology, 179–180

C

Canadian horseweed (*Conyza canadensis*), 366
Cancers, 156, 298, 345–346
Capsaicin (8-methyl-*N*-vanillyl-6-nonenamide), 222
Carrageenan, 259
β-caryophyllene, 296
Catechin, 95, 156
Cavitation phenomenon, 168
Centor Score, 245
Chicory (*Cichorium intybus*), 364–366
Chili pepper (*Capsicum* spp), 222
Chinese camellia (*Camellia sinensis* L.), 95
Chinese medicine, 48, 55, 110, 120, 305–306
Chi-square statistical method, 181
Cholinergic hypothesis, 230
Cinnamon (*Cinnamomum verum* J. Presl), 95
Citrus aurantium, 222
Citrus plants, 248
Coccinia (*Coccinia grandis* L.), 95
Coconut water, fermented, 158
Coffee silverskin, 159
Cognitive disorders, 233
Coleus forskohlii, 222
Consortium for the Barcode of Life-Plant Working Group, 125
Consumption of functional foods, 4
Coptis chinensis Franch, 306
Corni fructus (*Cornus officinalis*), 302
Cosmeceutical agents. *see* Mushrooms (*Tremella fuciformis*)
Costus speciosus rhizome extracts, 247
4-Coumaric acid, 61
Cross-flow filtration, 164
Curcumin, 300
Currant (*Ribespull chelum* Turcz), 96
Cushing Syndrome, 204
Cyclic guanosine monophosphate (cGMP), 260
Cycloserine, 37

D

Dairy-based beverages, 142–143
  commercially available, 143
  drawbacks of, 143–144
Dandelion medicinal (*Taraxacum officinale* L.), 96
Diabetes, 345
  medicinal plants for, 292
  primary treatment for, 292
  types, 292
Diabetes mellitus (DM), 70
Diabetic peripheral neuropathy, 306
Diclofenac, 259–260
Dietary carbohydrates, 71
Dietary fiber, 153
Diffusion of functional foods, 4
Dimethylallyl diphosphate (DMAPP), 293
2,6-Dimethyl-3-amino-benzoquinone, 36
Dimocrpus longan Lour. pericarp, 300
2,2-Diphenyl-1-picrylhydrazyl (DPPH), 24, 126–127, 361
  scavenging capacity, 27–28
Diterpene Kaurenes, 324
Donepzil, 230

E

Echinacea spp., 238
Ecklonia cava, 302
Egyptian balanitis (*Balanitesa aegyptiaca* L.), 95
Emulsifying agents, 173
Energy, lack of, 111
Energy drinks, 111–112, 144–146
  active compounds in, 112
  commercialized, 145
  health risks, 146
  using herbal extracts, 145–146
Ent-kaurenes
  antibacterial activity, 321–322
  anti-inflammatory activity, 323–324
  antitumor activity, 322–323
  applications of, 324–327
  biosynthesis of, 320–321
  definition, 318
  inhibition of glucose transporters by, 326–327
  as solubilizing agent, 326
  structure, 318

structure–activity relationship of, 319–320
in sweetener industry, 324–325
Epicatechin (EC), 95, 156
Epigallocatechin (EGC), 156
Epigallocatechin gallate (EGCG), 156
Erythritol, 32
Essential oils, 247–248
Etiology of disease
  biological factors, 203–205
  genetic abnormalities, 204
  metabolic conditions, 204–205
  obesity, 207
  psychological factors, 206
  socioeconomic factors, 206–207

F

Fenugreek, 306
Fenugreek (*Trigonella foenum-graceum* L.), 95
Flavonoids, quantification of, 23–24, 27
Flax (*Linum usitatisumum* L.), 96
Folin Ciocalteu assay, 126–127
Free-radical DPPH scavenging capacity, 24
Free radicals damage, 340–341
Fructose 1,6-biphosphatase, 296–297
Fruit juices, 158
Functional beverages, 4
  acceptability and availability of, 174–181
  flavor retention and release, 182
  interaction with environment, 184
  light-sensitive ingredients, 185
  quality and safety issues in, 182–184
  sensory quality of, 173–174
  shelf life of, 182–184
  stability of, 184
  storage and packaging, 184–186
  texture and stability, 170–173
  use of emulsifiers, 173
Functional coffee, 159
Functional foods, 111, 138
  definition, 138–140
  development of, 141–142
  nutritional and health claims, 140–141
  types of, 142–143
Functional ingredients, 141, 170
  antioxidants, 151–152
  nutraceuticals, 152–153
  vitamins, 154–155

Fungal amylase, 82
Fusobacterium necrophorum, 244

G

Galantamine, 230
Galega officinalis L., 96
Gallic acid, 298
Gallocatechin (GC), 156
Garcinia cambogia, 219
Garlic (*Allium sativum* L.), 96, 238
Gasotransmitters, 260
Gelling agents, 172
Gentisic acid, 61
Ginseng (*Ginkgo biloba*), 238
Ginseng (*Panax ginseng* C.A. Meyer), 118–119, 146
  as an antistress and antioxidant agent, 146
  anticarcinogenic activity, 123
  barcoding regions, 125–126
  for blood sugar regulation, 124
  on cardiovascular system, 124
  DNA barcoding, 124–125
  DNA extraction, 125
  gas chromatography-mass spectrometry (GC-MS) analysis, 128–130
  ginsenosides of, 121–122, 124, 126
  history, 120
  immune system effects, 123–124
  inhibitory effects on CNS, 122–123
  pharmacological and medicinal benefits, 121–122
  phenolic compounds and antiradical activity of, 122
  production and geographical distribution, 119
  root growth and shape, 119
  traditional uses, 120–121
α-1,4-glucan-4-glucanohydrolase, 71
Glucose 6-phosphase, 296–297
Glucose transporters (GLUTs), 296, 300, 326
α-glucosidase, 70, 86, 296
Green tea (*Camellia sinensis* (L.) Kuntze), 112–113, 156, 222
  anticancer activity, 115
  antimicrobial activity, 114–115
  barcoding regions, 125–126
  caffeine content in, 126, 128
  DNA barcoding, 124–125

DNA extraction, 125
gas chromatography-mass spectrometry (GC-MS) analysis, 128–129
history, 113–114
HPLC chromatogram of, 127
names, 112
pharmacological effects, 114–115
plant, 113
production and geographical distribution, 113
uses, 114
Green tea polyphenol (–)-epigallocatechin-3-gallate (EGCG), 301
Group A *Streptococcus* (GAS), 244, 245, 247
Guayaeva (*Psidium guajava* L.), 95
Gut–brain axis, 211
neuroendocrine axis, 212–213
neuronal axis, 212

H

"Hard to press" fruits, 166
Haritaki (*Terminalia chebula* Retz.), 96
Heart disease, 344–345
Heliopsis Longipes (*A. Gray*) S.F. Blake, 249–250
analgesic effect of, 255–257
as an anti-infective agent, 254
antihyperalgesic effect of, 259–260
anti-inflammatory properties, 257–259
antimicrobial activity, 277–278
antioxidant activity, evaluation of, 263–264, 272–273
barcoding profile, 266–267, 279–280
biological properties, 252
cytotoxic activity, 273–275
DNA extraction, 266, 278–279
ethanolic and dichloromethane extracts, 261
ethanolic extracts, 262
extraction and analysis methods for, 262–263, 267–268
general description, 250
methanol and chloroform extracts, 261
morphology, 250
oral use, 252
PCR analysis, 267
phytochemical profiling of, 268–271
total phenolic compounds, quantification of, 263, 272
toxicity of, 261–262
in treatment of fungal infections, 252
vasodilation, 260–261
vasorelaxation, 261
Hepatic gluconeogenesis, 296–297
Herbal drinks, 148–150
types and benefits, 149
Herbal medicine, 4–5, 7, 21, 99, 110–111, 236, 239, 246–247
to treat pharyngitis, 247
1,3,5-Hexatriene, 36
High hydrostatic pressure (HHP) technology, 162
High-intensity pulsed electric field (HIPEF), 158
High-pressure processing (HPP) technology, 161–163
effect on sustaining ascorbic acid level, 163
in food and beverage processing, 163
horizontal and vertical types, 162
inactivation of microorganisms and enzymes, 163
in yogurt processing, 163
Hippocrates, 4
Hoodia gordonii, 219
Human umbilical vein endothelial cells (HUVECs), 307
Huntington's disease, 234
Hydrocolloids, 171–172
4-Hydroxybenzoic acid, 61
Hyperglycemia, 71
Hypogonadism, 204
Hypothalamic lesions, 204
Hypothyroidism, 204

I

Indian mango (*Mangifera indica* L.), 95
Inflammation, 332
acute, 332
cell phase, 337–338
cells involved in, 334
chronic, 332
cytokines involved in, 335
diseases related to, 338–340
mechanism of, 333–338
as a response of immune system, 332–333
silent phase, 336–337
treatments of, 346
vascular phase, 337

Index 385

Insulinoma, 204
Isopentenyl diphosphate (IPP), 293

J

Jambolana or yambolan (*Syzygium cumini* L.), 95
Jamun landraces (*Syzygium cumini* L.), 298
Just About Right (JAR) scale, 177, 179

K

Kaffir-lime (*Citrus hystrix* L.), 96
Kratom (*Mitragyna inermis* L.), 95

L

Labeled Affective Magnitude (LAM), scale, 177–179
Lactase nonpersistence, 144
Lactose intolerance, 143
Lemon balm (*Melissa officinalis* L.), 6, 95
 barcoding amplification, 23
 chemical composition of, 28–37
 clinical applications, 21–22
  antibacterial and sedative activity, 18
  antioxidant properties, 22
 common names, 17
 constituents of, 19
  infusion essential oil, 20
 DNA extraction from, 22–27
 in ethnomedical systems, 18
 gas chromatography–mass spectrometry (GC–MS) analysis, 24, 30–32
 phytochemical compounds in, 18–20
 toxic aspects of, 20–21
Lettuce (*Lactuca sativa*), 368–369
Lingonberry (*Vaccinium uliginosum* L.), 96
Lipids, 342–343

M

Maceration, 260
Maltodextrin/glucose polymers, advantages of, 146
Mannitol, 129–130
Maple syrup, 159
Marigold (*Calendula officinalis*), 361–363
Matcha, 151
Medicinal plants, 4, 96–98, 98, 236–238, 238, 295
 for diabetes, 292
Melanocyte-stimulating hormone α (α-MSH), 212
Membrane filtration technologies, 163–166
 for juice clarification, 165
 standardization of protein content, 164
 types, 164
Metabolic disorders, 204–205, 298
2-Methoxy-4-vinylphenol, 26
1-Methyl, 2,4-dimethyl-2-oxazoline-4-methanol, 36
Mexican traditional medicine, 248–249
Microfiltration, 165
Microwave-assisted extraction, 167
Microwave heating technology, 166–168
Microwave hydrodiffusion and gravity (MHG), 166–167
Microwave pasteurization, 167
Minerals, 157–159
Mirabilis jalapa L., 306
Monogenic obesity, 204
Moringa oleifera, 222
Mushrooms (*Tremella fuciformis*), 48–50
 acidic heteroglycans, 51
 as adjuvant drug, 55
 anti-aging and skin whitening effect, 61–63
 anti-inflammatory and antitumor activities, 54
 antioxidant activity of, 60–61
 antitumor properties, 56–57
 aqueous, 62
 carboxymethylated, 63
 on cell viability, 60
 chemical modification of TFPS, 52
 expression of SIRT1, 62
 free-radical degradation, 60–61
 functional groups in, 61
 glycosidic linkages, 51–52
 health benefits, 47
 hypoglycemic activity of, 57–58
 as immunomodulatory agent, 55–56
 neuroprotective action, 58–60
 oxidative-reduction degradation, 61
 against oxidative stress, 47–48
 polysaccharide extracts of, 47, 50–54
  extraction method, 52–54
 in promoting immune functions, 63
 protective properties of, 62

significance of, 46–47
stimulation of peripheral nerve regeneration, 58
structural variability of TFPS, 50–51
on subjective cognitive impairment (SCI), 59
on trimethylamine-induced learning and memory deficits, 59

N

Nanodrop® spectrophotometer, 22
Neuroendocrine axis, 212–213
Neuronal axis, 212
Neuropeptide Y (NPY), 213
Nonivamide (*N*-nonanoyl-4-hydroxy-3-methoxybenzylamine), 222
Nonstarch polysaccharides, 153
Nutraceuticals, 152–153
Nutty lotus (*Nelumbo nucifera* Gaertn.), 95

O

Obesity, 202, 208
  definition, 203
  epidemiology, 208–209
  etiology of, 207
  herbal products, 218–223
  nonpharmacological treatments, 218–223
  pathophysiology, 209–211
  pharmacological treatments, 216–217
Obesogenic environment, 203
9,12,15-Octadecatrien-1-ol,(Z,Z,Z), 26
7-*O*-galloyl-d-sedoheptulose, 302
Oleander (*Nerium oleander* L.), 96
Opuntia ficus-indica root, 98
Oropharyngeal sprays, 246
Orthosiphon stamineus ethanolic extract, 298
Oxidative stress, 341–344
  proteins and, 343–344
Oximatrine, 307

P

Palmitic acid (PA), 130
Palm-shaped adansonia (*Adansonia digitata* L.), 95
Panax species, 118
Parkinson's disease, 234

Passionflower (*Passiflora incarnata* L.), 6
  alkaloids in, 10, 13
  application, 7–8
  barcoding amplification, 23
  BZF compound, 14
  characteristics, 9
  chemical composition of, 28–37
  clinical applications, 13–15
    antiasthmatic action, 14
    anticonvulsant properties, 14
    antidepression and anti-anxiety effects, 15–16
    anti-inflammatory response, 16–17
    antioxidant activity, 15
    antitumoral activity, 14
    antitussive action, 14
    anxiolytic activity, 13
    for insomnia, 16
  DNA extraction from, 22–27
  ethnopharmacology, 9–10
  flavonoids in, 12
  gas chromatography–mass spectrometry (GC–MS) analysis, 24, 34–36
  genus, 7–8
  glycosides in, 10–11
  inter- and intraspecies dissimilarity among species, 9
  neuromuscular relaxing effects of, 13
  phytochemical compounds in, 10–11
  structural types, 10–11
  toxicity, 11–13
Patanga (*Caesalpinia sappan* L.), 96
Pectin methyesterase (PME), 167–168
Peppermint tea, 150
Pharyngitis, 244, 245
  alternative and complementary medicine for, 246–280
Phellinus igniarius phenolic extract, 300
Phenolic compounds, 297–298
  mechanisms of action, 298–304
Phenolic compounds, quantification of, 23, 27
Phytochemicals, 153
Phytoconstituents, 218–219
  with α-amylase enzyme inhibitory activity, 85–98
Pink catharanthus (*Catharanthus roseus* L.), 95
Plant-based drugs, 5

Polyphenols, 153
Polysaccharides, 46
Polyunsaturated fatty acids (PUFAs), 153
Pomegranate juice, 165
Pomegranate (*Punicagranatum* L.), 160
Prebiotics, 157
Principal component analysis (PCA), 181
Probiotics, 157
    commercially available, 158
Proopiomelanocortin (POMC), 212
Protein tyrosine phosphatase intracellular (PTP1B), 297
Protein tyrosine phosphathase 1B (PTP1B), 299
p-Synephrine, 222
Pumpkin, 306
2-Pyrrolidinemethanol, 36

Q

Qualitative descriptive analysis (QDA), 177, 181

R

Rambutan (*Nephelium lappaceum* L.), 113
    barcoding regions, 125–126
    caloric content of, 118
    content of essential oils, 118
    DNA barcoding, 124–125
    DNA extraction, 125
    fatty acids and triglycerides in, 118
    history, 116–117
    pharmacological effects, 117–118
    plant structure, 113
    production and geographical distribution, 116
    traditional uses, 117
    vitamin content, 118
Reactive oxygen species (ROS)
    oxidative stress and, 341–344
    physiological functions of, 341
    sources of, 341
Ready-to-drink tea, 150–151
Resorcinol, 30
Reverse osmosis, 164, 165–166
Rheumatic fever (RF), 244–245
Rheumatic heart disease (RHD), 244245
Rhodiola (*Rhodiola rosea* L.), 96

Rivastigmine, 230
Rosemary (*Rosmarinus officinalis* L.), 95
Rosmarinic acid, 20
2(*R*),3(*S*)-1,2,3,4-Butanetretol, 30

S

Saliva, 83
    oral hygiene and, 83
Salivary α-amylase, 83–84
    quantity and enzymatic activity of, 84
    role in starch digestion, 84
Salvia officinalis ethanolic extract, 247
Sassafras tea, 150
Sensorial test, 180
Siam weed (*Chromolaena odorata*), 363–364
Sodium-glucose type 1 (SGLT-1) cotransporter, 296
Solanum nigrum, 301
Sonication technology, 168–169
Soxhlet extraction, 262–263
    with gas chromatography, 263
Soy-based beverages, 147–148
Sports and performance drinks, 146–147
    commercially available, 147
    differences, 147
St. John's wort (*Hypericum perforatum*), 238
Starch, 73–75
Stevia Rebaudiana, 325
Streptococcal pharyngitis, 244–245
Superfruit extract, 159–160

T

Tacrine, 230
Tamarind (*Tamarindus indica* L.), 95
Taurine, 145
Terpenes, 292–294, 297
    biosynthesis of, 293
    mechanisms of action, 294–297
    structure, 293–294
    use, historical evidence, 294
Thermosonication treatment, 169
Thickening agents, 171
Tinospore cordy (*Tinospora cordifolia* (Thunb.) Miers), 96
Traditional medicine, 110

Trailing Eclipta/Bhangra (*Eclipta prostrata*), 367–368
Trigonelline, 306
Tulasi (*O. tenuiflorum* L.), 96
Type 2 diabetes treatment, 68

U

Ultrafiltration, 164
Ultrasonication, 262

V

Valienamine, 86
Vegetable- and fruit-based functional beverages, 144
   commercially available, 145
   global market for, 144
   health benefits, 144
   health concerns, 144
Vitamin A, 152

Vitamin B, 154–155
Vitamin C, 152
Vitamin D, 155
Vitamin E, 152
Vitamin K, 155
Vitexin, 13
Volumetric heating, 166

W

Wagner–Meerwein rearrangement, 320
White mulberry (*Morus alba* L.), 95

Y

Yin-Yang, 110

Z

Zygophyllaceae family, 238